T0205912

Emerging Chemical Risks in the Work Environment

Occupational Safety, Health, and Ergonomics: Theory and Practice

Series Editor: Danuta Koradecka
(Central Institute for Labour Protection – National Research Institute)

This series will contain monographs, references, and professional books on a compendium of knowledge in the interdisciplinary area of environmental engineering, which covers ergonomics and safety and the protection of human health in the working environment. Its aim consists in an interdisciplinary, comprehensive and modern approach to hazards, not only those already present in the working environment, but also those related to the expected changes in new technologies and work organizations. The series aims to acquaint both researchers and practitioners with the latest research in occupational safety and ergonomics. The public, who want to improve their own or their family's safety, and the protection of heath will find it helpful, too. Thus, individual books in this series present both a scientific approach to problems and suggest practical solutions; they are offered in response to the actual needs of companies, enterprises, and institutions.

Individual and Occupational Determinants: Work Ability in People with Health Problems
Joanna Bugajska, Teresa Makowiec-Dąbrowska, Tomasz Kostka

Healthy Worker and Healthy Organization: A Resource-Based Approach
Dorota Żołnierczyk-Zreda

Emotional Labour in Work with Patients and Clients: Effects and Recommendations for Recovery
Edited by Dorota Żołnierczyk-Zreda

New Opportunities and Challenges in Occupational Safety and Health Management
Daniel Podgórski

Emerging Chemical Risks in the Work Environment
Małgorzata Pośniak

Visual and Non-Visual Effects of Light: Working Environment and Well-Being
Agnieszka Wolska, Dariusz Sawicki, Małgorzata Tafil-Klawe

Occupational Noise and Workplace Acoustics: Advances in Measurement and Assessment Techniques
Dariusz Pleban

Virtual Reality and Virtual Environments: A Tool for Improving Occupational Safety and Health
Andrzej Grabowski

Head, Eye, and Face Personal Protective Equipment: New Trends, Practice and Applications
Katarzyna Majchrzycka

Nanoaerosols, Air Filtering and Respiratory Protection: Science and Practice
Katarzyna Majchrzycka

Microbial Corrosion of Buildings: A Guide to Detection, Health Hazards, and Mitigation
Rafał L. Górny

Respiratory Protection Against Hazardous Biological Agents
Katarzyna Majchrzycka, Justyna Szulc, Małgorzata Okrasa

For more information about this series, please visit: https://www.crcpress.com/Occupational-Safety-Health-and-Ergonomics-Theory-and-Practice/book-series/CRCOSHETP

Emerging Chemical Risks in the Work Environment

Edited by
Małgorzata Pośniak

CRC Press is an imprint of the
Taylor & Francis Group, an **informa** business

[First] edition published [2020]
by CRC Press
6000 Broken Sound Parkway NW, Suite 300, Boca Raton, FL 33487-2742
and by CRC Press
2 Park Square, Milton Park, Abingdon, Oxon, OX14 4RN

Library of Congress Cataloging-in-Publication Data

Names: Pośniak, Małgorzata, editor.
Title: Emerging chemical risks in the work environment / edited by Małgorzata Pośniak.
Description: Boca Raton : CRC Press, 2020. | Series: Occupational safety, health, and ergonomics : theory and practice | Includes bibliographical references and index.
Identifiers: LCCN 2020006996 (print) | LCCN 2020006997 (ebook) | ISBN 9780367489885 (hbk) | ISBN 9780367507565 (pbk) | ISBN 9781003051152 (ebk)
Subjects: LCSH: Industrial toxicology. | Industrial safety.
Classification: LCC RA1229 .E44 2020 (print) | LCC RA1229 (ebook) | DDC 615.9/02--dc23
LC record available at https://lccn.loc.gov/2020006996
LC ebook record available at https://lccn.loc.gov/2020006997

ISBN 13: 978-0-367-48988-5 (hbk)
ISBN 13: 978-0-367-50756-5 (ebk)

Typeset in Times
by Deanta Global Publishing Services, Chennai, India

Contents

Preface

The rapid increase of cancer, reproductive system diseases, and child development disorders, as well as disorders of the endocrine system, leading to abnormal metabolism, obesity, diabetes, and the prevalence of hormone-dependent cancer due to occupational exposure to chemicals is a challenge for the people responsible for safety and health at work.

The emerging risks from occupational exposure to carcinogenic, mutagenic, reprotoxic substances and endocrine disruptors and engineering nanomaterials require systemic activities to reduce chemical risks in the workplace supported by researchers

Therefore, the purpose of this monograph is to provide full and up-to-date knowledge on the risks posed by substances of the abovementioned chemical groups. This book is intended for occupational health and safety experts, employers, and employees who have contact with the above mentioned chemical agendas at work. Readers will find valuable information enabling them to identify chemical hazards and carry out of occupational risks assessment and, in particular, to take appropriate control measures to reduce them.

The authors of this book believe that the information contained in this monograph will increase awareness and knowledge of the threats posed to workers by these dangerous agents, and the application in practice of the correct methods of chemical risk management will contribute to the reduction and, in some cases, elimination of the threats.

Małgorzata Pośniak

Acknowledgment

Wishing a special thank you to Katarzyna Czarnecka for translating this book into English.

Editor

Małgorzata Pośniak Central Institute for Labour Protection – National Research Institute, Departmentof Chemical, Aerosol and Biological Hazards, Czerniakowska 16, 00-701, Warsaw,Poland.

Małgorzata Pośniak is the head of the Department of Chemical, Aerosol and Biological Hazards in the Central Institute for Labour Protection – National Research Institute. She graduated from the Faculty of Pharmacy of the Medical University of Warsaw, Poland, and prepared her dissertation (PhD) at the Lodz Technical University of Technology, Poland.

Her main research activities are focused on the analysis of airborne chemical pollution at the workplace and occupational risks assessment. She has developed several dozen methods for determining the concentrations of hazardous chemicals in the workplace air to assess occupational exposure of workers. These methods are established in Poland as Polish Standards for the protection of air purity and are widely used by occupational hygiene laboratories.

She is an expert member of the Polish Interdepartmental Commission for MAC and MAI Values as well as of the Group for Chemical Agents. She is the editor of *Principles and Methods of Assessing the Working Environment* and associate editor of the *International Journal of Occupational Safety*. She is also a member of the Chromatographic Analysis Commission of the Polish Academy of Sciences and an expert in chemical agents in the European Agency for Safety and Health at Work.

Contributors

Tomasz Jankowski Central Institute for Labour Protection – National Research Institute, Departmentof Chemical, Aerosol and Biological Hazards, Czerniakowska 16, 00-701, Warsaw, Poland.

Tomasz Jankowski graduated from the Faculty of Chemical and Process Engineering of the Warsaw University of Technology, Poland. He earned a PhD in Technical Sciences from CIOP-PIB. Currently, he is working at the Aerosol, Filtration and Ventilation Laboratory in the Central Institute for Labour Protection – National Research Institute. He gained experience in the field of aerosol testing and determining the air parameters in workplaces, assessing occupational exposure to aerosols, conducting occupational risk assessment and the control of occupational risks. He is the author or co-author of 34 publications. He participated in 36 national and 5 international research projects and in over 180 works, such as expert opinions commissioned by institutions and enterprises.

Katarzyna Miranowicz-Dzierżawska Central Institute for Labour Protection – National Research Institute, Departmentof Chemical, Aerosol and Biological Hazards, Czerniakowska 16, 00-701, Warsaw, Poland.

Katarzyna Miranowicz-Dzierżawska graduated from the Medical University of Warsaw, Poland (Faculty of Pharmacy, specialization: pharmaceutical analysis). She prepared her dissertation (PhD): "The evaluation of interaction between chosen organic solvents" at the Collegium Medicum of the Jagiellonian University, Cracow, Poland.

She works at the Department of Chemical, Biological and Aerosol Hazards (Laboratory of Toxicology) in the Central Institute for Labour Protection – National Research Institute.

Her professional activities are currently focused on in vitro toxicity studies of industrial chemicals (cytotoxicity; apoptosis; oxidative stress; combined action of chemicals; etc.) She is a member of the Polish Society of Toxicology and Polish Association of Industrial Hygienists.

Przemysław Oberbek Central Institute for Labour Protection – National Research Institute, Departmentof Chemical, Aerosol and Biological Hazards, Czerniakowska 16, 00-701, Warsaw, Poland.

Przemysław Oberbek graduated from the Faculty of Chemistry, University of Warsaw, Poland. He earned a PhD in Materials Science and Engineering at the Warsaw University of Technology, Poland. He works at the Department of Chemical, Biological and Aerosol Hazards (Aerosols, Filtration and Ventilation Laboratory) in the Central Institute for Labour Protection – National Research Institute. His researches are focused on occupational exposure to engineering nanomaterials, and assessment of airborne dangerous substances in the workplaces air. He is an expert at the Polish Standard Committee in a nanotechnologies technical committee (PKN/KT 314), the European Technical Committee for assessment of workplace exposure to chemical and biological agents (CEN/TC 137 WG3), and the European Technical Committee for Nanotechnologies (CEN/TC 352 WG 1).

Jolanta Skowroń Central Institute for Labour Protection – National Research Institute, Departmentof Chemical, Aerosol and Biological Hazards, Czerniakowska 16, 00-701, Warsaw, Poland.

Jolanta Skowroń graduated from the Faculty of Pharmacy of the Medical University of Warsaw, Poland, and prepared her dissertation (PhD) at the Military Institute of Hygiene and Epidemiology, Warsaw, Poland. She works in Laboratory of Toxicology, a part of the Department of Chemical, Aerosol and Biological Hazards in the Central Institute for Labour Protection – National Research Institute. Her main activities are currently focused on toxicity studies of chemicals using in vitro methods and preparation of health-based documentation on occupational exposure limits for toxic substances. She is a secretary of Polish Interdepartmental Commission for MAC and MAI Values, an expert of Group for Chemical Agents, member of Polish Society of Industrial Hygienist, member of American Conference of Governmental Industrial Hygienist, editor of *Principles and Methods of Assessing the Working Environment* and a member of Scientific Committee for Occupational Exposure Limits UE.

Lidia Zapór Central Institute for Labour Protection – National Research Institute, Departmentof Chemical, Aerosol and Biological Hazards, Czerniakowska 16, 00-701, Warsaw, Poland.

Lidia Zapór graduated from the Faculty of Biology of the University of Warsaw, Poland. She earned a PhD in Biological Sciences at the Military Institute of Hygiene and Epidemiology, Warsaw, Poland. She is the head of the Laboratory of Toxicology in the Central Institute for Labour Protection – National Research Institute. Her main activities are focused on toxicity studies of industrial chemicals, including nanomaterials, using in vitro methods. She is a member of the Experts Group for Chemicals Agents of the Polish Interdepartmental Commission for MAC and MAI Values and a member of Polish Society of Industrial Hygienists.

Series Editor

Professor Danuta Koradecka, PhD, D.Med.Sc., and Director of the Central Institute for Labour Protection – National Research Institute (CIOP-PIB), is a specialist in occupational health. Her research interests include the human health effects of hand-transmitted vibration; ergonomics research on the human body's response to the combined effects of vibration, noise, low temperature and static load; assessment of static and dynamic physical load; and development of hygienic standards; as well as development and implementation of ergonomic solutions to improve working conditions in accordance with International Labour Organisation (ILO) conventions and European Union (EU) directives. She is the author of over 200 scientific publications and several books on occupational safety and health.

***

The "Occupational Safety, Health, and Ergonomics: Theory and Practice" series of monographs is focused on the challenges of the twenty-first century in this area of knowledge. These challenges address diverse risks in the working environment of chemical (including carcinogens, mutagens, endocrine agents), biological (bacteria, viruses), physical (noise, electromagnetic radiation) and psychophysical (stress) nature. Humans have been in contact with all these risks for thousands of years. Initially, the intensity of the risks was lower, but over time it has gradually increased, and now too often exceeds the limits of man's ability to adapt. Moreover, risks to human safety and health, so far assigned to the working environment, are now also increasingly emerging in the living environment. With the globalization of production and merging of labour markets, the practical use of the knowledge on occupational safety, health and ergonomics should be comparable between countries. The presented series will contribute to this process.

The Central Institute for Labour Protection – National Research Institute, conducting research in the discipline of environmental engineering, in the area of working environment, and implementing its results, has summarized the achievements – including its own – in this field from 2011 to 2019. Such work would not be possible without the cooperation of scientists from other Polish and foreign institutions as authors or reviewers of this series. I would like to express my gratitude to all of them for their work.

It would not be feasible to publish this series without the professionalism of the specialists from the Publishing Division, the Centre for Scientific Information and Documentation and the International Cooperation Division of our Institute. The challenge was also the editorial compilation of the series and ensuring the efficiency of this publishing process, for which I would like to thank the entire editorial team of CRC Press – Taylor & Francis.

***

This monograph, published in 2020, has been based on the results of a research task carried out within the scope of the second to fourth stage (2011–2019) of the Polish National Programme "Improvement of safety and working conditions" partly supported – within the scope of research and development – by the Ministry of Science and Higher Education/National Centre for Research and Development, and within the scope of state services – by the Ministry of Family, Labour and Social Policy. The Central Institute for Labour Protection – National Research Institute is the Programme's main coordinator and contractor.

1 Introduction

Małgorzata Pośniak

CONTENTS

1.1 INTRODUCTION

At the end of the twentieth century, chemical substances and their mixtures have become widespread in all areas of the global economy, and they are one of the most common factors that are harmful to human health in the work environment. They are also a systematically growing risk to the natural environment and to the health and life of the societies of all countries, both rich and highly industrialized countries, as well as those that are economically underdeveloped and have a low economic level. The world's society cannot imagine functioning without the use of chemical products in everyday life.

No industry could develop without chemicals. Chemistry has made people's lives easier, more pleasant and more comfortable. It is however important to bear in mind that many substances are the cause of serious diseases and the degradation of the environment.

Despite the considerable development of chemical technologies and progress in managing occupational risk caused by harmful chemicals and suspended particulates, in recent years, around the world, many of the problems associated with their effects on the health and lives of workers remain unresolved.

Chemical substances are hazardous throughout their entire life cycles, from their production, through processing and use, to their disposal. They are not only a source of environmental pollution, but a hazard to workers in the majority of workplaces. According to the US Environmental Protection Agency, more than 50 million chemicals circulate on the earth, of which only 85,000 are subject to a law on control [Krimsky 2017]. In 2018, 222.6 million tonnes of chemicals hazardous to human health were manufactured in the EU-28, including 15–18%, i.e., 33.4–40 million tonnes of carcinogens, mutagens and cancer-causing reprotoxic agents, along with those arising from occupational exposure. The production of industrial chemicals was largely concentrated in Western Europe [Eurostat 2018].

In the countries of the European Union, in 2017, occupational cancer diseases caused 106,300 deaths, which is 20 times more than the number of deaths caused by accidents at work. The International Labour Organisation however estimates that each year occupational exposure to carcinogens causes 666,000 deaths worldwide, which is twice as many as due to accidents at work. According to the World Health

Organization (WHO) and the International Agency for Research on Cancer (IARC), cancer kills 8.2 million people worldwide each year, and 14 million new cancer incidences are reported each year. It is estimated that before the year 2035, cancer mortality will increase by 78% and cancer incidence by 70% [Takala 2015, 2018].

The number of cases of diseases resulting from occupational exposure to substances harmful to reproduction or causing hormonal disorders is also systematically increasing. These disorders often appear even many years after exposure. It is estimated that in the European Union alone, more than 160 billion euros are being spent annually on treating diseases caused by endocrine disruptors and on supporting individuals who are unable to work due to health damage. The cost is approximately 1.28% EU Gross Domestic Products.

Infertility is at the forefront of the effects of contact with substances harmful to reproduction and is a growing concern for many developed and highly developed countries. As a result of its prevalence, infertility is considered by the World Health Organization (WHO) to be a social disease. According to the WHO data, approximately 60–80 million couples around the world permanently or temporarily face infertility. In highly developed countries, 10–12% of couples face infertility [Kumar and Singh 2015]. Risks for the reproductive health and maintenance of the hormonal balance of workers indicate that it is necessary to take measures to intensify and conduct further detailed epidemiological and toxicological studies, as well as the intensification of legislative and organizational activities to focus on these dangerous substances.

There is a particular need for studies to be conducted and scientifically proven data to be gathered on the combined effects of chemical substances due to a significant lack of results in the available literature. It is especially important due to the fact that workers are occupationally exposed to mixtures, not individual substances, and the combined effects might be significantly different from the effects of exposure to a single chemical.

Nanomaterials are new agents whose impact in the future is difficult to estimate at present. Specially designed nanomaterials (ENMs, engineered nanomaterials) are widely used, especially in construction, energy, transport, chemical, automotive, electronic, textile, telecommunications and cosmetics industries, as well as in medicine, environmental protection, agriculture and many others. Many nanomaterials have already reached very high production capacities. Statistics show that the rate of growth of nanomaterial production is increasing. Depending on the source, it is estimated that the global nanotechnology market could reach between approximately USD 11–55 billion in 2022, with an average annual growth rate of more than 22% since 2017 [Inshakova and Inshakov 2017]. Due to the systematically growing number of hazardous substances and workers exposed to them, more attention should be paid to the promotion of appropriate chemical safety management in the workplace, and above all to the promotion of effective methods to control risks to health and life posed by hazardous substances.

Special attention, in the process of occupational safety management, should be paid to newly introduced chemical substances and their mixtures and to those substances that cause an increase in hazards in the course of their production, processing and use, namely, the management of emerging chemical risks in the workplace.

Research conducted in recent years by European and global experts has shown that emerging chemical risks with regard to occupational health and safety in the future will be linked to the workers' exposure to newly introduced nanomaterials as well as substances that are carcinogenic, mutagenic and toxic to reproduction and endocrine disruptors.

At present, it is not possible to completely eliminate the use of chemical agents dangerous to workers' health. Therefore, the only solution that can, to some extent, reduce the risks associated with production, processing and use of hazardous agents is to encourage employers to make every effort to consciously and correctly manage chemical risks in the workplace.

REFERENCES

Inshakova, E., and O. Inshakov. 2017. World market for nanomaterials: Structure and trends. *MATEC Web Conf* 129:02013. doi: 10.1051/matecconf/201712902013.

Krimsky, S. 2017. The unsteady state and inertia of chemical regulation under the US Toxic Substances Control Act. *PLoS Biol* 15(12):e2002404. doi: 10.1371/journal.pbio.2002404.

Kumar, N., and A. K. Singh. 2015. Trends of male factor infertility, an important cause of infertility: A review of literature. *J Hum Reprod Sci* 8(4):91–196. doi: 10.1186/s12958-015-0032-1.

Statistic Explained. 2018. Chemicals production and consumption statistics. https://ec.europa.eu/eurostat/statistics-explained/index.php/Chemicals_production_and_consumption_statistics#Total_production_of_chemicals (accessed January 30, 2020).

Takala, J. 2015. Eliminating occupational cancer in Europe and globally. Working Paper 2015.10. Brussels: ETUI, European Trade Union Institute. https://www.etui.org/content/download/21462/179550/file/WP+2015-10-Eliminating+occupational+cancer+Web+version.pdf (accessed January 20, 2020).

Takala, J. 2018. *Carcinogens at work: A look into the future*. Vienna: Austrian EU Presidency.

2 Nanomaterials in the Work Environment

Lidia Zapór and Przemysław Oberbek

CONTENTS

2.1 INTRODUCTION

Currently, nanotechnology is the most rapidly developing interdisciplinary branch of science, which presents an opportunity to bring engineering to an entirely new level, allowing for impressive results for the economy in the near future. This technology is revolutionary for practically every branch of the industry. In the European Union, nanotechnology is considered a Key Enabling Technology (KET) of the twenty-first century. It is also often seen as an important factor in the fourth industrial revolution. Engineered nanomaterials (ENMs) or manufactured nanomaterials (MNMs) have found a number of applications in the fields of electronics, construction, energetics, transportation, and medicine; the chemical, automotive, textile, telecommunications, and cosmetics industries; environmental protection; and agriculture, food, and other industries. A number of nanomaterials, such as carbon nanotubes, carbon black, or oxide nanomaterials, are produced in very high capacities.

It is estimated that, depending on the source, the global nanotechnology market may reach 11–55 billion USD by 2022, with an average annual growth rate of 22% from 2017 (Compound Annual Growth Rate, CAGR ~22%) [Inshakova and Inshakov 2017]. The scale of production and the wide spectrum of applications provide an initial estimate of the professional groups potentially exposed to nanomaterials.

ENMs are structures that have at least one dimension in the 1–100 nm scale [EC 2018a]. The size of this magnitude influences a number of physicochemical properties of the ENMs, changing them considerably from their bulk counterparts, which can also cause different effects on living organisms. It is highly probable that due to their sizes, which are comparable to those of basic biological structures, ENMs can interact with cell components and cause health effects, which are difficult to estimate in both humans and animals, as well as changes in the environment. Due to this fact, ENMs are currently considered a new and emerging risk by international scientific bodies and organizations concerned with occupational health and safety. The magnitude of the risk is highly difficult to assess as of now, as it depends mainly on the results of epidemiological studies and progress in the area of toxicological studies, exposure assessment methodology, as well as organizational and legal solutions. The following chapter attempts to comprehensively deal with the aforementioned issues.

2.2 CHARACTERISTICS OF NANOMATERIALS

2.2.1 Terminology, Definitions, and Types of Nanomaterials

To give an understanding of how small the nanoworld is, examples from everyday life are shown: a sheet of paper is about 100 thousand nanometers thick, a strand of human hair measures between 80 and 100 nm, a strand of human DNA (deoxyribonucleic acid) has a diameter of 2.5 nm, and the covalent radius of a sulfur atom is 0.1 nm (102 picometers). Such is the scale at which nanotechnology operates.

The International Organization for Standardization (ISO), in its ISO/TS 80004-1:2015 document (Nanotechnologies – Vocabulary – Part 1: Core Terms), defines a nanomaterial as "a material with any external dimension in the nanoscale or having internal structure or surface structure in the nanoscale", where nanoscale is defined as the size range from approximately 1 to 100 nm. The generic term "nanomaterial" is inclusive of "nanoobject" and "nanostructured material". Viruses, single molecules, and even atoms can be found in the nanoscale, as shown in Figure 2.1.

In ISO/TS 80004-2:2015 (Nanotechnologies – Vocabulary – Part 2: Nanoobjects), the following terms were used to describe selected nanoobjects, according to their shape and nanoscale dimensions:

- *nanoobject* – a material with one, two, or three external dimensions in the nanoscale,
- *nanostructured material* – a material with internal or external nanostructures,

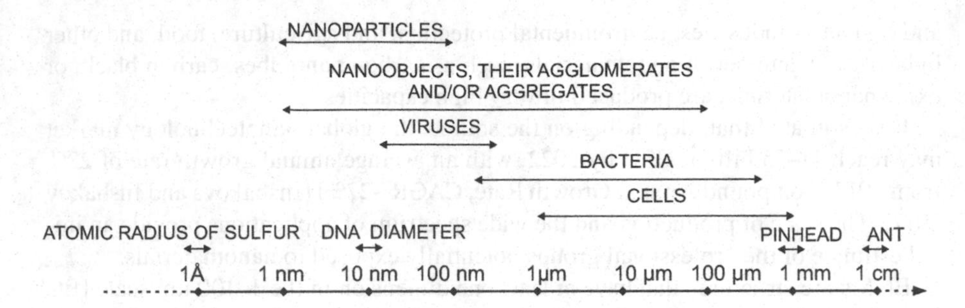

FIGURE 2.1 Scale from pico- to centimeters, including chosen objects.

- *nanoparticle* (NP) – a nanoobject having three external nanoscale dimensions,
- *nanoplate* – a nanoobject with one external nanoscale dimension and the remaining dimensions significantly larger (at least three times larger than the nanoscale dimensions; these dimensions do not necessarily have to be in the nanoscale),
- *nanofiber* – a nanoobject with two similar nanoscale dimensions and a third significantly larger dimension (similar external dimensions may differ by less than three times their value, and the significantly larger dimension may differ from them by more than three times),
- *nanotube* – a type of a hollow nanofiber,
- *nanorod* – a type of a full nanofiber,
- *nanocrystal* – a nanoobject with a crystalline structure.

Nanoobjects have the tendency to form sets of strongly bound particles with a total surface area smaller than the sum of the total surface areas of individual particles (aggregates) or sets of loosely bound particles with a total surface area similar to the sum of the total surface areas of individual particles (agglomerates). The definition of nanomaterials includes nanoobjects, their agglomerates and aggregates (NOAA).

Moreover, in the ISO/TS 80004-2:2015 standard, the source of the nanomaterials is taken into account:

- *engineered nanomaterial* (ENM) – a nanomaterial designed for a specific purpose or function,
- *manufactured nanomaterial* (MNM) – a nanomaterial produced specifically to have selected properties or composition,
- *incidental nanomaterial* – a nanomaterial produced unintentionally as a by-product of a process.

The term *ultrafine particle* (UFP) is also used for incidentally produced nanomaterials. As per the ISO/TR 27628:2007 definition (Workplace Atmospheres – Ultrafine, Nanoparticle and Nano-structured Aerosols – Inhalation Exposure Characterization and Assessment), it is a particle with a nominal diameter of 100 nm or less. UFP is used to describe both naturally occurring and incidentally produced particles (e.g.

in technological processes). UFPs have heterogeneous shapes, sizes, and chemical compositions.

Nanomaterials can also be classified according to the number of dimensions in the nanoscale:

- *zero-dimensional* – with three dimensions in the nanoscale (e.g. quantum dots or single nanoparticles),
- *one-dimensional* – with two mutually perpendicular dimensions outside the nanoscale (nanorods, nanotubes, and nanofibers),
- *two-dimensional* – with one dimension in the nanoscale (such as nanoplates or nanolayers), and
- *three-dimensional* – heterogeneous and homogeneous materials composed of nanosized crystals (e.g. multi-nanolayers or dispersion of nanoparticles) [Jeevanandam et al. 2018].

The most common nanomaterials in the work environment are carbon-based metals and their alloys, inorganic non-metallic compounds, nanoclays, nanopolymers and dendrimers, quantum dots, and nanocomposites.

The European Commission (EC) recommends the following definition of a nanomaterial: "a nanomaterial is a natural, incidental or manufactured material containing particles, in an unbound state or as an aggregate or as an agglomerate and where, for 50% or more of the particles in the number size distribution, one or more external dimensions is in the size range 1 nm–100 nm" [EC 2018a]. The EC also recommends treating nanoparticle aggregates and agglomerates (even with dimensions outside the nanoscale) as nanomaterials due to their nanostructure. The Commission Regulation (EU) 2018/1881 [EC 2018b] has introduced the term "nanoform". Fullerenes, graphene flakes, and single-walled carbon nanotubes with at least one dimension below 100 nm are also considered nanoforms.

2.2.2 Characteristic Properties of Nanomaterials

Changes in material properties after their structure is changed to the nanoscale result mainly from a larger ratio of surface atoms or ions to those inside the particle as well as quantum limitations of electrons in the nanoparticles. A characteristic property of nanomaterials is the large surface area of boundaries of separation. In nanoparticles or extended nanoobjects (such as nanorods, nanotubes, or nanofibers) the external surfaces are the boundaries of separation (interphase boundaries and grain boundaries), and in the case of nanocrystalline materials internal surfaces act as the boundaries of separation [Gleiter 2000; Padmanabhan and Gleiter 2014].

Consequently, nanomaterials have a larger specific surface area than their microscale counterparts. This results in, among others, an increase in catalytic activity, the number of defects and stress in the crystal lattice, chemical reactivity, tendency to agglomerate, absorption and adsorption capabilities, and higher mechanical strength [Dreaden et al. 2012; Xie et al. 2012; Barrak et al. 2019; Gajanan and Tijare 2018].

The use of engineered nanomaterials is becoming increasingly common, both at the workplace and in everyday life. Apart from a range of potential benefits, the use of nanomaterials can also endanger human health. Nanoobjects in the free state or released from aggregates and agglomerates (e.g. during their degradation and dissolution) have dimensions much smaller than the sizes of living cells, as shown in Figure 2.1. This means that they can easily pass through biological membranes, and this property is used in medical applications, e.g. targeted therapies, drug delivery, and cancer therapies. However, uncontrolled, incidental absorption of nanoobjects by biological structures that are in constant contact with the external environment, such as skin, eyes, or the airways, remains a problem.

2.3 TYPES OF NANOMATERIALS AND THEIR MAIN APPLICATIONS

ENMs have found a range of applications in the chemical, textile, cosmetics, electronics, sports, and automotive industries as well as in the fields of medicine, energetics, biotechnology, environmental protection, telecommunications, and transportation [Sahu 2016; Inshakova and Inshakov 2017]. Thanks to nanotechnology, a wide range of unique products are already available, e.g. extremely durable construction materials, compact computers with immense data processing capabilities, quantum computers, thin optical fibers, dichroic glass, high-efficiency microchips, carbon nanotube antennae that collect light in the visible spectrum, super-slippery coatings, walkways and non-woven textiles, air purification systems, specialized biomaterials and biomarkers, as well as drug carriers. The following section is a review of the types of nanomaterials and their applications [EC 2012; EPA 2017; Dreaden et al. 2012; Gajanan and Tijare 2018].

2.3.1 Carbon-Based Nanomaterials

This class of nanomaterials includes mainly nanofibers and carbon nanotubes, fullerenes, carbon black, and graphene. Carbon nanofibers are almost entirely composed of stretched carbon structures, chemically similar to graphite. Their high mechanical strength is due to a highly organized chemical and geometric structure. These materials are also infusible, chemically resistant, as well as resistant to wear and sudden changes in temperature. Carbon nanofibers have mainly found applications as a dispersive phase material in nanocomposites (carbon-metal, carbon-ceramic, carbon-polymer, and carbon-carbon) or composite materials in which nanostructured materials constitute the dispersive phase. They are also used in light constructions, lithium-ion batteries, fuel cells, vibration damping materials, filtration fabrics, and fuel lines.

Carbon nanotubes (CNTs) are composed of one or more layers of carbon atoms, rolled up in a hexagonal lattice analogous to that in graphite, constituting either long single-wall carbon nanotubes (SWCNTs) or multi-wall carbon nanotubes (MWCNTs). They exhibit high electrical and thermal conductivity and a high strength-to-weight ratio. These materials have found applications in, among others, disk drives and automobile fuel lines. They are also used as polymer additives in paints, fuel cells, electrodes, electrolytes, and battery membranes.

Fullerenes are molecules composed of an even number of 60 or more carbon atoms, creating a polycyclic system (as a kind of nanotubes closed at both ends).

Their applications are chiefly in the biomedical field (as contrast agents and in drug delivery systems).

Carbon black is a black powder containing 80–95% amorphous carbon. The majority of these particles are in the 1–100 nm range as well as in the agglomerated state. The material is used mainly for the production of tires, pigments, inks, anti-static fillers, decorative fibers, electrodes, and carbon brushes.

Graphene exhibits both the properties of metals and semiconductors but is not, in fact, classified under any of these categories. It is, however, an excellent conductor of heat and electricity. Graphene is also characterized by low resistivity, high opacity, and high tensile strength. Its current applications include integrated circuits, transistors, transparent conductive electrodes, solar and fuel cells, and a range of sensors and filters, e.g. for the sorption of heavy metal ions from contaminated waters.

2.3.2 Metals and Metal Alloys

The most popular, currently produced metal nanoparticles are gold, silver, iron, copper, and titanium. The most common metal alloys are platinum and palladium. Gold nanoparticles are most often used in medical applications such as diagnostics. They are also used in optics, solar cell technology, lubricants, catalytic converters, sensors, and special coatings.

Nanosilver is mainly used for its antibacterial, antifungal, and antiviral properties (products typically containing nanosilver are used in hospital textiles, dressings, containers for contact lenses, sports clothing, odorless undergarments, cosmetics, etc.). Other commercially available metals often used as nanomaterials are, among others, platinum nanoparticles, platinum and palladium alloys (in electronics and chemical catalysis and copper nanopowder for printed electronics and inks), iron nanoparticles (magnetic recording tapes), and titanium alloys (medical implants and materials used in the automotive, aviation, and space industries).

2.3.3 Inorganic Non-metallic Nanomaterials

The most common commercially available inorganic and non-metallic nanomaterials are synthetic amorphous silica, titanium dioxide, zinc oxide, aluminum oxide, aluminum hydroxide, iron oxides, cerium dioxide, and zirconium dioxide. Various nanoforms of synthetic amorphous silica have found applications in sectors such as textiles, leather, paper, cosmetics, food, electronics, and construction as well as in the production of detergents, paints, and varnishes. They are also used to reinforce elastomers (mainly tires), shoes, and cable shields, for crude oil refining, and as drying agents to protect products during transportation and storage. Titanium dioxide nanoparticles are good UV radiation filters. TiO_2 nanoparticles have special electrical, photocatalytic, and antimicrobial properties. This material is used in sunscreens, plastic and metal coatings, varnishes for maintenance, solar cells, catalysts, protective coatings, the so-called self-cleaning products (windows, cements, tiles, hospital textiles), deodorants, and air purification systems.

Zinc nano-oxide is a colorless and effective UV filter but in a different spectrum than TiO_2. It is used as an active agent in self-cleaning products, sunscreens, varnishes, ceramics, and electronics, as well as in liquid crystal displays and solar cells.

Aluminum oxide nanoparticles increase the coatings' wear resistance and are used as fillers in polymers and tires as well as in protective glasses and scratch-resistant windows, floors, bar code scanners, precise optical components, catalysts, ceramic filtration membranes, and flame retardants.

Iron oxide nanoparticles have found a number of applications as components of pigments, polishes, catalysts, fuel cells, oxygen sensors, and optoelectronic equipment as well as in water purification systems and the remediation of soil and ground waters.

Cerium (IV) nano-oxide has special optical properties due to which it has found applications in optical, electrooptical, microelectronic, and optoelectronic devices. It is used to polish silicon wafers and glass surfaces as an anticorrosive material and is also used as a catalytic diesel fuel additive.

Zirconium nano-dioxide is used in optical connectors, fuel cells, lithium-ion batteries, catalysts, ceramic membranes, cements, dental fillers, dentures, fluorescent lamps, and polishes. This material is sintered in the powder form of ceramic materials with unique properties, chiefly high fracture toughness.

2.3.4 Nanoclays

Nanoclays are nanoparticles of layered mineral silicates, such as kaolinite, bentonite, montmorillonite, hectorite, and halloysite. Nanoclays have found applications as components of tires, paints, inks, greases, polymer nanocomposites, cosmetics, and drug carriers.

2.3.5 Nanopolymers and Dendrimers

There are currently commercially available polymer nanofilms, nanotubes, nanowires, nanorods, and nanoparticles. Polymer nanoparticles are polymer units used in the nanoscale. They are used in drug carriers and as fillers in polymer composite matrices. Nanostructured polymer films are used as coatings, e.g. in biomedical applications. Polymer nanotubes, nanowires, and nanorods have potential applications in sensors and in micromechanical, electronic, magnetic, optic, and optoelectronic devices.

Currently, work is being carried out on the development of conductive textiles, wearables (wearable devices), and "intelligent fibers", which change their properties depending on the environmental conditions.

Dendrimers are a distinct group characterized by specific polymer structures. They are fractal-like, with a regular, branched structure and a high specific surface area. Dendrimers have found applications in drug concentration controllers, diagnostic tests, liquid diffraction grids, lasers and light-emitting diodes, catalysts, or semi-permeable membranes.

2.3.6 Quantum Dots

Quantum dots are crystalline semiconductors with dimensions from 2 to 10 nm, and their electrical properties depend on the shape and size of individual crystals.

Quantum dots are particles small enough to have their properties significantly changed after adding a single electron. The fluorescence of quantum dots can be controlled by modifying their size and geometry. Because of their absorption spectrum tuning and a high extinction coefficient, quantum dots are already used, or are planned to be used, in photovoltaic devices, lasers, LED diodes, photodetectors, sensors, and single-photon sources. Due to properties resulting from their nanometric dimensions, they are more stable and precise as markers in medical diagnostics than organic dyes used to date. Plans are underway to use them in drug carriers, for tracking viruses in the body, and in single-electron transistors. They are most commonly obtained from indium phosphide, cadmium selenide, and sulfide.

2.3.7 Nanocomposites

Nanomaterials can constitute the dispersive phase in composites, particularly in a polymer matrix. The sheer number of possible combinations of materials and their constituents, as well as their weight ratios in the final products, makes it highly complicated to single out particular nanocomposites for consideration in this monograph. They also significantly differ in material properties, making it impossible to list those most important for health safety. However, it is important to take note of nanocomposites as a group of materials. They are mainly used in the production of high-strength sport and construction materials, self-cleaning surfaces, conductive polymers, and bone implants.

2.4 IMPACT OF NANOMATERIALS ON HUMAN HEALTH

2.4.1 Effects of Specific Properties of Nanomaterials on Toxicity

Nanomaterials can have toxic effects on the human body depending on their chemical nature and physical properties, the most significant of which are size and shape of the particles, their surface area, the state of aggregation and agglomeration, solubility, surface charge, surface modifications, crystalline structure, etc. The properties of substances with particles in the nanoscale are different from their bulk counterparts in many respects. Nanoparticles have a relatively low mass, an extensive specific surface area (the outer area of a solid substance in relation to the mass of the substance (m^2/g)), varying chemical reactivity, a larger capacity for oxidation, a different surface charge, and varying solubility in liquids.

A number of these properties may influence the behavior of the nanoparticles in living organisms. Nanogold is an apt example. In the bulk form, gold is a yellow metal, which does not interact with biological materials; however, nanogold is purple or red, depending on the grain size, and it easily binds to proteins. This is why nanogold can be seen as both an anticancer medicine and as a potentially toxic substance.

2.4.1.1 Particle Size

A particle with a size in the order of 10^{-9} m is considered to be on the molecular level. To illustrate, the diameter of a single atom is of the order of 0.1 nm; many pathogenic viruses are about 100 nm in size, an average bacterium is 2.5 μm long, the diameter of red blood cell is approximately 7 μm, and the diameter of a pulmonary alveolus is approximately 400 μm.

The basic components and structures of living cells are also in the nanometer range: the diameters of the human DNA and protein molecules are approximately 2.5 nm and 5–50 nm, respectively, and the thickness of cell membranes does not exceed several tens of nanometers, as shown in Figure 2.1. The size of ENM particles allows them to freely cross cell membranes due to which they can interact with the components of living cells and disrupt their function. It is an important fact when considering nanomaterial toxicity that there are virtually no barriers for the spread of particles smaller than 10 nm in the human body. Therefore, ENMs can cross natural barriers in the body, such as the blood-brain barrier, placental barrier, lungs, skin, and intestines. Particle size is also the decisive parameter when it comes to their absorption, distribution, and deposition; it also plays a role in dermal absorption (as described in Section 2.4.2.2).

Nanoparticles with sizes below 5 nm are considered the most dangerous due to their ability to cross into the cell nucleus and interact with genetic material (DNA).

2.4.1.2 Specific Surface Area

The fragmentation of a substance into the nanoscale causes an increase in its specific surface area, subsequently causing an increase in chemical activity, especially the catalytic properties, influencing a number of biochemical processes inside the cell – mainly the redox processes, generating free radicals – and, in consequence – oxidative stress. Surface reactivity is considered to be a common feature of all nanomaterials, which plays a key role in their mechanism of action. Animal studies led to a hypothesis that an extensive specific surface area in combination with the low solubility of the ENMs can be a critical factor in lung inflammation [Braakhuis et al. 2014; ECETOC 2013; Oberdörster 1995].

2.4.1.3 Shape

Shape is a highly significant parameter influencing ENM toxicity, especially if they are absorbed via inhalation. Rigid fiber ENMs, fulfilling the WHO criteria for inhalable fibers, length > 5 μm, diameter < 3 μm, aspect ratio (ratio of length to diameter) > 3:1, and the so-called HARNs – High Aspect Ratio Nanomaterials, are especially hazardous. This category includes not only carbon nanotubes (CNTs) but also a number of nanosized metal structures (nickel oxide nanowires and titanium dioxide nanofibers) and graphene nanoplatelets. Fibrous nanomaterials can damage macrophages, causing the disruption of the phagocytosis process, as well as show a high capacity for fibrogenesis (the formation of excess fibrous connective tissue), leading to cancerous changes (mesothelioma), similar to the cases of exposure to asbestos fibers. Short fibers can be absorbed by macrophages, but the long and rigid ones damage them mechanically, inducing the release of proinflammatory mediators (this process is called "frustration" of macrophages). Tangled or twisted fibers are not considered rigid [Catalán et al. 2016].

Basically, particle shape influences the membrane formation processes during in vivo endocytosis or phagocytosis. It has been observed that endocytosis of spherical nanoparticles is quicker in comparison to elongated nanoparticles or fibers. More importantly, spherical nanoparticles are relatively less toxic. The toxicity of spherical (granular) nanomaterials is highly dependent on their chemical composition and water solubility (biopersistence).

2.4.1.4 Dissolution

Dissolution is a measure of biodurability of ENMs; it is dependent on their chemical and physical properties (size and surface area) as well as on the suspension medium, including its ionic strength, pH, and temperature. Biodurability, defined as the ability to resist chemical and biochemical alteration, is a significant contributor to biopersistence, regarded as the ability of a material to persist in the body in spite of physiological clearance mechanisms. Dissolution and biodurability are both important parameters in the risk assessment of nanomaterials [Utembe et al. 2015].

Nanomaterials with good solubility in water (solubility > 100 mg/l) lose their nano-specific behavior, and their toxicity depends, as in the case of conventional substances, mainly on their chemical composition. They can, however, exhibit greater toxicity than their parent substances due to their extensive specific surface area, allowing for the release of a larger number of ions causing free radical reactions [WHO 2017 and literature therein].

Poor water solubility determines nanomaterial biopersistence or their ability to remain in the organism for an extended period of time. Biopersistence is considered a basic property of nanomaterials, inducing the formation of changes in the lungs both by granular particles and fibers [Sellers et al. 2015; Gebel et al. 2014; Braakhuis et al. 2016]. Nanomaterials, such as Poorly Soluble Particles (PSP) and Granular Biopersistent Particles (GBP), with large specific surface areas, can weaken the macrophage efficiency in removing solid particles via phagocytosis. In consequence, the alveolar clearance process is impaired, and the particles are deposited in the pulmonary alveoli for extended periods of time, which leads to the so-called lung particle overload. The accumulated alveolar macrophages release proinflammatory cytokines due to oxidative stress, which causes inflammation in the lungs, proliferation of the alveolar epithelium, metaplasia, and, in consequence, fibrosis and malignancy. Such carcinogenic activity has been observed in cases of rats exposed to titanium dioxide (TiO_2) [Oberdörster 1995; NIOSH 2011; ECETOC 2013].

Apart from their water solubility, the bioavailability of the ENMs also influences their fat solubility. Human airways are covered with a pulmonary surfactant – a lipoprotein complex consisting of lipids, phospholipids, and proteins (mainly specific surfactant proteins but also albumins and immunoglobulins). Apart from the aerodynamic size of the ENMs, it is their lipid affinity that can cause deposition in the lungs. Lipophilic ENMs with a high affinity to lung surfactants are an even greater burden on the lungs than hydrophilic ENMs [Wohlleben et al. 2016].

2.4.1.5 Surface Chemistry (Surface Charge)

Surface morphology and functionalization, as well as the surface charge of ENMs are important parameters influencing their toxicity. ENMs with a cationic charge have been shown to have a more significant cytotoxic effect, disrupt cell membrane integrity, as well as cause mitochondrial and lysosomal damage; it is easier for them to cross mammalian cell membranes and the blood-brain barrier, and they interact with genetic material in the cell to a greater extent than ENMs with an anionic or neutral charge. Therefore, they exhibit higher toxicity [Gatoo et al. 2014; Havrdova et al. 2016]. Surface charge also plays a role in nanoparticles crossing the layer of the gastrointestinal mucus. Additionally, this physical parameter of the ENMs seems to be responsible for possible changes in the shape and size of the particles due to the creation of aggregates and agglomerates, influencing the processes of ENMs crossing into the cells.

Surface charge also influences ENMs' interactions with proteins in the body. Such interactions are based on the formation of strongly bound complexes, seen under a microscope as a "corona". The protein corona (biocorona) around the nanoparticles is spontaneously created the moment they start interacting with the biological environment. The formation of the biocorona influences the surface properties of the particles as well as their size, and, therefore, their bioavailability [Borisova 2018]. The protein corona can also change the ENMs' charge and influence a number of metabolic processes. The process of its formation might have a profound influence in prognosing ENM toxicity, e.g. due to the modulation of the absorption process and the elimination of nanoparticles in the lungs. Konduru et al. [2017] hypothesize that nanoparticle biokinetics in the lungs depends on the composition of proteins in the corona. However, the exact influence of corona is still up to further assessment.

ENMs, especially of organic origin, such as carbon nanotubes, are often functionalized, which influences their physicochemical properties (e.g. hydrophilic-hydrophobic, acid-base) and their electrochemical and catalytic properties. In the case of carbon nanotubes, toxicity is often influenced by surface modifications and contaminants, e.g. by metallic catalysts. Similarly, in the case of modified aluminosilicates, e.g. montmorillonite, it was shown that the cytotoxic effect is more dependent on the function group toxicity than on the properties stemming from their nanoscale dimensions [Zapór and Zatorski 2011].

2.4.1.6 Mechanisms of ENMs Toxicity

To date, in vivo, in vitro, and in silico studies concerning the influence of ENMs' specific properties on their toxicity allowed for the determination of toxic mechanisms at the cellular level. It is currently believed that ENMs can cause undesirable effects in the cells via the following mechanisms:

- Oxidative stress paradigm (oxidative damage)

  Oxidative stress, i.e. the capability of the ENMs to overproduce reactive oxygen species (ROS), causes inflammatory response, DNA damage, protein denaturation, or lipid peroxidation. These biological effects can be influenced by physicochemical properties of the ENMs (i.e. size and shape of the particles, surface area, surface reactivity, etc.) [Donaldson et al. 2003; Moller et al. 2010].
- Fiber paradigm (a mechanism similar to the damage caused by fibers)

  This mechanism can be ascribed to nanofibers, in particular, those with high rigidity and biopersistence, fulfilling the WHO fiber criteria and nanofibers with a high aspect ratio (HARNs) [Donaldson et al. 2010]. Theoretically, any single fiber can cause genotoxic effects, which may lead to cancer development.

- Genotoxic damage

  Gene damage has a fundamental significance in the origin of cancers and a number of chronic diseases. When present in sex cells, they can cause genetic diseases or affect reproduction. Published data show that ENMs can cause both direct and indirect damage to genetic material inside the cells. The key indirect mechanism is oxidative stress [Donaldson et al. 2003],

- Release of toxic ions

  The mechanism, characteristic of metals and metal oxides, is sometimes called the "trojan horse" effect. After entering the cells via endocytosis and being transported to the lysosomes, metal particles in the acidic environment begin to release toxic ions, leading to cell death [Braakhuis et al. 2014]. There is also limited evidence that the release of metal ions (e.g. from ENMs containing Ag, TiO_2, and ZnO) in an acidic environment can cause their toxic effects both after inhalation and gastrointestinal exposure [Pietroiusti et al. 2017].

The influence of properties specific to nanomaterials on their toxicity is significant and well-documented, and according to the Organisation for Economic Co-operation and Development (OECD) guidelines, the following parameters should be taken into account in ENM risk assessment: chemical composition (% additives; % contaminants), initial particle size, particle size distribution, shape, surface area, and surface charge (Zeta potential), surface morphology (crystallinity and amorphicity), water solubility (hydrophilicity/lipophilicity), dispersivity, reactivity, redox potential, adsorption capacity, and dustiness. Moreover, it is advisable to study the agglomeration, aggregation, biopersistence, and protein biocorona as factors influencing ENMs' biokinetics. Guidelines concerning the range and methodology of the studies are cyclically published in OECD environments and in health and safety publications (Series on the Safety of Manufactured Nanomaterials). It is also important to note that ENMs' physical parameters can also be subject to change during their life-cycle and in different biological environments. For example, ZnO nanoparticles are soluble in the low pH conditions of the digestive tract (stomach) but biostable in the neutral pH of the respiratory tract. This is why it is important to study ENMs' biokinetics [Braakhuis et al. 2014; OECD 2018; 2019].

2.4.2 Absorption and Translocation of Nanomaterials in the Human Body

Generally speaking, the inhalation route has the greatest significance in occupational exposure to all chemicals, including nanomaterials. However, as nanomaterials present in the air sediment quickly and can be deposited on working surfaces or protective clothing, dermal and gastrointestinal exposure may also be significant [Pietroiusti et al. 2017; 2018].

2.4.2.1 Inhalation Exposure

The transport of aerosol particles, and their absorption and deposition in the airways, depends both on the physicochemical properties of the ENMs and the physiology of the respiratory tract. As shown in Figure 2.2, in the human respiratory tract, there are three distinct functional areas, with different structures and mechanisms of particle deposition and elimination:

- extrathoracic region (nasopharynx) – oral cavity, nasal cavity, pharynx, and larynx,
- tracheobronchial region – trachea, bronchi, bronchioli, and terminal bronchioli,
- pulmonary region – respiratory bronchioli, alveolar ducts, and alveoli.

The respiratory tract from the trachea to the tertiary bronchi is covered with mucus, forming a protective layer, in which particles are internalized and removed as a result of mucociliary clearance through coughing or swallowing. In this region, clearance depends on the number of cilia and ciliary beat frequency and on the mucus quality and quantity [Bierkandt et al. 2018; Fernández Tena and Casan Clarà 2012].

The lungs consist of bronchi and pulmonary alveoli (300–500 million), blood, and lymphatic vessels. Each alveolus is covered with lung capillaries, forming an

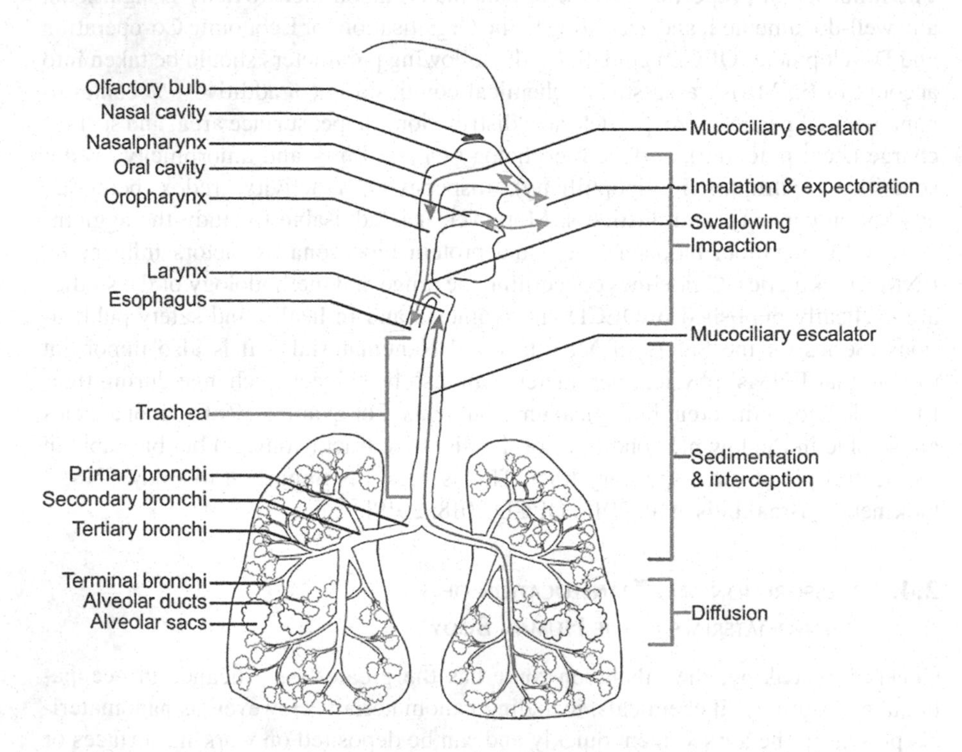

FIGURE 2.2 Human respiratory system scheme (Reprinted with permission from Bierkandt et al. 2018. The impact of nanomaterial characteristics on inhalation toxicity. *Toxicol Res* 7:321–346).

extensive network of over 280 billion, which constitutes the gas exchange region with an area of 50–130 m^2. The pulmonary alveoli walls are built from two types of epithelial cells, type I and II pneumocytes. Tight junctions between the epithelial cells along with the pulmonary surfactant, a mucous lining of the lung alveoli, act as an additional barrier for particle transport. Particles capable of crossing this barrier enter the circulatory system or are phagocyted by alveolar macrophages [Geiser 2010].

Physical properties of the particles – their size, size distribution, and shape – are of key importance in their deposition in particular sections of the airways. Oberdörster et al. [1994] introduced a model for nanoparticle absorption in the lungs, which is cited in a number of scientific papers and toxicological reports. It has been shown that the smallest particles, about 1 nm in size, basically do not reach the lung alveoli, because 90% of them are deposited in the upper respiratory tract (nasopharynx), and only 10% in the tracheobronchial tree. Five nanometer particles are equally deposited in particular areas of the respiratory tract (about 30% in each area), but 20 nm particles are deposited in about 50% in the lung alveoli.

The extended duration of nanoparticle deposition in the lungs facilitates their penetration into the epithelial cells of the respiratory tract and later entry into the circulatory and lymphatic systems. It was demonstrated that nanoparticles with dimensions around 50 nm enter the alveolar epithelial cells via passive diffusion and accumulate in the cytoplasm, but the ~100 nm nanoparticles enter via clathrin-dependent endocytosis and are accumulated in the endosomes [Thorley et al. 2014]. Larger particles are removed via phagocytosis by macrophages populating the lung alveoli.

The lung mononuclear phagocyte system, composed of, among others, macrophages, is naturally equipped to eliminate environmental contaminants, penetrating the respiratory tract. However, animal studies have shown that the effectiveness of this process in the case of nanoparticles might be significantly smaller than the process of elimination of the larger particles. For example, in rats exposed to TiO_2 inhalation, the half-life of nanoparticle retention in the pulmonary alveoli was 117 days for particles sized 250 nm and up to 541 days for the 20 nm particles [Oberdörster et al. 1994].

Particles not removed by the macrophages can pass through to the circulatory system. The presence of nanoparticles in the bloodstream has been determined after inhalation exposure of test animals to nearly each of the studied ENMs, e.g. polystyrene nanoparticles, a number of metals, fullerenes C60, cerium dioxide, titanium dioxide, and carbon black [Sarlo et al. 2009; Schleh et al. 2012; Kreyling et al. 2002, 2014; Naota et al. 2009; Geraets et al. 2012]. ENMs can be distributed through the organism via the circulatory system. The target organs in which nanoparticles were most often temporarily deposited were the liver, spleen, kidneys, hearth, testicles, and the brain [SCENHIR 2014; Geraets et al. 2012].

The efficiency of the translocation of inhalatory ENMs to the circulatory system and internal organs was estimated to be really small, generally not exceeding 4% of the dose, which has been established based on gold, iridium, silver, carbon ENMs, and cerium dioxide [Kreyling et al. 2002; 2014; Schleh et al. 2012; SCENHIR 2014; Li et al. 2016].

2.4.2.2 Dermal Exposure

Skin is the largest human organ, with a surface area of about 1.5–2 m^2. As shown in Figure 2.3, it is composed of three layers, from the outermost layer: the epidermis, dermis, and subcutaneous tissues. The epidermis is composed of five layers: stratum corneum, stratum lucidum, stratum granulosum, stratum spinosum, and stratum basale. The stratum corneum is composed of tightly connected terminally differentiated cells (corneocytes) and is a natural barrier of the body. This barrier can be easily weakened by, for example, mechanical damage (abrasions), inflammation, exposure to UV radiation, or microbial colonization. Experimental studies to date show that only very small particles, below 4 nm, are capable of entering the surface channels between corneocytes and pass through undamaged skin. Flexible ENMs, such as liposomes and micelles, are an exception, as they are capable of penetrating undamaged skin even with sizes above 4 nm. The most likely way of nanoparticle absorption is through the hair follicles, sebaceous glands, and sweat glands. This is the most effective way of absorption for particles in the 4–20 nm range. The penetration of particles in the 20–45 nm range has only been observed when the skin is damaged, while non-soluble particles above 45 nm did not even pass through damaged skin [Larese Filon et al. 2015; 2016; Wang et al. 2018].

In the process of ENM dermal absorption, their chemical properties play a significant role. A particular risk is connected to the ENMs of metallic origins or carbon ENMs with metallic residues, because metal ions may be released in contact with

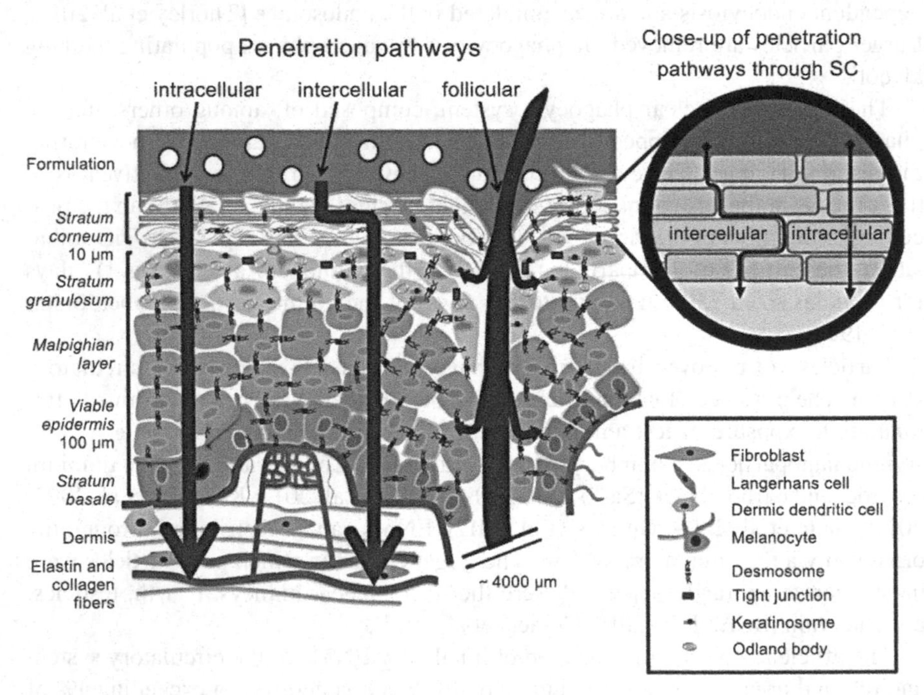

FIGURE 2.3 Skin penetration of chemicals (Reprinted with permission from Bolzinger et al. 2012. Penetration of drugs through skin, a complex rate-controlling membrane. *Current Opinion in Colloid & Interface Science* 17:156–165).

their suspensions or emulsions or the wet environment of the skin. It has been shown that metallic nanoparticles are mainly absorbed through hair follicles. Moreover, the follicles can be a reservoir for metallic nanoparticles. Hair follicles constitute over 0.1% of the total area of the skin, which makes this route of exposure significant. Long-term release of ions can increase the risk of allergic contact dermatitis in the case of ENMs containing sensitizing metals, such as palladium and cobalt [Larese Filon et al. 2015; Wang et al. 2018].

The physiological state of the skin is an important factor in risk assessment. A damaged skin barrier enables the absorption of the ENMs and increases the risk of their penetration into the dermis. In animal studies, free migration of titanium dioxide, zinc oxide, and silica nanoparticles between the deeper layers of the skin has been shown, however, mainly in the case of disorders of the epidermis. The process has not been observed for healthy skin [Larese Filon et al. 2015; Wang et al. 2018].

Larese Filon et al. [2015] have reviewed both in vivo and in vitro studies concerning the probability of hazards occurring after dermal exposure in occupational environments. Despite the available data on dermal absorption of metallic nanoparticles (FeO, Ag, ZnO, Au, Pt, Rh), the authors have stated that the risk of absorption is negligible due to the small amount of absorbed substances. Nanoparticles of sensitizing metals (Ni, Co, Pd) may pose a risk, but it is not clear whether they are absorbed in the form of particles or ions. It seems that quantum dot exposure might be particularly dangerous. It was shown that quantum dots under 12 nm could cross over epithelial layers to the dermis and subsequently enter into the bloodstream and accumulate in the internal organs. Studies on undamaged guinea pig skin, exposed to CdSe quantum dots approximately 7 nm in diameter for 24 h, have shown that they crossed into deeper skin layers [Ryman-Rasmussen et al. 2006]. Exposure to quantum dots containing cadmium needs to be particularly controlled due to high risk connected to the toxic and carcinogenic effects of this element.

As part of the works conducted by the European Scientific Committee on Consumer Safety (SCCS), various commercially available ENMs have been evaluated from the safety perspective of dermal exposure. In the case of some nano-SiO_2 forms, it has been proven that they can penetrate through the outer layers of the skin and reach living parts of the epidermis or even the dermis. Nano-SiO_2 has also been found in the dendritic cells [SCCS 2015]. However, no skin penetration by nano-TiO_2 has been demonstrated [SCCS 2013]. TiO_2 nanoparticles were usually found only in the outermost layers of the cornified layer of the epidermis.

In occupational exposure conditions, dermal absorption can be a significant route of exposure for workers coming into contact with nanomaterials in the form of suspensions, solutions, water-oil emulsions, and gels (colloids). Absorption through the skin may be facilitated by particular working conditions, such as humidity and pressure. Potential adverse effects may involve both local exposure symptoms, present in the skin, as well as systemic effects, taking place after dermal absorption or accidental ingestion after transferring the ENMs by hand-to-mouth contact.

The International Organization for Standardization (ISO) has developed a technical specification, ISO/TS 21623:2017 (Workplace exposure – Assessment of dermal exposure to nanoobjects and their aggregates and agglomerates (NOAA)), where a systematic approach to the assessment of potential occupational risk linked to the exposure to nanoobjects, their aggregates and agglomerates (NOAA) is described, resulting from the production and use of nanomaterials and nanotechnology. In the overall assessment, it has been stated that data concerning dermal exposure to ENMs is controversial. Despite a number of conflicting results, it seems that the dermal absorption of nanoparticles is possible, but occurs on a very small scale. Numerous in vivo studies have not shown nanoparticle absorption through undamaged skin. This hypothesis might yet be revised as analytical methods of dermal exposure assessment are further developed [Wang et al. 2018].

2.4.2.3 Oral Exposure

Oral exposure to nanomaterials in the work environment can take place in the case of direct hand-to-mouth contact or through ingestion of aerosol grains in the form of air contaminants or coughed up as the respiratory tract is naturally cleaned as described in ISO/TS 21623:2017.

The importance of gastrointestinal ENM absorption is supported by studies on animals exposed to nanomaterials through inhalation. After exposing rats to nano-CeO_2, the largest amount of ENMs was found in the feces (71–90%), followed by the lungs (18%), and urine and other internal organs (4–6%) [Li et al. 2016].

The rate of nanoparticle absorption in the gastrointestinal tract depends on the size of the particles and their electrical charge. The particles with a diameter of 14 nm cross the mucous layer covering the intestinal epithelium in 2 minutes, particles with a diameter of 415 nm particles, in 30 min, and particles with a diameter of 1,000 nm do not cross the mucus barrier. Positively charged particles are captured by the negatively charged mucus layer; however, negatively charged particles diffuse and reach the enterocytes and are transported to the internal organs [Hoet et al. 2004].

The efficiency of the ENMs absorption process is relatively low. Pietroiusti et al. [2017] show examples of studies on rodents, in which the absorption of amorphous silica (7 nm or 10–25 nm) after a 24- or 84-day exposure was only 0.25%. In turn, in another 10-day study, 500 nm TiO_2 particles, administered by gavage, were collected in amounts ranging from 0.11% in the stomach to 4% in the large intestine.

Data regarding the nanoparticle distribution in the organism after oral administration to test animals indicate that the process involves a number of organs and is dependent on both the size of the particles and the dosage. Similarly, in the case of silver nanoparticles (8, 20, 80, 110 nm), their presence was found in all the examined organs: liver, spleen, kidneys, heart, brain, lungs, testicles, and thymus. The highest contents of nano-Ag were noted in the liver, spleen, and kidneys [SCENHIR 2014]. In the case of titanium dioxide (25, 80 nm), it was observed that the larger particles were mainly retained in the liver, and the smaller particles in the liver, spleen, kidneys, and lungs [NIOSH 2011].

2.4.3 Potential Toxic Effects of Nanomaterials in Humans

Potential health effects stem from the unique properties of the ENMs. Due to the diversity of physicochemical properties of the ENMs, the effects they may cause in the human body are difficult to foresee. Based on the existing data, mainly from in vivo studies, it can be concluded that adverse health effects linked to exposure to ENMs can be located in their entry points – the respiratory tract as well as, in particular, target organs.

2.4.3.1 Effects on the Respiratory System

The largest body of information on the adverse effects of the ENMs on the respiratory system was provided by in vivo studies of titanium dioxide, metal oxides, and ENMs containing carbon (carbon nanotubes, carbon black, graphene oxides, and fullerenes). The effects included, among others, inflammation, granuloma formation, and lung fibrosis [NIOSH 2011; 2013a].

Pulmonary tissue inflammation has also been observed after the exposure of test animals to fibrous ENMs, with high aspect ratio nanomaterials (HARNs), as well as biopersistent ENMs (PSP and GBP). Such nanomaterials tend to accumulate in the lungs for extended periods of time, may impair macrophage function, and lead to the disruption in the functioning of the respiratory system and pulmonary fibrosis (fibrogenesis). The disruption of macrophage functioning, apart from slowing down the process of particle elimination from the lungs, may lead to the release of inflammatory reaction mediators, allergic reactions, and activation of the immune system [Geiser 2010; Poh et al. 2018]. Fibrogenesis might contribute to carcinogenic changes (mesothelioma). The process was particularly noted for rigid fibers, fulfilling the criteria set for fibers by the WHO [2017].

As described in Section 2.4.1.1, one of the mechanisms of lung inflammation in laboratory animals (mainly rats) is the hypothesis of "lung overload" with poorly soluble particles. Lately, the "lung particle overload" theory as a mode of ENMs action in the human body is questioned, mainly due to significant interspecies differences in the kinetics of inhaled particles in rats compared to other rodent species (mice, hamsters), and primates (monkeys) and humans [Bevan et al. 2018; Warheit et al. 2016].

There are currently no sufficient data to support the existence of harmful effects of the ENMs to the human respiratory system. One of the theories indicating mechanisms of ENM toxicity assumes its basis to be the similarity of nanomaterials to the biological effects of ultrafine fractions of environmental dust (with diameters equivalent to nanoparticle sizes, 1–100 nm). Epidemiological data (based on e.g. Diesel exhaust or smog) show that this fraction is responsible for tissue inflammation, leading to respiratory and circulatory system disorders [Stone et al. 2017]. Studies carried out on workers exposed to carbon black seem to confirm similar mechanisms of the action of the ENMs [Zhang et al. 2014] and nano titanium dioxide [Zhao et al. 2018], which have shown that continuing inflammatory responses and related oxidative stress caused changes in lung functioning. In workers exposed to nano-TiO_2, *serum surfactant protein D* – a biomarker of chronic obstructive pulmonary disease (COPD) – has also been found [Zhao et al. 2018]. In turn, in workers occupationally exposed to carbon nanotubes and nanofibers, including MWCNT, fibrosis biomarkers have been identified in the blood and saliva [Beard et al., quoted after Schulte et al. 2019].

The health effects scheme of ENMs, based on the mechanisms for ultrafine dust, is shown in Figure 2.4.

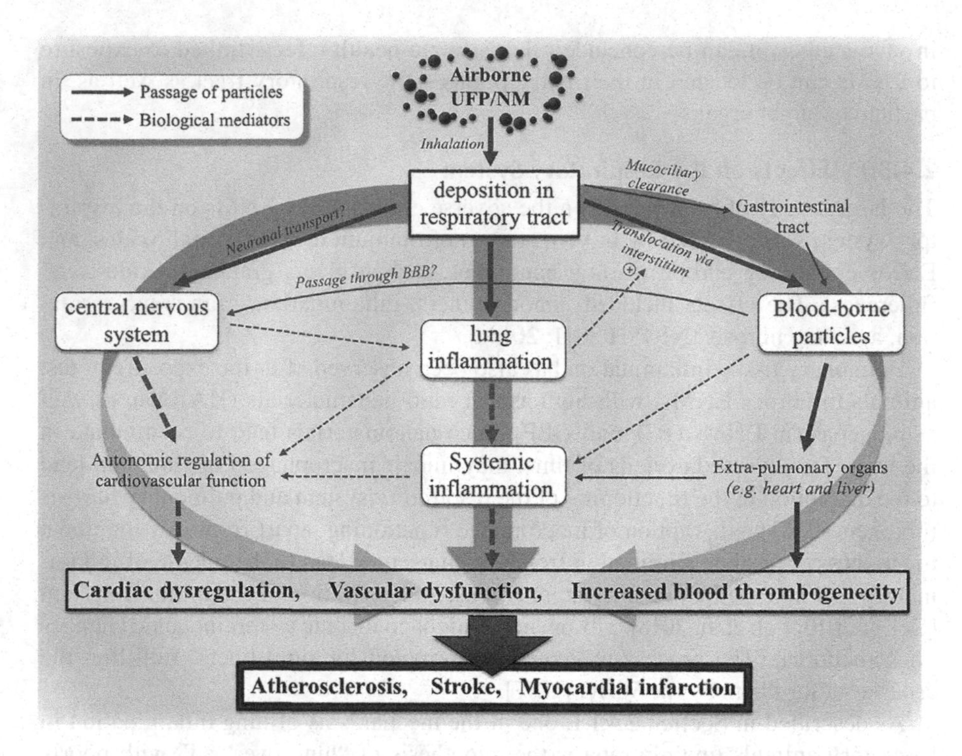

FIGURE 2.4 Nanoparticle health effects after inhalation exposure (Reprinted with permission from Stone et al. 2017. Nanomaterials versus ambient ultrafine particles: An opportunity to exchange toxicology knowledge. *Environ Health Persp* 125(10):106002).

2.4.3.2 Effects on the Cardiovascular System

Epidemiological studies confirm that environmental exposure to ultrafine dust leads to cardiovascular diseases [Pope et al. 2004]. By analogy, it is suspected that ENMs can have similar effects. In workers exposed to titanium dioxide nanoparticles or carbon nanotubes (MWCNTs), a heightened level of cardiovascular disease biomarkers was detected, intercellular adhesion molecule 1 (ICAM-1) and vascular cell adhesion molecule 1 (VCAM-1) [Zhao et al. 2018; Kuijpers et al. 2018], which confirms the aforementioned hypothesis.

Animal studies provide more data regarding the ENMs' effects on the cardiovascular system. In animals exposed to metal nanoparticles, carbon nanotubes, and fullerenes, nanoparticle interactions with red and white blood cells, platelets, or endothelial cells, as well as particle accumulation in the heart are shown [Nemmar et al. 2004]. Rossi et al. [2019] have shown that in studies on rats the direct effects of titanium dioxide nanoparticles on heart cells include oxidative stress, the release of proinflammatory cytokines, and pro-fibrotic gene expression, leading to irreversible structural changes in the organ. These changes led to arrhythmia and heart failure.

2.4.3.3 Specific Target Organ Toxicity after Repeated Exposure

The ENMs' capability of translocating or moving from the place of deposition to other organs with blood and lymph can result in changes in the organs and systems.

ENM translocation has been observed in animals after both inhalation and oral exposure. After oral administration of Ag, ZnO, TiO_2, SiO_2, and CNT nanoparticles, inflammatory changes in the small intestine and damage to the liver, spleen, and kidneys were observed [Pietroiusti et al. 2017]. After intratracheal and intravenous administration of nano-Au (1.4 nm, 18 nm), low Au levels (<0.2% of the administered dose in mass) have been found in the uterus. After inhalation exposure to nanoparticles of metals or metal oxides (e.g. Mn, Mo, Ag, Fe, Zn, Cu, Au, Al., Ti, and Ce), it has been shown that they can accumulate not only in the airways, but also migrate to the liver, spleen, heart, kidneys, brain, lymphatic glands, reproductive cells, and undergo bioaccumulation [Geraets et al. 2012; Pietroiusti et al. 2017]. In most cases, ENMs' penetration into the systemic circulation was restricted to several percentages of the total dose for all routes of exposure [Kreyling et al. 2014; Gulson et al. 2012].

There is currently no precise literature data on the ENMs' biokinetics in humans. Even though the extrapolation of animal studies data suggests that there is a low probability of acute toxic effects presenting themselves, it is possible that in persons subjected to chronic exposure (as in the work environment), there is a higher risk of disease linked to the accumulation of low doses of biopersistent ENMs [WHO 2017].

2.4.3.4 Neurotoxic Effects

Nanoparticles present in the bloodstream can cross the blood-brain barrier (BBB) and be deposited in the brain, leading to neurodegenerative and cognitive changes [Borisova 2018; Węsierska et al. 2018]. Neurological changes, including pathologic changes, in the brain have been observed, independent of the route of administration, in rats exposed to gold, copper, silver, and aluminum nanoparticles (50–60 nm), as well as in mice exposed to iron nanooxide. The nanoparticles' capability of crossing the blood-brain barrier can cause key problems in environmental neurotoxicology, but this property of the nanoparticles is considered promising in nanoneurotechnology, as ENMs can be used in drug delivery, imaging, and treatment [Borisova 2018].

In inhalation exposure, there is a risk of the nanoparticles migrating through the neuronal route – via the olfactory nerve endings in the nose, through the nervous system, and finally to the brain. This route of ENM translocation does not involve the BBB. Size-dependent particle migration through the nervous system to the brain has been observed in cases of animal inhalation exposure to TiO_2, MnO, Ag, CdTe, ZnO, and Co [Wang et al. 2017; Borisova 2018; SCENHIR 2014 and literature there in]. Size-dependent iron, silver, and gold nanoparticle bioaccumulation has been observed in rat brains after inhalation exposure to nanoparticles of these metals [Borisova 2018]. Some researchers believe that this route of ENM uptake is smaller in humans than in animal models, because the human olfactory epithelium constitutes 5–10% of total nasal mucous membranes in comparison with 50–55% in rats [Fröhlich and Salar-Behzadi 2014; Tian et al. 2019]. On the other hand, the neuronal absorption route of approximately 30 nm particles is clinically confirmed based on data concerning the penetration of viruses into the brain (e.g. lymphocytic choriomeningitis, polio, or herpes viruses) [Tian et al. 2019].

2.4.3.5 Developmental Toxicity

Experimental studies indicate that ENMs' developmental toxicity cannot be excluded. Even though data on ENMs' developmental toxicity is still limited and insufficient for risk assessment of pregnant women and the offspring, the probability of such effects is high, especially taking into account the fact that epidemiological studies have shown a strong correlation between the exposure to ultrafine atmospheric air particles and adverse effects on fetal development [Shah and Balkhair 2011].

ENMs administered by the inhalation, oral, and intravenous routes crossed the placental barrier in rodents and penetrated the fetus [Stapleton et al. 2018]. ENMs' developmental toxicity was observed after directly affecting the tissues of the fetus or by causing inflammatory reactions in mothers and disrupting their hormonal balance [Hougaard et al. 2010; Sun et al. 2013].

CNT or TiO_2 inhalation exposure of pregnant rodents caused fetal lesions and oxidative damage to the placenta [Qi et al. 2014; Pietroiusti et al. 2018]. In turn, MWCNT exposure caused inflammation in mothers, correlated with an increase in the incidence of birth defects [Fujitani et al. 2012]. Inhalation exposure of pregnant mice to nanosilica has also caused frequent pregnancy complications [Shirasuna et al. 2015].

Neurobehavioral disorders and damage to the nervous system (open field test) have been shown in the offspring of mothers after inhalation exposure to CNTs or TiO_2 [Hougaard et al. 2010; Jackson et al. 2013]. Similarly, the inhalation of carbon black (Printex 90) by pregnant mice induced neurodevelopmental changes in the offspring [Umezawa et al. 2018].

2.4.3.6 Effects on the Endocrine System

The hormonal (endocrine) system controls all the functions of the human (or other vertebrate) body on every stage of development. Disruptions in its functioning lead to a number of diseases, including cancers (hormone-related tumors of the nipple, prostate gland, testicles, and ovaries), and to reproductive disorders (fertility disorders, genital development disorders, fetal damage), as well as metabolic disorders (diabetes, obesity) [Priyjam et al. 2018].

Studies aiming to assess endocrine activity of chemicals, including nanomaterials, most often focus on potential interactions of xenobiotics with hormone receptors, with the most often assessed being estrogen, androgen, and thyroid receptors. Quantum dots, cadmium sulfides, fullerenes, and molybdenum trioxide nanoparticles are suspected of causing estrogenic effects and influencing thyroid function [Priyjam et al. 2018; Assadi et al. 2017]. Nanoparticles of ZnO, CeO_2, and TiO_2 caused metabolic disorders in the form of insulin secretion disorders [Iavicoli et al. 2013; Chen et al. 2018]. In turn, inflammatory reactions and cytokine secretion caused by a number of the ENMs (gold, CNT, TiO_2) can contribute to insulin resistance and the development of type 2 diabetes [Priyjam et al. 2018].

Another important mechanism of xenobiotic interference with the endocrine system is the effect on the functioning of the key enzymes involved in the synthesis and metabolism of endogenic steroids. Steroid hormones are one of the key factors in the regulation of vertebrate reproduction, as well as in a number of processes connected to growth and development [Larson et al. 2014]. ENMs' reprotoxicity seems to be supported by the results of in vivo and in vitro studies, in which morphological changes in the testicles and ovaries, damage to DNA in sperm cells, decrease in sperm mobility, and spermatogenesis and oogenesis disorders, as well as changes in the secretion of testosterone and 17β-estradiol by, among others, silver, copper, gold, titanium dioxide, zinc oxide, iron oxide, and palladium nanoparticles, carbon nanotubes, and fullerenes were observed [Iavicoli et al. 2013; Yang et al. 2017; Leso et al. 2018]. Moreover, quantum dot exposure can affect in vivo testicular function in mammals [Jain et al. 2012].

Hypothetic mechanisms of endocrine-disrupting effects were tied to oxidative stress, direct damage to DNA in sperm cells, modifications to the expression of genes coding proteins and enzymes responsible for the biosynthesis, metabolism and release of sex hormones, and the ability of the endocrine disruptors to bind to hormonal receptors [Iavicoli et al. 2013].

2.4.3.7 Effects on the Immune System

A review by Pietroiusti et al. [2018] indicates that interactions between nanoparticles and the immune system can involve both innate and acquired immune response. A number of in vitro and in vivo studies confirm that ENMs' exposure can lead to immune response, characterized by inflammasome activation and the induction of proinflammatory cytokines [Pietroiusti et al. 2018]. Induction of proinflammatory cytokines has also been observed in workers occupationally exposed to carbon black, TiO_2, and CNTs [Zhang et al. 2014; Zhao et al. 2018].

ENMs can induce immunotoxic effects, causing the death of immune cells or by influencing immune-specific signaling pathways [Dusinska et al. 2017]. ENMs' potential immunotoxic role can also involve immunosuppressive effects. Immune cells overloaded with nanoparticles can have disrupted defense mechanisms (they also undergo accelerated apoptosis and autophagy), which cause the host (human) to be more susceptible to bacterial infections [Pietroiusti et al. 2018].

The fact that nanomaterials are distributed to the lymph nodes indicates a possibility of modifying the immune response to bacterial and viral antigens, as well as other foreign proteins [Poh et al. 2018]. Moreover, ENMs can be the carriers of bacterial endotoxins, containing a highly immunizing lipopolysaccharide (LPS) [Li and Boraschi 2016].

2.4.3.8 Influence on the Human Microbiome

In recent years, a lot of attention has been focused on the issue of ENMs' influence on the human microbiome. The disruption of proper intestinal flora, known as dysbiosis, is linked to diseases such as colitis, inflammatory bowel disease, diabetes, and metabolic syndrome. The airways are also populated by microorganisms, and disturbances of their homeostasis lead to chronic diseases, e.g. chronic obstructive pulmonary disease, asthma, mucoviscidosis.

A number of ENMs used commercially and in consumer products are toxic to bacteria (Ag, TiO_2, ZnO, graphene, CeO_2, CNTs) [Poh et al. 2018]. Therefore, it is possible for the ENMs to disturb the microbiological balance after penetrating into the body – both in the intestines and lungs, leaving it more vulnerable to a number of diseases. This hypothesis is confirmed by the dose-dependent influence of nanosilver on the intestinal microbiome [van den Brule et al. 2016; Javurek et al. 2017]. It is necessary to determine whether the ENMs affect the human microbiome in order to assess both their occupational and environmental safety.

2.4.3.9 Genotoxic Effects

Genotoxic effects have also been observed in both in vitro and in vivo studies after exposure to ENMs, such as Ag, TiO_2, carbon black, CeO_2, CNTs, and other fibrous materials. The genotoxic effect mechanism involved the direct and indirect impact of the ENMs on DNA [ECETOC 2013; Pietroiusti et al. 2018; Dusinska et al. 2017]. Direct (primary) genotoxicity refers to DNA damage caused by direct physical interaction between the particles and DNA, and also to DNA damage in which reactive oxygen species (ROS) take part, in the absence of inflammation. Indirect (secondary) genotoxicity is induced by inflammation and refers to the effects of DNA damage induced by ROS and reactive nitrogen species (RNS), as well as other secondary inflammation mediators (cytokines, chemokines) released in response to ENMs' exposure.

2.4.3.10 Carcinogenic Effects

In the case of rigid fibrous ENMs, there is a risk that they might show properties similar to asbestos fibers (mesothelioma). However, a review prepared by the experts from the International Agency for Research of Cancer (IARC) was the basis for the position that there is only sufficient evidence only for a particular type of multi-walled carbon nanotubes, Mitsui 7 (MCWNT-7), which have been classified in the 2B group (possibly carcinogenic to humans). In the case of other MCWNT types (with similar sizes), it has been established that the evidence is limited; for the SWCNTs, however, no evidence of carcinogenic effects has been found and they were classified into Group 3 (not classifiable as to their carcinogenicity to humans) [IARC 2018]. However, the National Institute for Occupational Safety and Health (NIOSH) has found the effects of carbon nanotubes and nanofibers to be toxic regardless of their type (SWCNT, MWCNT) and degree of contamination [NIOSH 2013a]. Some of the most recent studies indicate that the risk of inducing cancerous processes is linked exclusively to long and rigid nanotubes and is not connected to tangled nanotubes [Catalán et al. 2016]. IARC has also ruled that there is sufficient scientific data to classify nano-TiO_2 and carbon black as substances possibly carcinogenic to humans (Group 2B) [IARC 2010]. Similarly, the Committee for Risk Assessment of the European Chemicals Agency has recommended, as of late, for TiO_2 to be classified as suspected of causing cancer when inhaled [ECHA 2017b], regardless of the particle size. TiO_2-induced cancers are a result of chronic lung inflammation caused by particle retention in the lungs of exposed animals (mainly rats), leading to lung particle overload. Alveolar macrophages, in response to lung particle overload, generate reactive oxygen species, which may act as carcinogens when interacting with the genetic material. According to the NIOSH experts, titanium dioxide is not a direct carcinogen, but has a secondary genotoxic effect linked to the size of the particles and their surface area [NIOSH 2011].

Potential carcinogenicity hazard is related to oxide nanomaterials and metals. A number of metals with a particle size of more than 100 nm (bulk form) show carcinogenic and mutagenic effects (nickel, cadmium, arsenic). Therefore, taking into account the high chemical reactivity of metal nanoparticles at the molecular level, it can be expected for them to show at least the same toxicity as the larger particles.

2.4.3.11 Epidemiological Studies

Epidemiological studies concerning the influence of the ENMs on human health are difficult to conduct, in part due to the fact that there is no single nanotechnology industry. Nanotechnology is usually used in many industrial sectors and the number of exposed workers in a single company can be rather small. As a result, it poses a technical difficulty to identify and assess the exposure of cohorts to particles with sizes appropriate for epidemiological studies.

According to Warheit et al. [2016], inflammation due to the retention of particles in the lungs, leading to cancer processes in animals, has no confirmation in humans. The analysis of epidemiological data of workers exposed to poorly soluble particles, such as titanium dioxide and carbon black, did not show a correlation between exposition to particles and lung cancer or other non-malignant respiratory system diseases [Warheit et al. 2016].

Studies carried out on 227 persons employed in 14 Taiwan facilities using nanomaterials (carbon nanotubes, titanium dioxide, silica dioxide, nanosilver, and other nanomaterials including nanoresins, nanogold, nanoclay, nanoalumina, and metal oxides) have shown a higher level of antioxidative enzymes and cardiovascular markers (fibrinogen, ICAM) in comparison to the control group [Liou et al. 2012]. Survey studies have been conducted in workers, checking for exposure symptoms. The main symptoms were irritation of the respiratory tract (coughing, sneezing) and exacerbated dermatitis, with cardiovascular symptoms (arrhythmia, angina pectoris) being present as well [Liao et al. 2014].

Schulte et al. [2019] reviewed the results of 27 epidemiological studies in workers exposed to ENMs most highly used in terms of tonnage: carbon black, synthetic amorphous silica, aluminum oxide, barium titanate, titanium dioxide, cerium dioxide, zinc oxide, carbon nanotubes/nanofibers, and silver. Even though, in most studies, no pathological changes have been found in workers, a number of biomarkers indicating the possibility of harmful effects in the respiratory and cardiovascular systems have been found. Inflammation and fibrosis markers were dominant.

2.5 STRATEGIES FOR THE GROUPING OF NANOMATERIALS APPROPRIATE TO TOXIC EFFECTS

Placing nanotechnology products on the market requires the identification of their particular hazards for health and the environment. Toxicological studies, which are the basis of risk assessment, take a long time, are expensive, and contradict the social expectations in regard to the protection of experimental animals.

Taking into account the sheer diversity of nanomaterial forms, the costs of their assessment and registration can be so high as to prevent innovations in this area. For many years, alternative methods have been in development for the registration (classification and labeling) of chemicals. The basis of the alternative methods is in vitro studies on cells isolated from different organs (mostly mammalian). Another way is to implement in silico methods – a wide range of computational modeling techniques, allowing for conclusions to be drawn about different toxicity aspects based on the developed models, e.g. pharmacokinetics modeling (Physiologically Based Pharmacokinetic, PBPK) or Quantitative Structure-Activity Relationships (Q)SAR, allowing for the correlation between the structure, physicochemical properties, and toxic effects of the active substances to be established.

Drawing conclusions about the toxicity of a particular substance based on alternative methods is a multi-step process, in which different methodologies can be used, such as those based on the weight of evidence, which is the basis for Integrated Testing Strategy (ITS), the read-across approach based on analogy, or the grouping of substances [WHO 2017; EC 2018b].

ITS has the character of sequential analysis and is conducted in stages, from the evaluation of the completeness of existing toxicity data, through the proposals of using particular research methods (particularly the alternative methods), and, in the final stage – deciding as to the substance toxicity or the necessity of performing additional animal studies. The "weight of evidence" analysis is used to evaluate information based on the existing data. Such an approach allows to hold the study at any of the stages, should the obtained data be deemed sufficient for risk assessment and substance classification. ITS are developed in regard to each of the toxic effects of the assessed substance [Stone et al. 2014; Zapór 2016].

The grouping of substances and the read-across approach are based on the experimentally proven assumption that structurally similar substances have similar physicochemical properties, so they will have similar biological effects. Therefore, they can be treated as a "group" or a "category" of substances. The application of the "group" concept requires the physicochemical effects on human health and the environment as well as the environmental fate can be projected based on data concerning a reference substance from a given group through the interpolation of other substances, which allows for the anticipation of health risks with no further studies [OECD 2014; EC 2018b]. The grouping of ENMs basing on toxic properties should facilitate toxicological risk assessment.

Even though the grouping strategy is widely used and effective in the evaluation of bulk substances, solutions for nanomaterials are still being developed [ECHA 2014; OECD 2014].

In recent years, a number of research projects were realized to develop the bases of material grouping strategies, such as ITS-NANO (Intelligent Testing Strategy for Engineered Nanomaterials), NANoREG (A Common European Approach to the Regulatory Testing of Manufactured Nanomaterial), MARINA (Managing Risk of Nanomaterials), NANOMILE (Engineered nanomaterial mechanisms of interactions with living systems and the environment: a universal framework for safe nanotechnology), GUIDEnano (Assessment and mitigation of nano-enabled product risks on human and environmental health), and DF4nanoGrouping (Decision-making framework for the grouping and testing of nanomaterials). A comparison of these strategies has been shown in several review articles in recent years [Landvik et al. 2018; Oomen et al. 2018; Lamon et al. 2019; Zapór 2016].

In the ENMs' grouping concepts, developed within the framework of those and a number of different projects, several aspects related to the specific structure and properties of the ENMs are taken into account. These concepts are developed mainly in terms of ENMs' exposure risk assessment in humans, but can also be used for ecotoxicity assessment [Oomen et al. 2015; Landvik et al. 2018; Hund-Rinke et al. 2018].

2.5.1 Criteria for Grouping Nanomaterials for Hazard and Risk Assessment

Most often, in different ENMs grouping concepts, three parameters are taken into account: physicochemical properties of the ENMs, ENMs' behavior in the environment, and their Mode of Action (MoA) [Braakhuis et al. 2016; Gebel et al. 2014].

The assumption that the physicochemical properties determine the behavior of nanomaterials over their entire life cycle is the basis for the concept of the grouping of nanomaterials, developed within the project framework of ITS-NANO and MARINA, which has been adopted for the REACH registration purposes [ECHA 2016]. According to these concepts, nanomaterials can be classified into three basic groups, depending on their physical and chemical properties and their influence on exposure, toxicokinetics, toxicity, ecotoxicity, and environmental effects ("environmental fate"). These parameters have been defined as "nanomaterial functionality parameters": *What they are, What they do, Where they go*) [ECHA 2017a; Oomen et al. 2014; 2015; Sellers et al. 2015; Stone et al. 2014].

They include:

- Characterization ("what they are"): both physical and chemical identification in terms of composition, impurities, size and size distribution, shape and aspect ratio (including rigidity), surface characteristics (coating, chemistry, functionalization, surface charge), surface area, porosity, crystallinity, etc.
- Behavior (Fate) ("where they go"): biological (toxicokinetics, biodistribution) and environmental fate described by solubility (water solubility and rate of dissolution in relevant media), hydrophobicity, Zeta potential, dispersibility, dustiness, etc.
- (Re)activity ("what they do"): their reactivity, physical hazards (explosivity, flammability, potential for accidental ignition), biological reactivity (free radical production, catalytic activity, ion release), toxicodynamics, photoreactivity, etc.

It was recommended in the MARINA project that the grouping of nanomaterials should be supported by information pertaining to their kinetics (absorption, distribution, and biopersistence), exposure (exposure scenarios), and hazards (risk characterization, including early toxic effects in higher organisms). This information constitutes the so-called four pillars of strategy (MARINA Risk Assessment Strategy).

The first step in the proposed strategy is to group the nanomaterials with similar physicochemical properties and create subgroups with similar toxicokinetics and hazards, including potential exposure in the whole life cycle of the nanomaterial. The next step is to define the specific toxic effects (risk characterization) and supplement the missing data (read-across modeling is particularly important at this stage). This strategy is especially helpful for registrants assessing the risk of several nanoforms of a single nanomaterial [Oomen et al. 2014; 2015; Landvik et al. 2018].

In the European Centre for the Ecotoxicology and Toxicology of Chemicals (ECETOC), basing on research conducted within the DF4nanoGrouping project, a decision tree was developed for the grouping of nanomaterials in respect to potential risks. In this approach, both the properties of the material itself (*intrinsic properties*) and those dependent on the external environment (*extrinsic properties*), responsible for biophysical interactions of the particles with the environment. The environment is to be understood as the natural environment, body environment, as well as in vitro culture media and matrices in which nanomaterials are placed. The main criteria for grouping nanomaterials are the toxic effect mechanisms (*MoA – Mode-of-Action*), both on the cellular and systemic levels. Hence, this approach allows for the risk assessment of nanomaterials during their whole life cycle [Arts et al. 2015].

DF4nanoGrouping allows for a nanomaterial to be placed in one of four main groups:

- Group I – Soluble ENMs – water-soluble nanomaterials (solubility > 100 mg/l), non-biopersistent, for which chemical composition is more important for risk assessment than the nanostructure itself. Soluble nanomaterials lose their nano-specificity after dissolution in water or biological fluids and should be further considered as conventional substances. Further risk assessment in this group should be based on the read-across approach with respect to soluble substances with possible toxic effects, e.g. through ion release.
- Group II – Biopersistent fibrous ENMs – non-soluble, biopersistent, rigid with high aspect ratio, fulfilling the WHO criteria for inhalable fibers, which should be assessed depending on the potential toxicity of the fibers. Other fibers (HARNs) can constitute a subgroup depending on their solubility and biopersistence.
- Group III – Biopersistent non-fibrous ENMs (GPB), chemically passive – passive nanomaterials, biopersistent nanomaterials which do not fulfill criteria for fibers which:
  a) do not exhibit particular biological activity (low surface reactivity, resulting in a low capacity of free radical production < 10% Mn_2O_3 in FRAS {Free Radical Analytical System} or cytochrome c assay tests);
  b) are not potentially toxic (there are no active ingredients in the chemical composition); do not cause cellular effects at concentrations ≤10 μg/cm²; No Observed Adverse Effect Concentration (NOAEC) in short-term inhalation study (STIS) > 10 mg/m³;
  c) are not mobile (aggregate in environmental media), do not move from the site of contact (are not biodistributed), because they are captured by the mononuclear phagocyte system (MPS). Such materials are, e.g.,

respirable nanomaterials, granular, and biopersistent, and whose high concentrations cause toxic effects only in case of lung particle overload.
- Group III does not pose a significant risk to health.
- Group IV – Biopersistent non-fibrous ENMs, chemically active – active nanomaterials, which require further in-depth study. Materials included in this group are biopersistent, non-fibrous, in case of which there is a risk of toxic effects in low concentrations due to their particular structure (chemical composition, reactive surface, and the possibility of biophysical interactions).

Nanomaterial grouping is carried out in three stages. In the first tier, nanomaterials are assigned to main groups basing on their intrinsic physicochemical properties. The list of assessed parameters and methods for their testing is in compliance with OECD guidelines [OECD 2014]. In the second tier, extrinsic material properties are taken into consideration, including in vitro test results, e.g. solubility in biological fluids, reactivity (formation of free radicals), cytotoxicity, genotoxicity. The third-tier grouping includes the results of in vivo studies: assessment of specific toxic effects on a systemic level (organ burden, genotoxicity, reproductive toxicity, etc.), reversibility of adverse effects, biodistribution, biopersistence.

The DF4nanoGrouping is considered universal, because it connects a number of previously developed grouping concepts and allows for risk assessment of nanomaterials during their entire life cycle [Oomen et al. 2018].

Substance grouping and read-across approach are among the most often used alternative methods of acquiring data for the purposes of registrations submitted under the REACH regulation, allowing for the limitation of both expensive studies and quick identification of hazards and classification of various ENMs.

2.5.2 Grouping Strategies and the Work Environment

Grouping strategies find their applications not only in the classification of nanomaterials for registration purposes, but also for exposure and occupational risk assessment in the work environment. In the work environment safety assessment, apart from the physicochemical properties specific to nanomaterials, the route of exposure, toxicity mechanism, and biopersistence are of utmost importance [Sellers et al. 2015; Gebel et al. 2014; Braakhuis et al. 2016]. Biopersistence is defined as the capability of a nanomaterial to remain in the cells, tissues, and organs of the organism for extended periods of time. In the case of the inhalation route, nanomaterials with a lung half-life of more than 40 days are considered biopersistent [Arts et al. 2015].

One of the first suggested approaches to nanomaterial grouping was developed by the British Standard Institution (BSI) [BSI 2007]. Nanoparticles/nanomaterials were classified into four categories, taking into account physicochemical and toxic properties of their larger counterparts:

- fibrous nanomaterials, assuming that fiber is a particle with an elongation factor larger than 3:1 and a length of over 5,000 nm;

- nanomaterials whose larger (not nano) counterparts have carcinogenic or mutagenic effects, cause asthma, or are toxic for reproduction (Carcinogenic, Mutagenic, Asthmogenic, Reproductive – CMAR);
- non-soluble or poorly soluble non-fibrous nanoparticles not belonging to the CMAR category;
- soluble nanoparticles which cannot be classified into any of the categories listed above.

A similar approach was suggested by the National Institute for Occupational Safety and Health (NIOSH) [Kuempel et al. 2012]. According to NIOSH, nanomaterials can be classified into one of the following groups:

- particles with high solubility, the toxicity of which is connected to the release of ions;
- poorly soluble low toxicity particles;
- poorly soluble high toxicity particles;
- fibrous particles for which the toxicity is related to biopersistence and genotoxicity.

Bundesanstalt für Arbeitsschutz und Arbeitsmedizin, BAuA [Gebel et al. 2014], have classified nanomaterials into the following groups based on the risks they pose:

- soluble particles;
- granular biopersistent particles without specific toxicological properties, which can induce inflammatory responses in the lungs due to so-called lung overload effect;
- biopersistent nanomaterials with specific toxic effects (e.g. releasing toxic ions, with toxic function groups, showing catalytic activity);
- biopersistent fibrous materials.

The Institut für Arbeitsschutz der Deutschen Gesetzlichen Unfallversicherung (IFA) has proposed to divide nanomaterials into four categories:

- nanomaterials non-persistent in the environment, granular or water-soluble (solubility > 100 mg/l);
- nanomaterials persistent in the environment, granular (non-fibrous) with density > 6 000 kg/m^3;
- nanomaterials persistent in the environment, granular and nanofibrous with density < 6 000 kg/m^3, whose effects similar to asbestos can be excluded;
- rigid fibers, persistent in the environment, whose effects similar to asbestos, cannot be excluded.

The grouping strategy and the read-across approach have been included in the WHO recommendations [2017] and in the REACH regulation [EC 2018b], as tools helpful in ENMs' risk assessment. The technical guidelines as to grouping substances for classification/categorization and labeling are included in ECHA [2017a] and OECD [2019] publications.

According to the WHO experts from the Guideline Development Group (GDG) it is considered the best practice to classify MNMs into three groups: those with specific toxicity, those that are respirable fibers, and those that are granular biopersistent particles [WHO 2017].

GDG has carried out an assessment of the existing results of studies of chosen nanomaterials from the aforementioned groups toward their classification as per Globally Harmonized System of Classification and Labeling of Chemicals (GHS). The most common hazard classes ascribed to the ENMs are:

- specific target organ toxicity after repeated exposure (SWCNT, MWCNT, Ag, Au, SiO_2, TiO_2, CeO_2, ZnO);
- carcinogenicity (MWCNT, TiO_2);
- germ cell mutagenicity (SWCNT, MWCNT);
- serious eye damage (MWCNT);
- respiratory or skin sensitization (Ag).

2.6 OCCUPATIONAL EXPOSURE ASSESSMENT CRITERIA FOR NANOMATERIALS

Occupational exposure control is one of the most important measures to prevent health effects related to inhalation exposure to ENMs, and the basic measure to enable exposure control is the determination of an Occupational Exposure Limit (OEL) value. The OEL values can be determined based on health criteria, as the maximum concentration of a substance in the air at the work environment. This means that, based on existing knowledge of the effects of a substance on humans or animals, it is possible to determine a threshold concentration, below which exposure is considered safe and will probably have no adverse health effects, for example, Treshold Limit Values (TLV) set by industrial hygienists (ACGIH, EU countries). There are three types of exposure limits: time-weighted average (TWA) limit, short-term exposure limit (STEL), and peak or ceiling limit.

Other OELs also include consideration of the technological feasibility e.g. Recommended Exposure Limits (REL) used by NIOSH or economic feasibility e.g. Permissible Exposure Levels (PEL) used by OSHA in the United States. In the European Union under the REACH legislation a Derived No-Effect Level (DNEL) and, for certain groups of substances, a Derived Minimum Exposure Level (DMEL) are established.

As there is no single, globally accepted methodology for establishing the OEL values, differences in OELs in different countries may exist. Currently, no country has established legally binding rules based on OEL health criteria referring to nanomaterials. It is because of the great diversity of ENMs and the insufficient amount of toxicological study results. Moreover, the existing data does not provide unambiguous information as to the dose–response relationship, which is the basis for the establishment of OEL. There is also ongoing discussion on the methodology

TABLE 2.1
Nano Reference Value (NRV) for Chosen Categories of Nanomaterials [van Broekhuizen et al. 2012]

NM Category	NRV (TWA-8h)	Examples
Rigid fibers, persistent in the environment, in case of which effects similar to asbestos cannot be excluded.	0.01 fibers/cm^3	SWCNT or MWCNT, or metal oxide fibers
Persistent in the environment, granular with density > 6000 kg/m^3	20000 particles/cm^3	Ag, Au, CeO_2, CoO, Fe, FexOy, La, Pb, Sb_2O_5, SnO_2
Persistent in the environment, granular and nanofibers, in case of which effects similar to asbestos are excluded, with density < 6000 kg/m^3	40000 particles/cm^3	Al_2O_3, SiO_2, TiN, TiO_2, ZnO, $CaCO_3$, nanoclay, carbon black, C_{60} fullerene, dendrimers, polystyrene
Granular, non-persistent in the environment or water soluble (solubility > 100 mg /l).	OEL values as in the case of bulk substances, if established	Lipid particles in nanoemulsions, sucrose

of measurement and metrics, in which OEL values in relation to ENMs should be expressed – should they be based on mass, the number of particles, or ENMs surface area.

In the case of some nanomaterials, the manufacturers (e.g. Bayer, Nanocyl) or organizations, e.g. NIOSH or New Energy and Industrial Technology Development Organization in Japan (NEDO), suggest OEL values (Table 2.1). Works are also in progress on the application of the grouping and read-across approaches, which could be used when substance-specific data is missing [Mihalache et al. 2017; Schulte et al. 2018]. Groups or categories of ENMs, for which OEL values are proposed, are compatible with categories developed for hazard identification and risk assessment. In the case of the group approach, OELs concerning the ENMs groups are designated Occupational Exposure Band (OEB) [Mihalache et al. 2017; Schulte et al. 2018].

BSI has also suggested a pragmatic approach to the ENMs, i.e. using the so-called benchmark/dose level as the acceptable exposure level in exposure assessment [BSI 2007]. The reference substances for nano-OEL are bulk substances, for which OEL values have been established, noting that the benchmark dose is not based on health criteria. And so, the BSI has proposed the following Benchmark Exposure Levels (BEL):

- for fibrous nanomaterials (e.g. carbon nanotubes), fulfilling the WHO definition of a fiber – BEL = 0.01 fibers/cm3;
- for nanomaterials whose larger counterparts are Carcinogenic, Mutagenic, Asthmogenic, Reproductive (CMAR) – BEL = 0.1 × OEL (bulk);
- for non-soluble or poorly soluble and non-fibrous ENMs, not belonging to the CMAR group – 0.066 × OEL (bulk) or 20 000 particles/cm^3;
- for soluble ENMs, which cannot be placed into any other category – BEL = 0.5 × OEL (bulk).

In the case of nanomaterials, exposure assessment carried out based on substance mass concentration data does not fully reflect the level of exposure, because the small size and large specific surface area of nanomaterials cause dust with even low mass concentration and may contain a large number of nanoparticles with a large active surface. Therefore, reference values for nanomaterials Nano Reference Value (NRV) expressed in particle number concentration are being proposed. NRV values are not based on health criteria, but are limit values, exceeding of which should result in the implementation of appropriate exposure control measures. It is important to note that the suggested values can be subject to change as knowledge about nanomaterial toxicity advances. Exposure assessment based on measurements in the work environment in relation to the NRV values requires appropriate instruments, measurement methods, and measurement strategies to be used.

The European Agency for Safety and Health and Work (EU OSHA), followed by the Institut für Arbeitsschutz der Deutschen Gesetzlichen Unfallversicherung (IFA, Germany) and the National Institute for Public Health and the Environment in the Netherlands (RIVM), has accepted the division of nanomaterials into four categories along with the proposed NRV reference values, based on particle number concentration [van Broekhuizen et al. 2012].

Mihalache et al. [2017] reviewed the quality of exposure assessment and proposed OEL values (56 values in total), both for particular nanomaterials and particular categories (groups). They have shown large differences in the OEL values, especially for metals (100–300 times). The smallest differences in the OEL values have been noted for fibers [Mihalache et al. 2017].

In the Guidelines on Protecting Workers from Potential Risks of Manufactured Nanomaterials, WHO recommends the OEL limit values developed by the industry and research institutes to be used in worker exposure assessment, despite the differences in the values and methodology of their determination. It also stresses that the OEL suggested for the ENMs do not mean safe exposure levels, below which no adverse health effects arise [WHO 2017]. Exemplary OEL value differences for chosen ENMs have been shown in Table 2.2.

Derived no-effect level (DNEL) is the maximum level of human exposure to a substance produced or placed on the market in the European Union, under conditions defined in the REACH regulation. In estimating the DNEL values, among others, the following are taken into account: routes of exposure, time of exposure, and nature of effects (local and target organ effects, etc.). These values are shown in registration documents of chemical substance manufacturers as a part of chemical safety assessment. Examples of DNEL values for several ENMs are shown in Table 2.2

Abbreviations and acronyms used in Table 2.2: TWA – time-weighted average; OEL(PL) – occupational exposure limit (period limited – 15 years); REL – recommended exposure limits; NRV – nano reference values; DNEL – derived no-effect level; OEL – occupational exposure limit; NIOSH – National Institute for Occupational Safety and Health; NEDO – New Energy and Industrial Technology Development Organization in Japan; RIVM – National Institute for Public Health and the Environment in the Netherlands; IFA – Institute for Occupational Safety and Health of the German Social Accident Insurance, AGS – Committee on Hazardous Substances in Germany; EU – European Commission.

TABLE 2.2
Suggested Reference Values of Acceptable Concentration of Chosen Nanomaterials in the Work Environment [quoted after WHO 2017 and OECD 2018 without citing references]

Category of ENMs	OEL Name and Concentration	Study Reference
MWCNT (Baytubes)	8 h TWA = 0.05 mg/m^3	Bayer 2010 (Germany)
MWCNT (Nanocyl NC 7000)	8 h TWA = 0.0025 mg/m^3	Nanocyl 2009 (Belgium)
CNTs (SWCNT and MWCNT)	REL (8 h TWA) = 0.001 mg/m^3 (as a respirable elemental carbon)	NIOSH 2013 (US)
SWCNT and MWCNT	OEL(PL) = 0.03 mg/m^3	NEDO 2011 (Japan)
Fibres (non-entangled)	Acceptance level (default), respirable fraction = 0.01 fibers/cm^3	AGS 2013, (Germany)
MWCNT	DNEL = 0.02 mg/m^3 (acute inhalation, lung inflammation) DNEL = 0.004 mg/m^3 (acute inhalation, systemic immune effect)	EU 2010
MWCNT	DNEL = 0.034 mg/m^3 (chronic inhalation, lung inflammation); DNEL = 0.0007 mg/m^3 (chronic inhalation, systemic immune effect)	EU 2010
Fullerenes	DNEL = 0.044 mg/m^3 (acute inhalation)	EU 2010
Fullerenes	DNEL = 0.0003 mg/m^3 (chronic inhalation)	EU 2010
Fullerenes, C60	NRV = 40000 particles/cm^3	RIVM 2012 (Netherlands)
Fullerenes, C60	OEL(PL) = 0.39 mg/m^3	NEDO 2011 (Japan)
TiO_2 (21 nm)	DNEL = 0.017 mg/m^3 (chronic inhalation)	EU 2009
TiO_2 (< 100 nm)	REL (8 h TWA) = 0.3 mg/m^3	NIOSH 2011 (US)
TiO_2 (< 100 nm)	OEL(PL) = 0.61 mg/m^3	NEDO (Japan)
TiO_2 (< 100 nm)	OEL < 0.5 mg/m^3	AGS (Germany)
TiO_2 (< 100 nm)	NRV = 40000 particles/cm^3	RIVM 2012 (Netherlands)
Carbon black	8h TWA = 3.5 mg/m^3 (PEL)	HSE 2013 (UK); NIOSH 2007 (US)
Carbon black	NRV = 40000 particles/cm^3	RIVM 2012 (Netherlands)

Exposure to chemicals can also be assessed based on the measurement results of occupational exposure markers for a particular substance present in the blood or urine of the exposed worker. The results of current epidemiological studies do not provide a scientific basis enough to conduct biological monitoring for exposure to nanomaterials [Gulumian et al. 2016 and literature therein]. Some researchers indicate a possibility of detection of blood biomarkers, e.g. fibrotic processes. However, these data need to be validated.

Lung fibrosis markers (serum TGF-β1 and sputum KL-6), antioxidant enzymes, such as SOD and GPX, and cardiovascular markers, such as fibrinogen, ICAM, and interleukin-6, are possible biomarkers for medical surveillance of workers handling engineered nanomaterials [Liou et al. 2012; 2015].

NIOSH recommends health surveillance, mainly in the case of exposure to carbon nanotubes and nanofibers, in the form of screening tests, including spirometric tests and basic chest X-rays [NIOSH 2013a; 2013b].

2.7 NANOMATERIALS IN THE CONTEXT OF STANDARDS AND LEGISLATION

2.7.1 Standards and Legislation Regarding Nanomaterials and Nanotechnologies in the World

There is currently no legislation strictly regulating safe working conditions with nanomaterials. It is therefore a good practice to use standards and recommendations concerning the subject at hand. National and international (governmental and non-governmental) groups are engaged in drafting documents concerning nanotechnology and nanomaterials, as well as set standards and recommendations for workplace safety in relation to nanomaterials. For example, the Chinese Ministry of Industry and Information Technology (MIIT) has published guidelines for industries creating new materials, such as graphene.

Some of the international organizations, along with their Technical Committees (TC), which set (voluntary) standards and guidelines in using nanotechnology are:

- International Organization for Standardization, ISO/TC 229 – Nanotechnologies, ISO/TC 146/SC2 – Workplace atmospheres,
- American Society for Testing and Materials (ASTM), International Committee E56 – Nanotechnology,
- International Electrotechnical Commission (IEC), TC 113 – Nanotechnology standardization for electrical and electronic products and systems,
- Institute of Electrical and Electronics Engineers (IEEE), Nanotechnology Council,
- Organization for Economic Co-operation and Development (OECD), Working Party on Nanotechnology (WPN), and Working Party on Manufactured Nanomaterials (WPMN).

The OECD is an international organization specializing in the economy, bringing together 35 highly developed democratic countries, including the United States. Since 2005, it has been publishing a number of reports, manuals, and recommendations from the "Series on the Safety of Manufactured Nanomaterials". In 2010, OECD published e.g. recommendations concerning personal protection in a work environment involving the use of ENMs. In 2012, it published a document concerning risk assessment of manufactured nanomaterials – "Important Issues on Risk Assessment of Manufactured Nanomaterials". All documents from the series are available on the following website: www.oecd.org/envehs/nanosafety/publications-series-safety-manufactured-nanomaterials.htm (accessed January 10, 2020).

Nanotechnologies also belong to research priority areas, set out in the World Health Organization (WHO) agenda. An important WHO document, including general recommendations for safe handling of nanomaterials is "Guidelines on protecting workers from potential risks of manufactured nanomaterials" [WHO 2017].

2.7.2 Standards and Legislation Regarding Nanomaterials and Nanotechnologies in the United States

The most important regulatory authorities in the United States, linked to the use of nanomaterials in the workplace, are:

- The United States Environmental Protection Agency (EPA), which introduces national environmental protection standards and takes enforcement actions in this area (e.g. The Toxic Substances Control Act – TSCA, Resource Conservation and Recovery Act – RCRA itp.);
- The National Institute of Standards and Technology (NIST), which introduces standards and reference materials for nanoparticles;
- National Cancer Institute (NCI);
- National Nanotechnology Initiative (NNI), a US federal government program concerning scientific research, engineering, technology, and the development of projects in the nanoscale;
- American National Standards Institute (ANSI), which maintains a database of nanotechnology standards and accredits organizations developing standards.

The US Federal Government research concerning measurements in science and technology is carried out by NIST. Experts from this institution head the international ASTM Committee E56 on Nanotechnology. The United States Technical Advisory Group (TAG) represents the United States in the ISO committee 229 on Nanotechnology. TAG is also responsible for formulating positions and propositions of documents concerning nanotechnology to be sent to ISO. The United States also heads the Working Group ISO/TC 229 WG3: Health, Safety and Environmental Aspects of Nanotechnologies, with a representative of the National Institute for Occupational Safety and Health (NIOSH). NIOSH is a federal agency in the United States, responsible for conducting research and publishing recommendations concerning health and safety at the workplace.

Among federal and state regulations, guidelines, and health standards concerning nanomaterials, the following are of particular note:

- EPA "Technical FactSheet – Nanomaterials" 2017 – which includes general information on standards, nanomaterials, and safety in their manufacturing and use;
- EPA "Nanotechnology White Paper" 2007 – which includes the definition of nanotechnology, benefits of nanomaterials for the environment, risk assessment, recommendations, and responsible technology development [Morris and Willis 2007];

- according to the Toxic Substances Control Act (TSCA), a number of nanomaterials are considered as "chemical substances" subject to the Act. EPA has released a one-time report framework for nanomaterials which are chemicals present on the market;
- nanomaterials used as pesticides are subject to the Federal Insecticide, Fungicide, and Rodenticide Act (FIFRA). Should their use as a pesticide cause the nanomaterial to remain in food or animal feed, the maximal residue level according to the Federal Food, Drug and Cosmetic Act (FFDCA) needs to be established;
- other issues related to nanomaterials may be regulated under other programs, such as Clean Water Act (CWA) and Clean Air Act (CAA), Resource Conservation and Recovery Act (RCRA), and Comprehensive Environmental Response, Compensation, as well as the Liability Act (CERCLA).

Besides federal standards and regulations, state-wide and local operations are also carried out in the United States. An example is a formal letter sent in 2010 and 2011 by the California Department of Toxic Substances Control (CA DTSC), requesting manufacturers to provide information regarding the chemical and physical properties, including analytical methods used and other relevant information about the produced carbon nanotubes, nano metal oxides, nanometals, and quantum dots [DTSC 2013]. In 2006, the city of Berkeley, California, adopted the first local regulation developed specifically for nanomaterials, requiring all the companies producing or using nanomaterials to disclose current toxicological data, if available [Berkeley 2006].

2.7.3 Standards and Legislation Regarding Nanomaterials and Nanotechnologies in Europe

In the European Union, the most important regulatory authority in the field of standardization is the CEN, and the technical committees related to nanomaterials are: CEN/TC 352 – Nanotechnologies and CEN/TC 137 – Assessment of workplace exposure to chemical and biological agents.

In the European Union member states, legal regulations concerning human health and safety in regard to nanomaterials are included in the Regulation (EC) No 1907/2006 of the European Parliament and of the Council of December 18, 2006, concerning the Registration, Evaluation, Authorisation and Restriction of Chemicals (REACH) [EC 2006] and Commission Regulation (EU) 2018/1881 of December 3, 2018, amending Regulation (EC) No 1907/2006 as regards Annexes I, III, VI, VII, VIII, IX, X, XI, and XII to address nanoforms of substances [EC 2018b].

According to the REACH Regulation and the CLP Regulation (Classification, Labelling, and Packaging), [EC 2008], nanomaterials are subject to appropriate classification, labeling, and packaging. Just as every other chemical substance placed on the market (manufactured or imported in quantity exceeding 1 metric ton per year), nanoforms have to be registered in the European Chemicals Agency in Helsinki (ECHA). In 2010, a manual titled "Nanomaterials In IUCLID 5.2" was published on the ECHA website, in which practical instructions on how to include nanomaterials in the registration dossier http://iuclid.echa.europa.eu/download/documents/userman ual/IUCLID_User_Manual_Nanomaterials_v1.0.pdf (accessed January 10, 2020).

Commission Regulation (EU) 2018/1881 also introduces the obligation to carry out a risk assessment of substance nanoforms. In the case of nanomaterials, REACH requires the distributors to inform the recipients of the substance nanoform properties within the supply chain, enable the workers to obtain the information, when the substances are used or could pose a risk during the work process. The 2018/1881 Regulation shall apply from January 1, 2020.

The CLP Regulation stipulates that all substances in the nanoform must be labeled according to the CLP criteria in order to convey information concerning hazards linked to particular nanomaterials.

2.8 OCCUPATIONAL EXPOSURE TO NANOMATERIALS

People most exposed to contact and adverse effects of the ENMs are the workers employed in the manufacture, packaging, and transportation of nanomaterials; also affected are academicians and those employed in research institutes, as well as users and consumers of nanotechnology products.

As the respiratory system is the main route of exposure to nanoparticles for humans [SCENIHR 2007; 2009], in the case of risk assessment for workers working with particular nanomaterials, it is important to start with exposure assessment to airborne NOAA [Singh and Nalwa 2007; Dolez and Debia 2015]. Low mass and small dimensions of nanoparticles cause them to behave similarly to gases in the air. The smaller and the less agglomerated the nanoparticles are, the easier they are transported and float up with warm air. They can connect to larger objects, including microscale suspended particulate matter (e.g. inhalable and respirable fractions). Their movement in the air is determined by gravitational forces to a lesser extent than in the case of microparticles, and is much more strongly influenced by diffusion mechanisms of nanoparticle movement between air particles and their chaotic collisions. In non-ventilated areas, dispersed nanoparticles can remain in the air even after 10 days [Grassian 2008]. Therefore, the air concentration of nanoparticles can still be high even long after a particular process and is strongly dependent on the functioning of collective protective measures.

The method for occupational exposure assessment of airborne particles is currently strongly based – in comparison to the micrometric fractions – on the analysis of their chemical composition and measurement of mass concentration in the personal breathing zone (PBZ), defined as a zone within the radius of 0.3 m from the nose and mouth of the worker. Particle fractions are separated by their effects on human health (inhalable, thoracic, and respirable) and refer to the probability of the airborne particles crossing into separate anatomical regions of the respiratory system. Precise information has been included in the following standards: ISO 7708:1995 (Air quality – Particle size fraction definitions for health-related sampling) and ISO 13138:2012 (Air Quality – Sampling conventions for airborne particle deposition in the human respiratory system). More information about the choice of measurement strategies for the measurement of airborne dust, the selection of samplers, transportation and storage of samples can be found in, e.g., manuals of the American Conference of Governmental Industrial Hygienists [ACGIH 2001; Vincent 1999] and the British Health and Safety Executive [HSE 2014].

In the occupational exposure assessment of micrometric dust, their contents in the work environment are referred to established reference levels or other parameters, set in a particular country in respect to the air concentration limits of a hazardous substance, given in mass and standardized concentrations as per national regulations [Deveau et al. 2015]. In the case of most nanomaterials there are, however, no legally established reference levels or concentration limit values. Few propositions of reference values for permissible concentration levels of particular nanomaterials have been published in an article [Mihalache et al. 2017] and in the guidelines by WHO [2017]. They have also been shown in Section 2.6., Table 2.2.

The main reason for the lack of reference mass concentration levels is the almost negligible share of the nano fraction in the total mass of objects present in an aerosol. The mass of one 10 μm particle corresponds to the mass of a billion 10 nm particles. Therefore, the total mass of nanoparticles in an aerosol may be a small percentage of the total mass of particulates, but the number of nanoparticles may be over 95% of all the suspended particulates. Therefore, unlike the mass concentration, the number concentration of nanoparticles is often very high. Hence, when it comes to exposure assessment, air concentration of nanoobjects should be referred to their total area or number, and not to their mass. If the average nanoparticle size is below 500 nm, NOAA characteristics such as number concentration and total specific surface area (length/width ratio in the case of nanofibers) are considered to be exposure indicators more significant than mass concentration [Mark 2007; Maynard and Aitken 2007; Branche 2009]. This is also confirmed by epidemiological studies of the entire population, which indicate a correlation between ultrafine particle inhalation and health effects [Pope 2000; Wichmann and Peters 2000; Delfino et al. 2005; Ohlwein et al. 2019].

Occupational exposure assessment in work with nanomaterials is conducted, among others, based on available data on the nanomaterial, through the health hazard assessment (Section 2.4.) and the NOAA exposure assessment. In Document nos. 82, 90, and 91 of the "Series on the Safety of Manufactured Nanomaterials", the OECD has introduced tools for the selection of proper research methods for the measurement of physicochemical parameters of NOAAs and solving problems related to the lack of data on a particular ENM. Document no. 82 (Strategies, Techniques and Sampling Protocols for Determining the Concentrations of Manufactured Nanomaterials in the Air at the Workplace) provides information on the recommended strategies and measurement techniques, as well as aerosol sampling protocols. Document no. 90 (Physical-Chemical Decision Framework to Inform Decisions for Risk Assessment of Manufactured Nanomaterials) provides guidelines as to the recommended research methods for the measurement of key physicochemical parameters for the characterization and identification of particular nanomaterials, exposure and behavior of a nanomaterial in the human body and the environment, as well as for evaluating the risks to human health and the environment. Document no. 91 (Guiding Principles for Measurements and Reporting for Nanomaterials: Physical

Chemical Parameters) helps in the identification of appropriate methods for solving problems linked to the lack of data for nanomaterial safety assessment, improvement of scientific research quality, and promoting consistent data reporting. Document nos. 90 and 91 complement each other and should facilitate a better understanding of the behavior of nanomaterials in the environment and living organisms.

In government and standardization documents, scientific articles, and recommendations, two main approaches to exposure assessment can be distinguished:

- the determination of reference values for nanomaterials, based on established limits for their larger counterparts or based on the grouping of nanomaterials – detailed information on this approach in Section 2.6.);
- task-based measurements, based on comparing the results of concentration measurements from the studied tasks with average background nanoparticle concentration, Section 2.8.1.

British Standards Institution [BSI 2007] pioneered the first approach, based on pragmatic nanomaterial levels on reference values of their larger counterparts. This approach did not gain much popularity and was met with criticism from the scientific community [Hendrikx and van Broekhuizen 2013; Mark 2007; Kumar et al. 2010]. The BSI has also proposed an alternative approach, based on the recommended number concentration levels. A reference level, based on number concentration was proposed, with a value of 20,000 particles/cm^3 for non-soluble nanomaterials. This approach has been further adopted and refined by the Institute for Occupational Safety and Health of the German Social Accident Insurance (IFA DGUV) and the University of Amsterdam (UVA) [van Broekhuizen and Hendrikx 2012; Schumacher and Pallapies 2009]. The recommendations were supplemented with data concerning nanofibers (concentration of which should not exceed 0.01 fiber/cm^3). The remaining nanomaterials were divided into two groups according to density, and two reference levels have been established – 20,000 particles/cm^3 for particles with a density higher than 6 g/cm^3 and 40,000 particles/cm^3 for particles with a density lower than 6 g/cm^3. The reference values were established with regard to 8-hour long work shifts. More information on establishing Nano Reference Values is provided in Section 2.4.

A number of different types of ultrafine particles and nanoparticles in different states of agglomeration are present in everyday human environment. Apart from engineered nanomaterials (ENMs), produced to obtain particular properties, the following may be sources of nanomaterials [Jeevanandam et al. 2018]:

- incidental events (unforeseen leaks from technological processes, glove boxes, thermal processes, etc.),
- by-products of industrial processes (e.g. combustion, chemical production, exhaust gas emission, welding fumes),
- technological processes (grinding, crushing, sieving, transportation, mixing, sharpening, sanding, and polishing),
- natural nanomaterials (e.g. viruses, volcanic ash, organic particles, products of photochemical reactions in higher layers of the atmosphere).

Average background particle concentration was determined by Seipenbusch et al. [2008] to be between 1,000 and 10,000 particles/cm^3 in clean rooms. In the BSI standard and European Commission Recommendations [EC 2014; BSI 2007; Stockmann-Juvala et al. 2014], it was stated that the average concentration of nanoparticles and ultrafine particles in the so-called urban air background, with no nanomaterial production plants in the vicinity, is between 10,000 and 50,000 particles/cm^3. This range was used to develop reference levels of 20,000 and 40,000 particles/cm^3 for ENMs.

Due to the highly developed surface of the nanoparticles, their geometric area should also be considered in exposure assessment. The currently available measurement devices, working in real time or in near real time, do not directly measure this parameter, but can determine e.g. lung deposited surface area (LDSA). The LDSA concentration is considered to be a proper parameter for the determination of negative health effects of an aerosol. It has been shown in toxicological studies, that LDSA correlates with adverse health effects [Oberdörster et al. 2005; Geiss et al. 2016; Kuuluvainen et al. 2016; Marcus et al. 2016]. LDSA includes the deposition efficiency of various airborne objects in different regions of the lungs. A model for lung deposition of inhaled objects (which includes LDSA) was described and defined in relation to a reference worker [ICRP 1994].

2.8.1 Task-Based Measurements

Task-based measurements rely on the registration of NOOA air concentration (in real time or near real time; meters for this kind of measurement are discussed in Section 2.8.2.2.) during different work stages, worker activities, events, etc. (N_p, nanoparticles-process). Such a solution allows for the identification of the source or sources of emissions during a particular nanotechnological process, and the measurement results can also be used to assess the effectiveness of control measures, assure compliance with reference exposure level values, as well as detecting failures or unsealed parts of a system.

Due to the fact that the presence of background nanoparticles (N_b, nanoparticles-background) in the work environment is inevitable, the background needs to be included in the task-based measurements in order to mathematically distinguish it from nanoparticles emitted during work processes (N_p). The average nanoparticle concentration registered during a process or an event (N_p) is reduced by the average background concentration (N_b) or is compared to the average background nanoparticle concentration value (N_p/N_b). More information on background nanoparticle registration can be found in Section 2.8.3.2.

From the magnitude of the increase in a concentration level above the background level, the significance of this increase can be determined and the level of exposure estimated. For example, according to criteria proposed by E. Jankowska and co-workers [2016] from CIOP-PIB (Central Institute for Labour Protection – National Research Institute, Poland), worker exposure level can be determined on a four-level scale, using meters registering number or area concentration of particles in real or near real time.

The first lowest level of exposure occurs when the N_p/N_b ratio is below 1.1 (which means that the concentration value determined during an activity with NOAA, which differs less than 10% from the background level, is not significant). The second level of exposure is of the ratio 1.1–1.5, and the third – 1.5–2. The fourth highest level of exposure corresponds to N_p/N_b ratio higher than 2 – which means the nanoparticle concentration is double the average background ground level.

This four-level scale was developed to enable its use in a risk assessment method, proposed in the technical specification of ISO/TS 12901:2012/2014 (Nanotechnologies – Occupational risk management applied to engineered nanomaterials; Part 1: Use of the control banding approach and Part 2: Principles and approaches). The ISO document provides a body of information on exposure level assessment and risk assessment for given nanotechnological processes, e.g., based on decision trees for various technological processes. A comparison between the exposure level with exposure category (on a five-level scale from A to E, e.g., based on available literature health data) allows for the assessment of occupational exposure.

In task-based measurements, aerosol samples are also taken for further microscopic analysis, in order to assess NOAA sizes and shapes and to confirm the air presence of ENMs from a particular process – more on microscopic methods and sampling in Section 2.8.2.1.

During exposure assessment and emission from given processes or activities, all contextual information as well as data on air parameters (temperature, pressure, relative humidity, flow rate) should be noted. A time-activity diary (TAD) should be kept during the tests to note all relevant information, including the precise time of their occurrence, on the start and end of work, random events, worker activities, breaks, and activities during the process and related to the process. After gathering data on concentration vs. time, they can be compared to the TAD information.

Task-based exposure assessment is considered to be the most significant in the context of activities highly prone to changes, typical for the nanotechnology industry, which utilizes batch processing more often than continuous mass production [Seixas et al. 2003]. Task-based approach was described, e.g., by Ham et al. [2012].

2.8.2 Measurement Methods and Analytical Techniques Used for Occupational Exposure Assessment

There are a number of techniques enabling the characterization, detection, and monitoring of NOAA concentration to assess human exposure. No single method is however versatile and universal enough to encompass all possible parameters of a given nanomaterial – and the more characteristics there are, the more accurate the method needs to be, the more expensive, less mobile, and more difficult to use it becomes. The choice of analytical techniques and testing instruments depends on a number of factors, including:

- the purpose of the measurement (e.g. determination of size, mass, area, or number concentration of particles),
- the possibility of performing the test in a given area and required testing equipment mobility, type of sample (solid, liquid, suspension, aerosol),
- test duration,

- instrument limitations and environmental conditions (too high or too low temperature, relative humidity or pressure may damage the instruments),
- local limitations (no access to electricity in at the work station, the presence of an explosive atmosphere, which requires spark-free tools, etc.),
- the required amount of the sample,
- sample preparation type and the possibility of its destruction, as well as the cost of testing (working time of the laboratory, scientific and technical workers, consumables, etc.).

Simple instruments can only detect the presence of nanoparticles, while more advanced instruments can determine number concentration, size distribution, and area concentration of the particles. Testing methods can be divided into "offline" (laboratory analyses of samples, material characterization – Section 2.8.2.1.) and "online" (Section 2.8.2.2), meaning those conducted in real time or near real time. In nanoparticle testing, several complementary methods are often used, including combinations of "online" and "offline" methods, which allow for a fuller description of health hazards and the work environment.

2.8.2.1 Offline Measurements

The offline methods serve to characterize a previously taken sample of nanoparticle suspension, aerosol, or nanomaterial, used in a manufacturing site or a consumer-ready nanoproduct – mainly for occupational risk assessment, including the assessment of exposure to NOAA. Results of offline measurements are available after laboratory analysis.

2.8.2.1.1 Nanoparticles Sampling

In the work environment, samples to be analyzed are taken with adequate samplers, connected to air aspirators. Relatively inexpensive and simple instruments are, e.g., respirable fraction cyclones, cascade impactors, and Mini Particle Sampler (MPS). The working principle is very simple: the aspirator draws in the air and the air stream is let through a metal net or directed to a proper substrate, e.g. a membrane, filter, or porous carbon membrane. As for the more expensive and complicated instruments, it is worth mentioning the Nanometer Aerosol Sampler (NAS), a personal sampler produced by Particlever (formerly Nanobadge) for sampling in the breathing zone, German personal sampling system Personen Getragenes Probenahme system (PGP), The Personal Nanoparticle Respiratory Deposition (NRD) produced by Zefon International, Thermal Precipitator Sampler for SEM analysis, produced by RJ Lee Group (TPS), and comprehensive cascade impactor systems, including Quartz Crystal Microbalance (QCM) cascade system, and Electrical Low Pressure Impactor (ELPI). The time of sampling, type of substrate, number of samples, method of transportation to the laboratory and preparation, all depend on the type of the method and measurement strategy.

2.8.2.1.2 Chemical Analysis Methods Applied to Nanomaterials

Samples from the air in the work environment are often taken in trace amounts, which makes them insufficient for typical chemical analysis methods, such as FTIR,

BET, ICP-MS, and AAS. For this reason, microscopic methods, which do not require large amounts of material, have gained popularity in the NOAA analysis. The most common microscopes in nanomaterial analysis are the Electron Microscopes – Transmission and Scanning, which use a high energy electron beam to scan the surface or transmit throughout the specimen. Scanning Probe Microscopes (SPM) use a mechanical probe to test the sample topography, thanks to which the obtained image is in the form of a relief. Atomic Force Microscopy (AFM) is the most popular SPM method, allowing for imaging with a better resolution than Electron Microscopes. More details are provided in Table 2.3.

Other often-used offline chemical measurement methods are: dynamic light scattering, nanoparticle tracking analysis, and X-ray diffraction (Table 2.3.). The analysis of an air sample should involve more than one analytical method.

2.8.2.1.3 Other Methods

Other methods can also be used to analyze the chemical composition of nanomaterials. These methods, also commonly applied to their bulk counterparts, are: Nuclear Magnetic Resonance (NMR), Raman spectroscopy, Inductively Coupled Plasma Mass Spectrometry (ICP-MS), Inductively Coupled Plasma Optical Emission Spectrometry (ICP-OES), Inductively Coupled Plasma Atomic Emission Spectrometry (ICP-AES), X-ray Photoelectron Spectroscopy (XRF), Time-of-Flight Secondary Ion Mass Spectroscopy (TOF-SIMS), X-Ray Photoelectron Spectroscopy (XPS), Atomic Absorption Spectrometry (AAS), Fourier-Transform Infrared Spectroscopy (FTIR). To determine nanoparticle size as well as surface area and porosity of bulk material, the BET (Brunauer-Emmett-Teller) method is used, based on the analysis of gas adsorption isotherms. Field Flow Fractioning (FFF) is also used to analyze the size of particles in colloids and suspensions. This method uses the difference in mobility of particles of different sizes. The Tunable Resistive Pulse Sensing (TRPS) method, based on the analysis of resistance caused by the flow of nanoparticles through a membrane, allows for the determination of their volume and surface charge.

2.8.2.1.4 Gravimetric Analysis

Gravimetric analysis is a class of laboratory techniques used to determine the mass and concentration of a substance by measuring mass change. Gravimetric analysis of suspended particulate matter is a standard method used for the determination of mass concentration in the work environment and assessment of occupational exposure. Most commonly, large dust fractions are measured, such as the inhalable fraction (penetrating the upper respiratory tract due to the size of the particles) and the respirable fraction (penetrating the lower respiratory tract), or fine particles (PM2.5, which refers to particles with an aerodynamic diameter of 2.5 μm or less) and coarse particles (PM10, which refers to particles with an aerodynamic diameter of 10 μm or less).

Due to the small size of the nanoparticles and their negligible contribution to the total mass of the aspired dust, gravimetric methods cannot provide any results strictly relevant for nanometric fraction analysis. It is important to measure the respirable fraction in the case of nanotechnological processes in which studied nanoparticles

TABLE 2.3
Examples of Analytical Chemical Methods for Offline Measurements

Method	Description
Scanning Electron Microscopy (SEM)	Uses a high-energy electron beam to scan the sample surface and register electron backscattering. The sample has to remain in vacuum and its surface has to be electrically conductive or coated with a conductive material. SEM is used to assess the shape, state of agglomeration, size, and morphology of the particles.
Transmission Electron Microscopy (TEM)	A high-energy electron beam is used to transmit throughout the specimen. TEM allows for a better image resolution than SEM. Transmission through nanoparticles makes it possible to see objects behind or inside the particle.
Energy-dispersive X-ray spectroscopy (EDS, EDX, EDXS or XEDS)	A technique most often used in tandem with SEM and TEM, for qualitative assessment of chemical composition of the samples, EDS indicates the presence of elements heavier than carbon, which unfortunately makes it difficult to successfully identify organic compounds.
Atomic Force Microscopy (AFM)	During AFM testing, the forces between the tip on an AFM cantilever and the sample surface are measured. The test does not need to be carried out in vacuum; it is possible to conduct even in a liquid environment. The smallest measurable areas can have a dimension of several tens of nanometers, the largest—several tens of micrometers. AFM is used, among others, to assess the shape, size, and morphology of the particles, but additional parameters can also be tested for, depending on modes and configurations used – such as phase contrast, Young's modulus, magnetic domains, and conductivity.
Dynamic Light Scattering (DLS)	This technique uses dynamic scattering of light on particles suspended in liquid. Based on dynamic light scattering, DLS measures the speed of Brownian motion of particles in the liquid and refers it to particle size; however, viscosity of the tested material has to be known. It is also possible to assess particle size distribution and their state of agglomeration using this method. Measured particle sizes are in the range of 0.5–10 μm.
Nanoparticle Tracking Analysis (NTA)	The nanomaterial sample has to be suspended in liquid and placed on a non-transparent, metalized background. Brownian motions of the particles are visualized by an optical microscope. NTA serves to determine particle size, particle size distribution, and enables nanoparticle viewing in real time. Measured particles sizes are in the range of approx. 50–500 nm.
X-ray diffraction (XRD)	The XRD method is based on recording diffraction patterns of X-rays crossing a crystal at different angles and the resulting interactions of radiation with electron clouds of the atoms. Based on the diffraction patterns, a three-dimensional electron density map of a crystal unit cell is determined. XRD is used for the determination of phase composition, crystallite size, degree of crystallinity, and bulk material purity.

form large and stable agglomerates and aggregates in the air, and their size distribution is in the respirable dust range. Another reason to conduct such an analysis is the possibility of assessing the share of the nano fraction in larger dust fractions. In order to determine the respirable fraction, adequate respirable fraction samplers (e.g. large volume cyclones), as well as cascade impactors as per current national standards, are used.

2.8.2.1.5 Dustiness of Nanomaterials

Dustiness is a parameter used during the assessment of occupational exposure to manufactured nanomaterials, determined by the manufacturers and distributors of ENMs. According to the definition in standard EN 1540:2011 (Workplace exposure – Terminology), bulk material "dustiness is the propensity of a material to generate airborne dust during its handling". According to ISO/TS 12025:2012 (Nanomaterials – Quantification of nanoobject release from powders by the generation of aerosols) "dustiness is derived from the amount of dust emitted during a standard test procedure".

Dustiness is a relative parameter, depending not only on the physicochemical properties of the material, but also on the method used for its determination [Evans et al. 2014; Lidén 2006; Pensis et al. 2010]. For this reason, results can only be compared to those obtained using the same method [Hamelman and Schmidt 2004; Hjemsted and Schneider 1996].

Methods described in the EN 15051:2017 "Workplace exposure – Measurement of the dustiness of bulk materials – Part 1: Requirements and choice of test methods" standard are based on gravimetric analysis only, which is insufficient in the case of nanomaterial dustiness testing. Five standards for the determination of bulk nanomaterial dustiness were developed in the CEN: EN 17199-1:2019 – EN 17199-5:2019 (Workplace exposure – Measurement of dustiness of bulk materials that contain or release respirable NOAA and other respirable particles – Part 1: Requirements and choice of test methods, Part 2: Rotating drum method, Part 3: Continuous drop method, Part 4: Small rotating drum method, and Part 5: Vortex shaker method). Methods for the assessment of dustiness provide important data on the diffusion and persistence of particulate matter in the air – they, however, need large amounts of nanopowders. Their applications are mostly limited to mass-produced nanopowders.

Flammability and explosivity of bulk materials containing nanoobjects was discussed in the CEN/TS 17274:2018 guide.

2.8.2.2 Online Measurements

"Online" methods, or those working in real time (or near real time), are mainly used to monitor the NOAA concentration at potential exposure areas and are mostly so-called counter methods, also known as direct reading instruments (DRI). Due to the fact that results are obtained in short time periods and in a continuous manner (e.g. every second, every 15 seconds, every minute, etc.), they can be related to particular processes, activities, and tasks performed by the workers or devices working with a given nanomaterial.

Similarly, in the case of "online" testing instruments, there is no single ideal device, and due to the lack of air homogeneity, the measurement is often highly inaccurate. Nearly all counter methods are based on the same working principle: an internal pump aspires a given air volume and directs it towards a sensor. Conditions in which "online" instruments work are also similar and they often range from 0 to 40°C and up to 90% air humidity. The devices differ, however, in their size, weight, parameters, accuracy, measuring range, measuring method, and the type of sensor used.

Among the most significant disadvantages of counter methods are their relativity (which results in a discrepancy between values shown on two different instruments) and the lack of proper calibration methods [OECD 2017; Levin et al. 2016; Hsiao et al. 2016]. For instruments determining NOAA mass concentration in air based on the number of objects counted in a given air volume, another problem is the need for aerosol density to be known. It is, however, often impossible due to background nanoparticle content and lack of aerosol homogeneity (aerosol density is most often assumed to be 1.7 g/cm^3). Those nanoparticles will always be present in the sampled air. Adequate measurement strategies are being adopted to minimize the drawbacks of measurement methods and to maximize the amount of data and its quality (Section 2.8.3).

Instruments most commonly used for online measurements are: Condensation Particle Counter, Diffusion Size Classifier, Differential Mobility Analyser, Scanning Mobility Particle Sizer, Aerosol Particle Mass Analyzer, Fast Mobility Particle Sizer, Wide Range Aerosol Spectrometer, Electrical Low Pressure Impactor, Tapered Element Oscillating Micro Balance. Their descriptions and general working principles are shown in Table 2.4.

As air parameters, such as temperature, relative humidity, air flow, and pressure, can affect suspended or airborne NOAA, they should be monitored during the testing as supplementary measurements. Air parameters are relatively easily determined using thermo-hygro-barometers, flow meters, microclimate meters, anemometers, etc.

Examples of commercially available testing instruments for online measurements with their specifications are shown in Table 2.5.

2.8.3 Measurement Strategies

Occupational exposure assessment for NOAA is a difficult and demanding task. It is a significant challenge to select relevant parameters, illustrating the scale of exposure and its potential health effects, adequate testing methods including their limitations, the separation of background nanoparticles, non-homogeneous temporal and spatial aerosol concentration distribution, and the choice of measurement strategies [Brouwer 2012]. For the selection of parameters, it is recommended to determine occupational exposure in real time (or near real time), based on the number concentration or area concentration, along with monitoring of the average nanoparticle size. This subject has been discussed in detail in the European standard EN 16966:2019. This standard specifies the use of different measurement parameters of NOAA inhalation exposure during basic and complex assessments, as per the description in EN 17058:2019. The document presents the consequences of the choice of particle parameters used to express suspended NOAA inhalation exposure and the operating

TABLE 2.4
Brief List of Direct Reading Instruments for Airborne Nanoparticles

Instrument	Description
Condensation Particle Counter (CPC)	Uses the nanoparticle condensation technique in order to enlarge them to a size measurable by an optical detector, and then to determine nanoparticle number concentration.
Diffusion Size Classifier (DSC)	Used for the measurement of size and number of particles, based on electrical charging of aerosols. Small size and weight are advantages of DSC and this method can be used for personal measurements. Low accuracy is, however, a disadvantage.
Differential Mobility Analyser (DMA)	Used to classify charged nanoparticles based on their mobility in an electric field. It is possible to sort particles into fractions and determine their size distribution. DMA is a component of a larger particle analysis system.
Electrical mobility spectrometer (SMPS and FMPS)	Measures air particle size distribution in relation to their electrical mobility diameter using several electrometers with very low detection limits or by CPC.
Aerosol Particle Mass Analyzer (APM)	Used for the determination of mass, mass concentration, and density of nanoparticles. Particle classification is performed based on the balance between the centrifugal and electrostatic forces.
Wide Range Aerosol Spectrometer (WRAS)	The combination of two methods allows for the determination of number concentration and particle size distribution for a wide range of sizes. Optical spectrometer counts submicron and micrometric objects (PM10, PM2.5, PM1, inhalable, thoracic, and respirable fractions). Nanoparticles are registered electronically, as in DSC.
Electrical Low Pressure Impactor (ELPI)	Used for nanoparticle size distribution and number concentration measurements in relation to aerodynamic diameter. Contains a cascade impactor, to shelves of which particles are directed depending on their charge. Such shelves may be removed and used for further analysis, e.g. microscopic, chemical, or gravimetric.
Tapered Element Oscillating Micro Balance (TEOM)	Air aspirated into TEOM is directed on a conical element, the vibrations of which change in proportion to the mass of deposited particles. This allows for the measurement of mass concentration of micrometric and submicron particles in aerosol.

principles, as well as advantages and disadvantages of selected measurement techniques for the determination of a range of aerosol parameters. Measurement strategies, including the problems listed above, have been discussed by Kuhlbusch et al. [2011] and Asbach [2015].

A practical strategy for NOAA exposure assessment was proposed by the BSI [BSI 2010] and in the NEAT 2.0. approach of the American NIOSH institute [Eastlake et al. 2016]. As long and complex measurement campaigns are expensive and time-consuming, a practical tiered approach, encompassing different levels of assessment, was also proposed. Results of each level enable the decision to move to a higher level of assessment, risk mitigation based on the obtained information or to stop

TABLE 2.5
Examples and Specification of Instruments for Online Measurements

Instrument	Weight (kg)	Concentration Range	Particle Size Range (nm)	Time Resolution (s)
CPC 3007, TSI	1.7	0–10^6 #/cm^3	10–1000	10
DiSCmini, TESTO	0.67	10^3–10^7 #/cm^3	10–300	1
SMPS™ 3938, TSI	9.1	0–10^7 #/cm^3	1–1000	15
FMPS™ 3091, TSI	32	0–10^7 #/cm^3	5.6–560	1
APM 3602, Kanomax	11	0.001–565 femtograms	12–1008	–
MiniWRAS®, GRIMM	7.6	3×10^3–5×10^6 #/cm^3	10–3500	60
ELPI®+, Dekati	22	10–10000	6–10000	0.1

the assessment and document the results. Descriptions of such an approach are published in the EN 17058:2019 standard, the propositions of German research institutes [BAuA 2011; Methner et al. 2010a; 2010b], OECD document [2015] and in scientific publications [Collier et al. 2015]. The tiered approach is discussed in chapter 2.8.3.1.

In chapter 2.8.3.2, recommendations on background nanoparticle measurements are proposed, chapter 2.8.3.3 discusses contextual information to facilitate data analysis of workplace measurements, and chapter 2.8.3.4 provides an introduction to statistical analysis of measurement data.

2.8.3.1 Tiered Approach and Recommendations

Tiered approach guides the user through a multi-step process not only to perform a nanomaterial exposure assessment, but also for a complete risk assessment. According to the OECD [2015], the general steps of the tiered approach to risk assessment are divided into three main stages: (a) gathering information, (b) basic exposure assessment, and (c) expert exposure assessment.

The aim of Level 1 is to gather as much information as possible according to the established best health and safety practices for a given workstation, including activities at the workplace and the materials used. If it is found at the data analysis stage that nanoobject release and occupational exposure to NOAA cannot be excluded, the procedure shall continue to Level 2.

The aim of Level 2 is to perform a basic exposure or emission assessment. This level focuses on a simple approach to establishing whether ENM exposure is possible through (a) an easy-to-use, portable equipment (DRI) or (b) the application of up-to-date knowledge. It also includes laboratory or site measurements, which can be performed using different monitoring strategies, e.g. spatial or temporal changes in the concentration to assess the increase of NOAA concentration above the background nanoparticle concentration. If significant exposure still cannot be excluded, it is necessary to move to the next level, Level 3.

Level 3 requires more accurate, extensive measurements, consequently being the most labor-intensive and expensive. It is only required to move to Level 3 when Levels 1 and 2 have shown significant exposure or if the used materials are highly harmful. Level 3 assessment needs to be adapted to specific needs of the workplace and nanomaterial used.

2.8.3.1.1 Recommendations

Sources and concentrations of aerosols can change in space and time – it is therefore good practice to conduct a number of measurement campaigns, as well as to register data on the ventilation system and air parameters which may influence exposure characterization.

For personal exposure assessment it is necessary to use small, battery-powered samplers and meters installed on the worker's clothing in their PBZ. Personal meter measurement data is used to determine the nanoparticle concentration during a work shift, as well as the assessment of emission during particular processes or their stages. For the latter, results of concentration measurements as a function of time need to be correlated with TAD (Section 2.8.1). TAD is being kept during the measurements and includes information on the time of breaks and processes-related activities, worker activities, random events, etc. To identify potential sources of emission, larger recording equipment is needed – typically performing continuous concentration measurements, which can be correlated with the location of emission, ventilation setup, and particular work processes. Portable stationary test devices are also used for near-field measurements – the instruments are placed as close to the worker as possible during the entire duration of the processes. In case of work conducted over a larger area, the device can be moved (e.g. using a trolley), but maintaining a distance of 1 m or less from the worker, taking care to avoid restricting its motion.

Due to the nanoparticles' tendency to agglomerate and form stable objects even in micrometric sizes, it is recommended to sample a respirable fraction for gravimetric analysis (determination of mass concentration of this fraction). Aerosol samples should also be taken for microscopic analysis to assess the size and shape of the objects, their state of agglomeration, as well as chemical analysis to confirm whether a given nanomaterial is present in the air. Samples should be taken from near-field or PBZ.

Kuhlbusch and co-workers [2018] also recommend to:

- perform the measurements directly at the workstation or simulate the process in laboratory conditions;
- assess background nanoparticle concentration and attempt to minimize it;
- perform "online" aerosol testing, using several instruments to compensate for DRI limitations;
- sample the aerosol for further "offline" laboratory analysis and carry out supplementary measurements.

2.8.3.2 Measurements of Background Nanoparticles

In order to account for measurement data related to background nanoparticles (N_b), the average concentration for a given nanotechnological process (N_p) (or processes, work stages, tasks) can be registered in a cleanroom or another very clean area with tightly controlled conditions. This approach is, in practice, focused on the process only and does not reflect all the conditions in the work environment. Moreover, not every process can be moved to a laboratory. In this case, the remaining solution is to simulate the process, which introduces another element of uncertainty and further removes the results from the real conditions.

Another approach is to register the background nanoparticle level when no processes are in progress. Background nanoparticle level should be determined at the same place where exposure assessment is being conducted, but no nanoparticle source may be active. In order to fulfill this condition, background measurement is performed at least 30 min before the beginning of the working process and averaged. However, the longer the measurement is, the more information on background fluctuations (measured by standard deviation). As the presence of other people, other particle emission sources, machine operation, etc., actively increase nanoparticle concentration in the environment, background registration can be performed during non-working days (weekends, public holidays), when air is the cleanest. It is also recommended to perform additional background measurements after the studied process has ended. Average background concentration is either subtracted from the average NOAA concentration determined during the process or the process concentration to background concentration ratio is calculated (N_p/N_b). Methods for determination of the significance of the ratio of nanoparticle level from a process/work stage to background nanoparticle level have been shown, e.g., in Annex E to EN standard 17058:2019.

In the case of continuous processes, far-field measurements are carried out. For these measurements, a reference room (e.g. an adjacent one) needs to be chosen, where the given process is not being conducted and the conditions are similar to the studied area. In that case, background measurement can be carried out during the whole NOAA exposure assessment for a given nanotechnological process.

2.8.3.3 Contextual Information

During exposure assessment, all contextual information, as well as data on air parameters (temperature, pressure, relative humidity, flow rate), should be noted. A relevant questionnaire should include, among others, information on processes using nanomaterials or those, in which NOAA may be incidentally produced; the size and description of the room where nanotechnological processes are being carried out; exposure protection measures (e.g. local exhaust ventilation; personal protective equipment; and other control measures present at the place where measurements are carried out). It should also include data on the duration of the nanotechnological processes, the main and side sources of emission, environmental conditions at the workplace, known physicochemical parameters of a given nanomaterial, sampling equipment, offline and online exposure analysis equipment, the method of choosing the site and location for measurements, as well as work timing (including detailed information about events and activities performed during the processes). The set of contextual information should be in compliance with the NECID database [PEROSH 2019], introduced in Annex C to EN standard 17058:2019, preferable for research harmonization and gathering data on occupational exposure to NOAA.

2.8.3.4 Statistical Analysis of Exposure Measurement Results

From the point of view of statistical analysis, the NOAA concentrations determined during the measurements are a time series. Statistically significant differences in size distributions of "background" nanoparticles and NOAA emitted during processes can be determined using appropriate descriptive statistics methods, including geometric mean and deviation from number concentration geometric mean, particle surface area, and mass concentration. Parametric statistics is mainly used for testing the dependence between concentration levels and other factors, and the variance analysis (ANOVA) with Bonferroni post hoc test and Student *t*-test is used to determine the level of significance of the differences in concentration levels between given events/stages of the studied processes [Ham et al. 2012]. As the standard *t*-test is not the perfect choice for NOAA concentration time series analysis, Annex D to EN standard 17058:2019 recommends the use of the autoregressive integrated moving average (ARIMA) model [Klein Entink et al. 2011]. Measurement results from different real-time concentration assessment devices are compared using Pearson's correlation analysis [Ham et al. 2016].

2.8.4 Example of Measurement Results in the Work Environment

Below are two examples of the author's own studies on nanoparticle emission in the work environment. The first example (Section 2.8.4.1) describes conditions during covering special plates with titanium nano dioxide (TiO_2). The second example (Section 2.8.4.2) describes conditions at the workstation of an operator of a press, extracting steel forms in high temperature and cooled by oil containing graphite nanoparticles, in a continuous process. The examples provided aim to show the contrast between a process with short-term, controlled emission and one with long-term, high emission of nanoparticles and/or ultrafine particles to the surrounding air. Other examples are presented and described in greater detail, e.g. in this publication on inhalation exposure to nanomaterials [Oberbek et al. 2019].

2.8.4.1 Covering Plates with Spray Containing Nano TiO_2

The first example describes conditions during spray coating a protective layer of 1% aqueous suspension of titanium nano dioxide on special plates. Nano TiO_2 is used in this application due to its antibacterial properties and self-cleaning effect under UV light. The process was conducted manually and the plates were placed in a half-open laminar flow cabinet, which aspirated air into its interior. The worker had a coat, latex gloves, safety glasses, and a mask fitted with an FFP2 filter. The room, with dimensions of 5.0 × 8.7 × 4.6 m (200.1 m^3), was equipped with efficient mechanical ventilation. Only one worker was present during the process.

The meter marked DM-A for short was placed inside the cabinet and the meter marked DM-B was placed on the worker's clothing, inside their breathing zone. First, the background nanoparticle level was registered by the DM-B (in a clean room, before the processes started that day). Graphs obtained from meters DM-A and DM-B during the process of covering a single plate with the protective layer

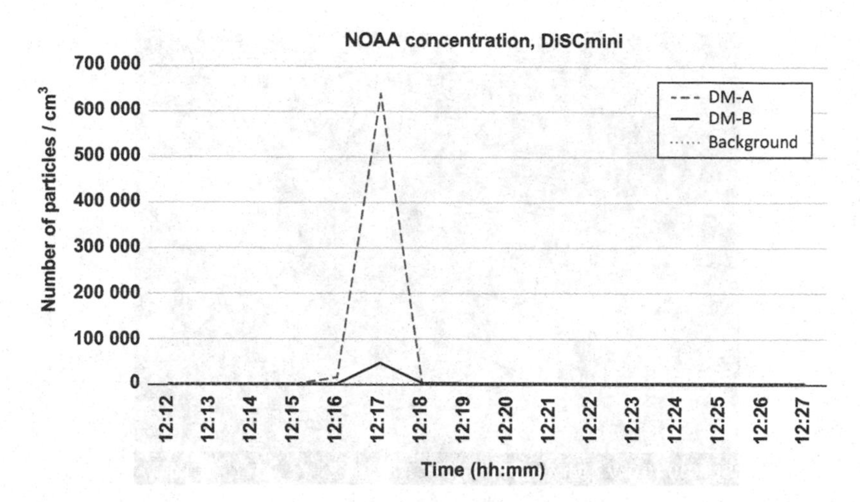

FIGURE 2.5 NOAA concentration measured in real time by the DM-A and DM-B meters during spray coating of special plates with aqueous suspension of titanium nano dioxide; results averaged over minutes.

are shown in Figure 2.5. The worker started the process by filling the gun reservoir with the TiO_2 suspension and connected the compressed air pump to the gun (no significant emission was noted during these activities). The main process of covering a single plate with protective layers lasted about a minute. Next, the worker waited for several minutes and exchanged the plate for another. To register the emission, two portable nanoparticle analyzers of DiSCmini (Matter Aerosol/Testo) were used.

Before the start of the process, the background nanoparticle level was relatively low, at 2,166 ±74 nanoparticles/cm³ with small fluctuations. Number concentrations, registered with the meters in real time, have shown a significant but short-lived increase in the concentration when the spraying process started. The DM-A inside the cabinet registered a surge up to approximately 640,000 nanoparticles/cm³ and the DM-B, installed on the worker's clothing and sampling air from their PBZ, registered a surge to approximately 48,000 nanoparticles/cm³. After a minute, the concentration dropped to the background nanoparticle level.

An aerosol sample was taken during the process for later microscopic analysis. Samples gathered on copper meshes were analyzed using the SEM (Hitachi SU8010) with an EDS spectrometer. A selected SEM image of the sample is shown in Figure 2.6. EDS spectrum of one of the objects selected during imaging is shown in Figure 2.7.

Spherical aggregates with a wide size distribution (70 nm to 1.5 μm) have been found in the aerosol. EDS analysis has shown the presence of titanium in the samples (peaks form oxygen and carbon are not taken into account due to significant presence of these elements in the air and the copper peak originates from the copper mesh).

The process was characterized by high emissions, but the increase in concentration was brief and in the long term, taking appropriate control measures, should not present a threat to the worker.

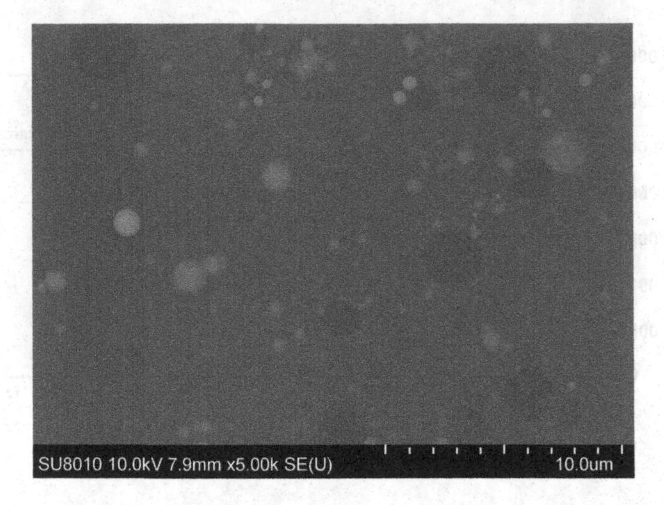

FIGURE 2.6 Selected SEM image of an aerosol sample taken during a process of spray coating special plates with aqueous suspension of titanium nano dioxide; 10 µm scale.

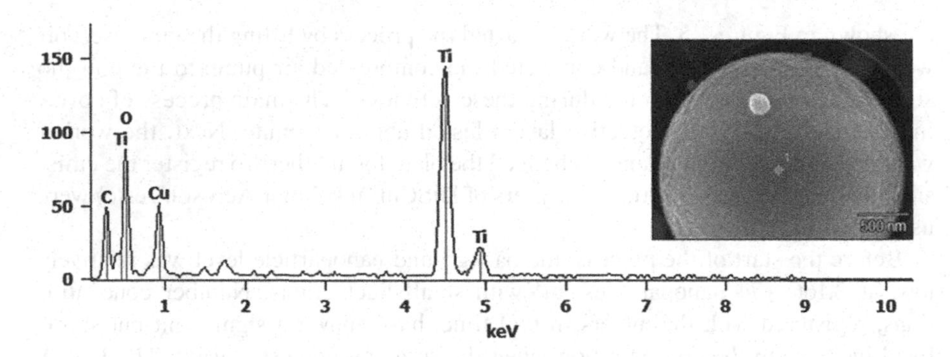

FIGURE 2.7 EDS spectrum of a chosen nanostructured object, 500 µm scale.

2.8.4.2 Emission of Graphite Nanoparticles and Ultrafine Metal Particles

The second example describes conditions during production of steel forms the workstation with a high-temperature press (~1,000°C) cooled by oil containing graphite nanoparticles.

The process was conducted in a continuous manner: three workers supervised the press operation during their work shift, visually inspecting the produced components. They occasionally turned the press off, replaced a component of the form and cleaned it. Most of their working time was spent in the operator cabin or outside the workstation. The workstation was placed in a large hall with dimensions of 37.1 × 83.5 × 13.3 m (41,201.4 m^3), equipped with hybrid ventilation. The press was 4.6 m wide and 6.3 m long.

To register the emission, two portable nanoparticle analyzers DiSCmini (Matter Aerosol/Testo) were used. The meter marked DM-A was placed near the press, approximately 1 m away from its chamber. The meter marked DM-B was placed near the operator cabin. Due to the continuous character of the process, the background nanoparticle level was registered in the closest worker social room (approx. 5 m from the workstation) with the DM-B. Figure 2.8 shows graphs obtained from the DM-A and DM-B meters in a chosen period of the press operation time.

The background nanoparticle concentration, determined in a room adjacent to the hall, was 10,984 ±709 nanoparticles/cm³. During the measurements (from 11:30 to 11:40 am), one of the workers supervised the press operation from the cabin, the second one was near the press control panel, near the DM-A. At 11:40 am, the main press chamber was opened and four metal elements were removed for inspection. At 11:47 am the press door was closed. At 12:10 pm, two of the workers opened the chamber again to refit the press. From 12:25 to 12:27 pm three workers installed a new mold in the press. At 12:26 pm, one of the workers flushed the interior of the chamber with a stream of water for one minute. At 12:29 pm, the next component was put in the mold and the chamber was closed. Number concentrations, registered in real time by the DM-A and DM-B meters, have shown high concentrations during the whole measuring time, with visible surges during activities carried out with the press door open. The average NOAA concentration, registered by the DM-A, was 75,995 ±19,062 nanoparticles/cm³ during the whole measurement time and 65,453 ±19,665 nanoparticles/cm³ for the DM-B. These values are higher than the expected NRV reference values for nanoparticles (the reference values are 20,000 or 40,000 nanoparticles/cm³, depending on nanomaterial density, more on this subject in Section 2.6., Table 2.1).

An aerosol sample was taken during the process, with the chamber door open, for later microscopic analysis. Samples gathered on copper meshes were analyzed using SEM (Hitachi SU8010) with an EDS spectrometer. A selected SEM image of the sample is shown in Figure 2.9. EDS spectrum of one of the objects selected during imaging is shown in Figure 2.10.

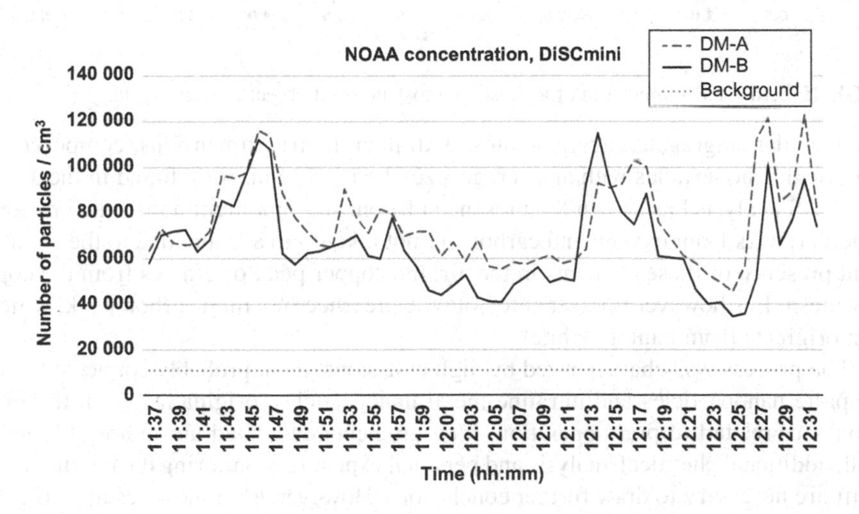

FIGURE 2.8 NOAA concentration measured in real time by the DM-A and DM-B meters during a chosen period of the press operation time; results averaged over minutes.

FIGURE 2.9 A selected SEM image of the aerosol sample, taken during activities with open press, cooled with oil containing graphite nanoforms; 10 μm scale.

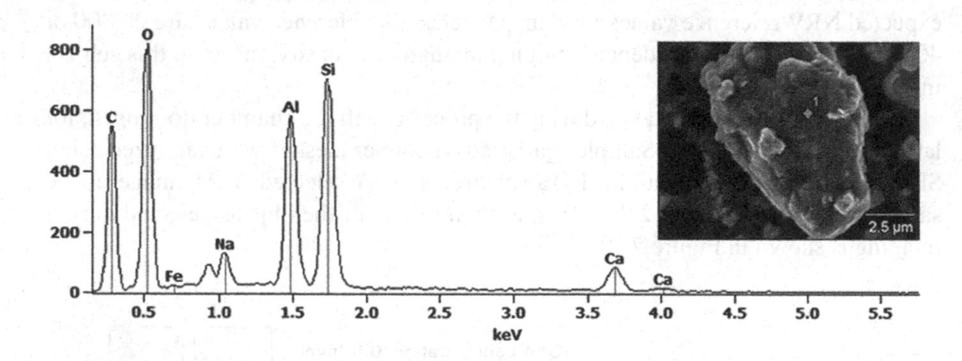

FIGURE 2.10 EDS spectrum of a chosen nanostructured object; 2.5 μm scale.

Irregular, aggregated agglomerates, with micrometric dimensions, composed of spherical nanoparticles with an average size of 55 ± 13 nm, were found in the aerosol. EDS analysis has shown the presence of iron, sodium, aluminum, calcium, and silicon (peaks from oxygen and carbon are not taken into account due to the significant presence of these elements in the air; the copper peak originates from the copper mesh. It is however important to note the presence of a high carbon peak, which can originate from nano graphite).

The process was characterized by high emissions, most probably connected with graphite nanoparticles and ultrafine metal nanoparticles, originating from thermal processes related to press operation. Measurements described above are only general; additional chemical analysis and personal exposure monitoring during the work shift are necessary to draw further conclusions. However, even those results indicate that the NOAA concentration near the press is maintained on a level above 40,000 nanoparticles/cm^3, and even more severe exposure is linked to work with the press open. In a case such as this, risk to workers is significant and should require a reaction from the employer and specialists for workplace health and safety. The process and the time workers spend near the press should be revised and additional control measures should be implemented.

2.9 SUMMARY

The dynamic growth of nanotechnology, visible in the high growth rate of both the number and tonnage of ENMs placed on the market, also means an increase of potential occupational exposure to nanomaterials at every stage of their life cycle, from research and development, through manufacturing, to waste treatment. This implies the necessity for rapid progress in the knowledge of the risks linked to ENMs and occupational exposure assessment. It is important to understand that in the ENMs' hazard identification, the term: "nanoparticle, nanoobject, nanomaterial" encompasses a vast variety of materials with different chemical structures and composition, as well as their shape and size. Specific properties of the ENMs, resulting from their dimensions, are a challenge for toxicological studies, exposure assessment methodology, and ensuring safety of their use. This literature review indicates that significant progress has been made in these areas in the recent years.

Harmonization of methods for the assessment of ENMs' physicochemical properties and the development of integrated strategies for the assessment of their toxicity have provided the basis for the development of the ENMs' grouping criteria, which can be used for their registration (classification and labeling), as well as for the assessment of health risk and occupational exposure. Progress in measurement techniques allows for the determination of ENMs' air concentration, which is a basis for the development of indicative/limit values of ENMs exposure, which can be used for the assessment of the work environment. Currently available research methods and measurement strategies allow for a comprehensive assessment of occupational risks linked to nanomaterials. However, they are not commonly applied, even in places where nanomaterials are specifically designed and produced. Moreover, the application of harmonized measurement procedures is necessary for exposure assessment. Large differences in suggested OEL values can influence decisions on occupational risk management (choice of risk control measures).

Available toxicological data, coming from animal studies and few results of epidemiological studies indicate that the use of precautionary measures while working with the ENMs is strongly grounded. Harmful effects linked to the possibility of ENMs interaction with the respiratory system, nervous system, carcinogenicity, reproductive toxicity, and endocrine activity may have a magnitude difficult to predict. It is therefore necessary to intensify toxicological studies in this area, as well as studies on the mechanisms of ENMs' toxicity. Data on bioavailability and toxicokinetics (absorption, distribution in the body, metabolism, excretion) of ENMs is especially desired. An interesting subject matter, related not only to the work environment, is the possible interaction of ENMs with the human microbiome, which can have an immense impact on human physiology and unknown clinical consequences.

It is also necessary to intensify epidemiological studies and the search for exposure biomarkers. As long as gaps in the knowledge on ENM toxicity are not filled, the materials will pose a new and emerging risk in the environment, including the work environment – a risk which needs to be actively prevented and controlled.

REFERENCES

ACGIH [American Conference of Governmental Industrial Hygienists]. 2001. *Air sampling instruments for evaluation of atmospheric contaminants*. 9th ed. Cincinnati, OH: American Conference of Governmental Industrial Hygienists (ACGIH).

Arts, J. H. E., M. Hadi, M. A. Irfan et al. 2015. A decision-making framework for the grouping and testing of nanomaterials (DF4nanoGrouping). *Regul Toxicol Pharmacol* 71(2) (Suppl):S1–S27. doi: 10.1016/j.yrtph.2015.03.007.

Asbach, C. 2015. Exposure measurement at workplaces. In: *Nanoengineering: Global approaches to health and safety issues*, ed. P. I. Dolez, 523–555. Amsterdam: Elsevier.

Assadi, F., M. Mohseni, K. Dadashi Noshahr et al. 2017. Effect of molybdenum nanoparticles on blood cells, liver enzymes, and sexual hormones in male rats. *Biol Trace Elem Res* 175(1):50–56. doi: 10.1007/s12011-016-0765-5.

Barrak, H., T. Saied, P. Chevallier et al. 2019. Synthesis, characterization, and functionalization of ZnO nanoparticles by N-(trimethoxysilylpropyl) ethylenediamine triacetic acid (TMSEDTA): Investigation of the interactions between Phloroglucinol and ZnO@TMSEDTA. *Arab J Chem* 12(8):4340–4347. doi: 10.1016/j.arabjc.2016.04.019.

BAuA [Bundesanstalt für Arbeitsschutz und Arbeitsmedizin], IUTA, BG RCI, VCI, DGUV (IFA), TUD. 2011. Tiered approach to an exposure measurement and assessment of nanoscale aerosols released from engineered nanomaterials in workplace operations. https://www.baua.de/DE/Angebote/Publikationen/Kooperation/Nanomaterialien-2.pdf%3F__blob%3DpublicationFile%26v%3D4 (accessed January 10, 2020).

Berkeley Municipal Code. 2006. Section 12.12.040 filing of disclosure information and section 15.12.050 quantities requiring disclosure. Council of the City of Berkeley. Ordinance No. 6,960-N.S. www.cityofberkeley.info/citycouncil/ordinances/2006/6960.pdf (accessed January 10, 2020).

Bevan, R. J., R. Kreiling, L. S. Levy et al. 2018. Toxicity testing of poorly soluble particles, lung overload and lung cancer. *Regul Toxicol Pharmacol* 100:80–91. doi: 10.1016/j.yrtph.2018.10.006.

Bierkandt, F. S., L. Leibrock, S. Wagener et al. 2018. The impact of nanomaterial characteristics on inhalation toxicity. *Toxicol Res* 7(3):321–346. doi: 10.1039/c7tx00242d.

Bolzinger, M. A., S. Briançon, J. Pelletier et al. 2012. Penetration of drugs through skin, a complex rate-controlling membrane. *Curr Opin Colloid Interface Sci* 17(3):156–165. doi: 10.1016/j.cocis.2012.02.001.

Borisova, T. 2018. Nervous system injury in response to contact with environmental, engineered and planetary micro- and nano-sized particles. *Front Physiol* 9:728. doi: 10.3389/fphys.2018.00728.

Braakhuis, H. M., A. G. Oomen, and F. R. Cassee. 2016. Grouping nanomaterials to predict their potential to induce pulmonary inflammation. *Toxicol Appl Pharmacol* 299:3–7. doi: 10.1016/j.taap.2015.11.009.

Braakhuis, H. M., M. V. D. Z. Park, I. Gosens et al. 2014. Physicochemical characteristics of nanomaterials that affect pulmonary inflammation. *Part Fibre Toxicol* 11(1):18. doi: 10.1186/1743-8977-11-18.

Branche, C. M. 2009. Approaches to safe nanotechnology: Managing the health and safety concerns associated with engineered nanomaterials. DHHS (NIOSH) Publ. No.

2009-125. https://www.cdc.gov/niosh/docs/2009-125/pdfs/2009-125.pdf (accessed January 10, 2020).

Brouwer, D. 2012. Control banding approaches for nanomaterials. *Ann Occup Hyg* 56(5):506–514. doi: 10.1093/annhyg/mes039.

BSI [Bristish Standards Intitution]. 2007. Nanotechnologies, Part 2: Guide to safe handling and disposal of manufactured nanomaterials. PD 6699-2:2007. http://www3.imperial.ac.uk/pls/portallive/docs/1/34683696.PDF (accessed January 10, 2020).

BSI [Bristish Standards Intitution]. 2010. Nanotechnologies: Guide to assessing airborne exposure in occupational settings relevant to nanomaterials. PD 6699-3:2010. https://shop.bsigroup.com/ProductDetail/?pid=000000000030206667 (accessed January 10, 2020).

Catalán, J., K. M. Siivola, P. Nymark et al. 2016. In vitro and in vivo genotoxic effects of straight versus tangled multi-walled carbon nanotubes. *Nanotoxicology* 10(6):794–806. doi: 10.3109/17435390.2015.1132345.

Chen, Z., Y. Wang, X. Wang et al. 2018. Effect of titanium dioxide nanoparticles on glucose homeostasis after oral administration. *J Appl Toxicol* 38(6):810–823. doi: 10.1002/jat.3589.

Collier, Z. A., A. J. Kennedy, A. R. Poda et al. 2015. Tiered guidance for risk-informed environmental health and safety testing of nanotechnologies. *J Nanopart Res* 17(3):155. doi: 10.1007/s11051-015-2943-3.

Delfino, R. J., C. Sioutas, and S. Malik. 2005. Potential role of ultrafine particles in associations between airborne particle mass and cardiovascular health. *Environ Health Perspect* 113(8):934–946. doi: 10.1289/ehp.7938.

Deveau, M., C. P. Chen, G. Johanson et al. 2015. The global landscape of occupational exposure limits: Implementation of harmonization principles to guide limit selection. *J Occup Environ Hyg* 12(Suppl):S127–S144. doi: 10.1080/15459624.2015.1060327.

Dolez, P. I., and M. Debia. 2015. Overview of workplace exposure to nanomaterials. In: *Nanoengineering: Global approaches to health and safety issues*, ed. P. I. Dolez, 427–484. Amsterdam: Elsevier.

Donaldson, K., F. A. Murphy, R. Duffin et al. 2010. Asbestos, carbon nanotubes and the pleural mesothelium: A review of the hypothesis regarding the role of long fibre retention in the parietal pleura, inflammation and mesothelioma. *Part Fibre Toxicol* 7:5. doi: 10.1186/1743-8977-7-5.

Donaldson, K., V. Stone, P. J. A. Borm et al. 2003. Oxidative stress and calcium signalling in the adverse effects of environmental particles (PM10). *Free Radic Biol Med* 34(11):1369–1382.

Dreaden, C., A. M. Alkilany, X. Huang et al. 2012. The golden age: Gold nanoparticles for biomedicine. *Chem Soc Rev* 41(7):2740–2779. doi: 10.1039/C1CS15237H.

DTSC [Department of Toxic Substances Control]. 2013. Chemical call-in/nanotechnology. https://dtsc.ca.gov/dtsc-website-archive/chemical-call-in-nanotechnology (accessed January 10, 2020).

Dusinska, M., J. Tulinska, N. El Yamani et al. 2017. Immunotoxicity, genotoxicity and epigenetic toxicity of nanomaterials: New strategies for toxicity testing? *Food Chem Toxicol* 109(1):797–811. doi: 10.1016/j.fct.2017.08.030.

Eastlake, A. C., C. Beaucham, K. F. Martinez et al. 2016. Refinement of the nanoparticle emission assessment technique into the nanomaterial exposure assessment technique (NEAT 2.0). *J Occup Environ Hyg* 13(9):708–717. doi: 10.1080/15459624.2016.1167278.

EC [European Commission]. 2006. Regulation EC No 1907/2006 of the European Parliament and of the Council of 18 December 2006 concerning the Registration, Evaluation, Authorisation and Restriction of Chemicals (REACH). *Off J Eur Union*. Document 02006R1907-20140410. https://eur-lex.europa.eu/legal-content/EN/TXT/?uri=CELEX%3A02006R1907-20140410 (accessed January 10, 2020).

EC [European Commission]. 2008. Regulation EC No 1272/2008 of the European Parliament and of the Council of 16 December 2008 on classification, labelling and packaging of substances and mixtures, amending and repealing Directives 67/548/EEC and 1999/45/EC, and amending Regulation (EC) No 1907/2006. *Off J Eur Union*. Document 32008R1272. https://eur-lex.europa.eu/eli/reg/2008/1272/oj?eliuri=eli:reg:2008:1272:oj (accessed January 10, 2020).

EC [European Commission]. 2012. Types and uses of nanomaterials, including safety aspects. COM(2012):572. https://eur-lex.europa.eu/LexUriServ/LexUriServ.do?uri=SWD:2012:0288:FIN:EN:PDF (accessed January 10, 2020).

EC [European Commission]. 2014. Guidance on the protection of the health and safety of workers from the potential risks related to nanomaterials at work. http://ec.europa.eu/social/BlobServlet?docId=13087&langId=en%20 (accessed January 10, 2020).

EC [European Commission]. 2018a. Definition of a nanomaterial. https://ec.europa.eu/environment/chemicals/nanotech/faq/definition_en.htm (accessed January 10, 2020).

EC [European Commission]. 2018b. Commission Regulation EU 2018/1881 of 3 December 2018 Amending Regulation (EC) No 1907/2006 of the European Parliament and of the Council on the Registration, Evaluation, Authorisation and Restriction of Chemicals (REACH) as regards Annexes I, III, VI, VII, VIII, IX, X, XI, and XII to address nanoforms of substances. *Off J Eur Union*. Document 32018R1881. https://eur-lex.europa.eu/legal-content/PL/TXT/?qid=1545225466697&uri=CELEX%3A32018R1881 (accessed January 10, 2020).

ECETOC [European Centre for Ecotoxicology and Toxicology of Chemicals]. 2013. Poorly soluble particles/lung overload. Technical Report No. 122. http://www.ecetoc.org/wp-content/uploads/2014/08/ECETOC-TR-122-Poorly-Soluble-Particles-Lung-Overload.pdf (accessed January 10, 2020).

ECHA [European Chemicals Agency]. 2014. Human health and environmental exposure assessment and risk characterisation of nanomaterials: Best practice for REACH registrants. http://echa.europa.eu/documents/10162/5399565/best_practices_human_health_environment_nano_3rd_en.pdf (accessed January 10, 2020).

ECHA [European Chemicals Agency]. 2016. Usage of (eco)toxicological data for bridging data gaps between and grouping of nanoforms of the same substance: Elements to consider. https://echa.europa.eu/documents/10162/13630/eco_toxicological_for_bridging_grouping_nanoforms_en.pdf (accessed January 10, 2020).

ECHA [European Chemicals Agency]. 2017a. Guidance on information requirements and chemical safety assessment. Appendix R.6-1 for nanomaterials applicable to the guidance on QSARs and grouping of chemicals. Version 1.0. https://echa.europa.eu/documents/10162/13564/appendix_r6-1_nano_draft_for_committees_en.pdf/cb821783-f534-38cd-0772-87192799b958 (accessed January 10, 2020).

ECHA [European Chemicals Agency]. 2017b. Committee for risk assessment RAC opinion proposing harmonised classification and labelling at EU level of titanium dioxide. https://echa.europa.eu/documents/10162/682fac9f-5b01-86d3-2f70-3d40277a53c2 (accessed January 10, 2020).

EPA [United States Environmental Protection Agency]. 2017. Technical fact sheet – nanomaterials. https://www.epa.gov/sites/production/files/2014-03/documents/ffrrofactsheet_emergingcontaminant_nanomaterials_jan2014_final.pdf (accessed January 10, 2020).

Evans, D. E., L. A. Turkevich, C. T. Roettgers et al. 2014. Comment on comparison of powder dustiness methods. *Ann Occup Hyg* 58(4):524–528. doi: 10.1093/annhyg/met086.

Fernández Tena, A., and P. Casan Clarà. 2012. Deposition of inhaled particles in the lungs. *Arch Bronconeumol* 48(7):240–246. doi: 10.1016/j.arbres.2012.02.003.

Fröhlich, E., and S. Salar-Behzadi. 2014. Toxicological assessment of inhaled nanoparticles: Role of in vivo, ex vivo, in vitro, and in silico studies. *Int J Mol Sci* 15(3):4795–4822. doi: 10.3390/ijms15034795.

Fujitani, T., K. Ohyama, A. Hirose et al. 2012. Teratogenicity of multi-wall carbon nanotube (MWCNT) in ICR mice. *J Toxicol Sci* 37(1):81–89.

Gajanan, K., and S. N. Tijare. 2018. Applications of nanomaterials. *Mater Today Proc* 5(1, Part 1):1093–1096. doi: 10.1016/j.matpr.2017.11.187.

Gatoo, M. A., S. Naseem, M. Y. Arfat et al. 2014. Physicochemical properties of nanomaterials: Implication in associated toxic manifestations. *BioMed Res Int* 2014:498420. doi: 10.1155/2014/498420.

Gebel, T., H. Foth, G. Damm et al. 2014. Manufactured nanomaterials: Categorization and approaches to hazard assessment. *Arch Toxicol* 88(12):2191–2211. doi: 10.1007/s00204-014-1383-7.

Geiser, M. 2010. Update on macrophage clearance of inhaled micro- and nanoparticles. *J Aerosol Med Pulm Drug Deliv* 23(4):207–217. doi: 10.1089/jamp.2009.0797.

Geiss, O., I. Bianchi, and J. Barrero-Moreno. 2016. Lung-deposited surface area concentration measurements in selected occupational and non-occupational environments. *J Aerosol Sci* 96:24–37. doi: 10.1016/j.jaerosci.2016.02.007.

Geraets, L., A. G. Oomen, J. D. Schroeter et al. 2012. Tissue distribution of inhaled micro- and nano-sized cerium oxide particles in rats: Results from a 28-day exposure study. *Toxicol Sci* 127(2):463–473. doi: 10.1093/toxsci/kfs113.

Gleiter, H. 2000. Nanostructured materials: Basic concepts and microstructure. *Acta Mater* 48(1):1–29. doi: 10.1016/S1359-6454(99)00285-2.

Grassian, V. H. 2008. *Nanoscience and nanotechnology: Environmental and health impacts.* Hoboken, NJ: Wiley.

Gulson, B., H. Wong, M. Korsch et al. 2012. Comparison of dermal absorption of zinc from different sunscreen formulations and differing UV exposure based on stable isotope tracing. *Sci Total Environ* 420:313–318. doi: 10.1016/j.scitotenv.2011.12.046.

Gulumian, M., J. Verbeek, C. Andraos et al. 2016. Systemic review of screening and surveillance programs to protect workers from nanomaterials. *PLoS ONE* 11(11):e0166071. doi: 10.1371/journal.pone.0166071.

Ham, S., N. Lee, I. Eom et al. 2016. Comparison of real time nanoparticle monitoring instruments in the workplaces. *Saf Health Work* 7(4):381–388. doi: 10.1016/j.shaw.2016.08.001.

Ham, S., C. Yoon, E. Lee et al. 2012. Task-based exposure assessment of nanoparticles in the workplace. *J Nanopart Res* 14(9):1–17. doi: 10.1007/s11051-012-1126-8.

Hamelman, F., and E. Schmidt. 2004. Methods for characterizing the dustiness estimation of powders. *Chem Eng Technol* 27(8):844–847. doi: 10.1002/ceat.200403210.

Havrdova, M., K. Hola, J. Skopalik et al. 2016. Toxicity of carbon dots: Effect of surface functionalization on the cell viability, reactive oxygen species generation and cell cycle. *Carbon* 99:238–248. doi: 10.1016/j.carbon.2015.12.027.

Hendrikx, B., and P. van Broekhuizen. 2013. Nano reference values in the Netherlands. *Gefahrstoffe reinhalt luft* 73(10):407–414. https://www.gefahrstoffe. de/libary/common/X075en.pdf (accessed January 10, 2020).

Hjemsted, K., and T. Schneider. 1996. Dustiness from powder materials. *J Aerosol Sci* 27(Suppl 1):S485–S486. doi: 10.1016/0021-8502(96)00315-1.

Hoet, P. H., I. Brüske-Hohlfeld and O. V. Salata 2004. Nanoparticles – known and unknown health risks. *J Nanotechnology* 2(1):12. doi: 10.1186/1477-3155-2-12.

Hougaard, K. S., P. Jackson, K. A. Jensen et al. 2010. Effects of prenatal exposure to surface-coated nanosized titanium dioxide (UV-Titan): A study in mice. *Part Fibre Toxicol* 7:16. doi: 10.1186/1743-8977-7-16.

HSE [Health and Safety Executive]. 2014. General methods for sampling and gravimetric analysis of respirable, thoracic and inhalable aerosols. *Health Saf Lab MDHS* 14/4. http://www.hse.gov.uk/pubns/mdhs/pdfs/mdhs14-4.pdf (accessed January 10, 2020).

Hsiao, T. C., Y. C. Lee, K. C. Chen et al. 2016. Experimental comparison of two portable and real-time size distribution analyzers for nano/submicron aerosol measurements. *Aerosol Air Qual Res* 16(4):919–929. doi: 10.4209/aaqr.2015.10.0614.

Hund-Rinke, K., K. Schlich, D. Kühnel et al. 2018. Grouping concept for metal and metal oxide nanomaterials with regard to their ecotoxicological effects on algae, daphnids and fish embryos. *NanoImpact* 9:52–60.

IARC [International Agency for Research on Cancer]. 2010. *Carbon black, titanium dioxide and talc*, 43–191. Lyon, France: International Agency for Research on Cancer.

IARC [International Agency for Research on Cancer]. 2018. *Some nanomaterials and some fibres*. Lyon, France: International Agency for Research on Cancer. https://monographs.iarc.fr/wp-content/uploads/2018/06/mono111.pdf (accessed January 10, 2020).

Iavicoli, I., L. Fontana, V. Leso et al. 2013. The effects of nanomaterials as endocrine disruptors. *Int J Mol Sci* 14(8):16732–16801. doi: 10.3390/ijms140816732.

Inshakova, E., and O. Inshakov. 2017. World market for nanomaterials: Structure and trends. *MATEC Web Conf* 129:02013. doi: 10.1051/matecconf/201712902013.

ICRP. 1994. Respiratory tract model. SAGE Publications Ltd, *ICRP* 24:1–11. doi: 10.1016/0146-6453(94)90004-3 (accessed January 10, 2020).

Jackson, P., S. Halappanavar, K. S. Hougaard et al. 2013. Maternal inhalation of surface-coated nanosized titanium dioxide (UV-Titan) in C57BL/6 mice: Effects in prenatally exposed offspring on hepatic DNA damage and gene expression. *Nanotoxicology* 7(1):85–96. doi: 10.3109/17435390.2011.633715.

Jain, M. P., F. Vaisheva, and D. Maysinger. 2012. Metalloestrogenic effects of quantum dots. *Nanomedicine* 7(1):23–37. doi: 10.2217/nnm.11.102.

Jankowska, E., P. Sobiech, and B. Kaczorowska. 2016. *Zalecenia w zakresie oceny ryzyka i do profilaktyki technicznej w odniesieniu do narażenia na nanoobiekty emitowane do powietrza podczas produkowania i stosowania nanomateriałów*. Warszawa: Centralny Instytut Ochrony Pracy – Państwowy Instytut Badawczy.

Javurek, A. B., D. Suresh, W. G. Spollen et al. 2017. Gut dysbiosis and neurobehavioral alterations in rats exposed to silver nanoparticles. *Sci Rep* 7(1):2822. doi: 10.1038/s41598-017-02880-0.

Jeevanandam, J., A. Barhoum, Y. S. Chan et al. 2018. Review on nanoparticles and nanostructured materials: History, sources, toxicity and regulations. *Beilstein J Nanotechnol* 9:1050–1074. doi: 10.3762/bjnano.9.98.

Klein Entink, R. H., W. Fransman, and D. H. Brouwer. 2011. How to statistically analyze nano exposure measurement results: Using an ARIMA time series approach. *J Nanopart Res* 13(12):6991–7004. doi: 10.1007/s11051-011-0610-x.

Konduru, N. V., R. M. Molina, A. Swami et al. 2017. Protein corona: Implications for nanoparticle interactions with pulmonary cells. *Part Fibre Toxicol* 14(1):42. doi: 10.1186/s12989-017-0223-3.

Kreyling, W. G., S. Hirn, W. Moller et al. 2014. Air–blood barrier translocation of tracheally instilled gold nanoparticles inversely depends on particle size. *ACS Nano* 8(1):222–233.

Kreyling, W. G., M. Semmler, F. Erbe et al. 2002. Translocation of ultrafine insoluble iridium particles from lung epithelium to extrapulmonary organs is size dependent but very low. *J Toxicol Environ Health A* 65(20):1513–1530.

Kuempel, E. D., V. Castranova, C. L. Geraci et al. 2012. Development of risk-based nanomaterial groups for occupational exposure control. *J Nanopart Res* 14:1029. doi: 10.1007/s11051-012-1029-8.

Kuhlbusch, T. A., C. Asbach, H. Fissan et al. 2011. Nanoparticle exposure at nanotechnology workplaces: A review. *Part Fibre Toxicol* 8:22. doi: 10.1186/1743-8977-8-22.

Kuhlbusch, T. A. J., S. W. P. Wijnhoven, and A. Haase. 2018. Nanomaterial exposures for worker, consumer and the general public. *NanoImpact* 10:11–25.

Kuijpers, E., A. Pronk, R. Kleemann et al. 2018. Cardiovascular effects among workers exposed to multiwalled carbon nanotubes. *Occup Environ Med* 75(5):351–358. doi: 10.1136/oemed-2017-104796.

Kumar, P., A. Robins, S. Vardoulakis et al. 2010. A review of the characteristics of nanoparticles in the urban atmosphere and the prospects for developing regulatory controls. *Atmos Environ* 44(39):5035–5052. doi: 10.1016/j.atmosenv.2010.08.016.

Kuuluvainen, H., T. Rönkköa, A. Järvinen et al. 2016. Lung deposited surface area size distributions of particulate matter in different urban areas. *Atmos Environ* 136:105–113. doi: 10.1016/j.atmosenv.2016.04.019.

Lamon, L., K. Aschberger, D. Asturiol et al. 2019. Grouping of nanomaterials to read-across hazard endpoints: A review. *Nanotoxicology* 13(1):100–118. doi: 10.1080/17435390.2018.1506060.

Landvik, N. E., V. Skaug, B. Mohr et al. 2018. Criteria for grouping of manufactured nanomaterials to facilitate hazard and risk assessment: A systematic review of expert opinions. *Regul Toxicol Pharmacol* 95:270–279.

Larese Filon, F., D. Bello, J. W. Cherrie et al. 2016. Occupational dermal exposure to nanoparticles and nano-enabled products, Part 1: Factors affecting skin absorption. *Int J Hyg Environ Health* 219(6):536–544. doi: 10.1016/j.ijheh.2016.05.009.

Larese Filon, F., M. Mauro, G. Adami et al. 2015. Nanoparticles skin absorption: New aspects for a safety profile evaluation. *Regul Toxicol Pharmacol* 72(2):310–322.

Larson, J. K., M. J. Carvan III, and R. J. Hutz. 2014. Engineered nanomaterials: As emerging class of novel endocrine disruptors. *Biol Reprod* 91(1):20. doi: 10.1095/biolreprod.113.116244.

Leso, V., L. Fontana, A. Marinaccio et al. 2018. Palladium nanoparticle effects on endocrine reproductive system of female rats. *Hum Exp Toxicol* 37(10):1069–1079. doi: 10.1177/0960327118756722.

Levin., M., O. Witschger, S. Bau et al. 2016. Can we trust real time measurements of lung deposited surface area concentrations in dust from powder nanomaterials? *Aerosol Air Qual Res* 16(5):1105–1117. doi: 10.4209/aaqr.2015.06.0413.

Li, Y., and D. Boraschi. 2016. Endotoxin contamination: A key element in the interpretation of nanosafety studies. *Nanomedicine* 11(3):269–287.

Liao, H. Y., Y. T. Chung, C. H. Lai et al. 2014. Sneezing and allergic dermatitis were increased in engineered nanomaterial handling workers. *Ind Health* 52(3):199–215.

Lidén, G. 2006. Dustiness testing of materials handled at workplaces. *Ann Occup Hyg* 50(5):437–439. doi: 10.1093/annhyg/mel042.

Liou, S. H., C. S. Tsai, D. Pelclova et al. 2015. Assessing the first wave of epidemiological studies of nanomaterial workers. *J Nanopart Res* 17:413. doi: 10.1007/s11051-015-3219-7.

Liou, S. H., T. C. Tsou, S. L .Wang et al. 2012. Epidemiological study of health hazards among workers handling engineered nanomaterials. *J Nanopart Res* 14(8):878–892.

Marcus, L., O. Witschger, S. Bau et al. 2016. Can we trust real time measurements of lung deposited surface area concentrations in dust from powder nanomaterials? *Aerosol Air Qual Res* 16(5):1105–1117. doi: 10.4209/aaqr.2015.06.0413.

Mark, D. 2007. Occupational exposure to nanoparticles and nanotubes. In: *Nanotechnology: Consequences for human health and the environment*, eds. R. E. Hester, and R. M. Harrison, 50–80. Cambridge: Royal Society of Chemistry.

Maynard, A. D., and R. J. Aitken. 2007. Assessing exposure to airborne nanomaterials: Current abilities and future requirements. *Nanotoxicology* 1(1):26–41. doi: 10.1080/17435390701314720.

Methner, M., L. Hodson, A. Dames et al. 2010a. Nanoparticle Emission Assessment Technique (NEAT) for the identification and measurement of potential inhalation exposure to engineered nanomaterials, Part A. *J Occup Environ Hyg* 7(3):127–132. doi: 10.1080/15459620903476355.

Methner, M., L. Hodson, A. Dames et al. 2010b. Nanoparticle Emission Assessment Technique (NEAT) for the identification and measurement of potential inhalation exposure to engineered nanomaterials, Part B: Results from 12 field studies. *J Occup Environ Hyg* 7(3):163–176. doi: 10.1080/15459620903508066.

Mihalache, R., J. Verbeek, H. Graczyk et al. 2017. Occupational exposure limits for manufactured nanomaterials: A systematic review. *Nanotoxicology* 11(1):7–19. doi: 10.1080/17435390.2016.1262920.

Moller, P., N. R. Jacobsen, J. K. Folkmann et al. 2010. Role of oxidative damage in toxicity of particulates. *Free Radic Res* 44(1):1–46.

Morris, J., and J. Willis. 2007. Nanotechnology white paper. U.S. Environmental Protection Agency. EPA 100/B-07/001. Washington, DC: Science Policy Council, U.S. Environmental Protection Agency. https://www.epa.gov/sites/production/files/2015-01/documents/nanotechnology_whitepaper.pdf (accessed January 10, 2020).

Naota, M., A. Shimada, T. Morita et al. 2009. Translocation pathway of the intratracheally instilled C60 fullerene from the lung into the blood circulation in the mouse: Possible association of diffusion and caveolae-mediated pinocytosis. *Toxicol Pathol* 37(4):456–462.

Nemmar, A., M. F. Hoylaerts, P. H. Hoet et al. 2004. Possible mechanisms of the cardiovascular effects of inhaled particles: Systemic translocation and prothrombotic effects. *Toxicol Lett* 149(1–3):243–253.

NIOSH [National Institute for Occupational Safety and Health]. 2011. *Occupational exposure to titanium dioxide*. Cincinnati, OH: Department of Health and Human Services. https://www.cdc.gov/niosh/docs/2011-160/pdfs/2011-160.pdf (accessed January 10, 2020).

NIOSH [National Institute for Occupational Safety and Health]. 2013a. *Occupational exposure to carbon nanotubes and nanofibers*. Cincinnati, OH: Department of Health and Human Services. https://www.cdc.gov/niosh/docs/2013-145/pdfs/2013-145.pdf (accessed January 10, 2020).

NIOSH [National Institute for Occupational Safety and Health]. 2013b. *Current strategies for engineering controls in nanomaterial production and downstream handling processes*. Cincinnati, OH: Department of Health and Human Services. https://www.cdc.gov/niosh/docs/2014-102/pdfs/2014-102.pdf (accessed January 10, 2020).

Oberbek, P., P. Kozikowski, K. Czarnecka et al. 2019. Inhalation exposure to various nanoparticles in work environment: Contextual information and results of measurements. *J Nanopart Res* 21(11):222. doi: 10.1007/s11051-019-4651-x.

Oberdörster, G. 1995. Lung particle overload: Implications for occupational exposures to particles. *Regul Toxicol Pharmacol* 21(1):123–135.

Oberdörster, G., J. Ferin, and B. E. Lehnert. 1994. Correlation between particle-size, in-vivo particle persistence, and lung injury. *Environ Health Perspect* 102(Suppl 5):173–179.

Oberdörster, G., E. Oberdörster, and J. Oberdörster. 2005. Nanotoxicology: An emerging discipline evolving from studies of ultrafine particles. *Environ Health Perspect* 113(7):823–839. doi: 10.1289/ehp.7339.

OECD [Organisation for Economic Co-operation and Development]. 2014. Grouping of chemicals: Chemical categories and read-across. https://www.oecd.org/env/ehs/risk-assessment/groupingofchemicalschemicalcategoriesandread-across.htm (accessed January 10, 2020).

OECD [Organisation for Economic Co-operation and Development]. 2015. Harmonized tiered approach to measure and assess the potential exposure to airborne emissions of engineered nano-objects and their agglomerates and aggregates at workplaces. ENV/JM/MONO(2015)19. http://www.oecd.org/officialdocuments/publicdisplaydocumentpdf/?cote=env/jm/mono%282015%2919&doclanguage=en (accessed January 10, 2020).

OECD [Organisation for Economic Co-operation and Development]. 2017. Strategies, techniques and sampling protocols for determining the concentrations of manufactured nanomaterials in air at the workplace. ENV/JM/MONO(2017)30. http://www.oecd.org/officialdocuments/publicdisplaydocumentpdf/?cote=env/jm/mono(2017)30&doclanguage=en (accessed January 10, 2020).

OECD [Organisation for Economic Co-operation and Development]. 2018. Investigating the different types of risk assessments of manufactured nanomaterials. ENV/JM/MONO(2018)24. http://www.oecd.org/officialdocuments/publicdisplaydocume ntpdf/?cote=env/jm/mono(2018)24&doclanguage=en (accessed January 10, 2020).

OECD [Organisation for Economic Co-operation and Development]. 2019. Physical-chemical decision framework to inform decisions for risk assessment of manufactured nanomaterials. ENV/JM/MONO(2019)12. http://www.oecd.org/officialdocuments/public displaydocumentpdf/?cote=env/jm/mono(2019)12&doclanguage=en (accessed January 10, 2020).

Ohlwein, S., R. Kappeler, M. K. Joss et al. 2019. Health effects of ultrafine particles: A systematic literature review update of epidemiological evidence. *Int J Public Health* 64(4):547–559. doi: 10.1007/s00038-019-01202-7.

Oomen, A. G., E. A. Bleeker, P. M. Bos et al. 2015. Grouping and read-across approaches for risk assessment of nanomaterials. *Int J Environ Res Public Health* 12(10):13415–13434.

Oomen, A. G., P. M. Bos, T. F. Fernandes et al. 2014. Concern-driven integrated approaches to nanomaterial testing and assessment: Report of the NanoSafety Cluster Working Group 10. *Nanotoxicology* 8(3):334–348.

Oomen, A. G., K. G. Steinhäuser, E. A. J. Bleeker et al. 2018. Risk assessment frameworks for nanomaterials: Scope, link to regulations, applicability, and outline for future directions in view of needed increase in efficiency. *NanoImpact* 9:1–13. doi: 10.1016/j.impact.2017.09.001.

Padmanabhan Anantha, K., and H. Gleiter. 2014. On the structure of grain/interphase boundaries and interfaces. *Beilstein J Nanotechnol* 5:1603–1615. doi: 10.3762/bjnano.5.172.

Pensis, I., J. Mareels, D. Dahmann et al. 2010. Comparative evaluation of the dustiness of industrial minerals according to European standard EN 15051, 2006. *Ann Occup Hyg* 54:204–216. doi: 10.1093/annhyg/mep077.

PEROSH [Partnership for European Research in Occupational Safety and Health]. 2019. Nano Exposure & Contextual Inf Database (NECID). https://perosh.eu/research-projects/perosh-projects/necid (accessed January 10, 2020).

Pietroiusti, A., E. Bergamaschi, M. Campagna et al. 2017. The unrecognized occupational relevance of the interaction between engineered nanomaterials and the gastro-intestinal tract: A consensus paper from a multidisciplinary working group. *Part Fibre Toxicol* 14(1):47. doi: 10.1186/s12989-017-0226-0.

Pietroiusti, A., H. Stockmann-Juvala, F. Lucaroni et al. 2018. Nanomaterial exposure, toxicity, and impact on human health. *WIREs Nanomed Nanobiotechnol* 10:e1513. doi: 10.1002/wnan.1513.

Poh, T. J., N. A. B. M. Ali, M. M. Aogáin et al. 2018. Inhaled nanomaterials and the respiratory microbiome: Clinical, immunological and toxicological perspectives. *Part Fibre Toxicol* 15(1):46. doi: 10.1186/s12989-018-0282-0.

Pope, C. A., 3rd. 2000. Epidemiology of fine particulate air pollution and human health: Biologic mechanisms and who's at risk? *Environ Health Perspect* 108(Suppl 4):713–723. doi: 10.1289/ehp.108-1637679.

Priyam, A., P. P. Singh, and S. Gehlout. 2018. Role of endocrine-disrupting engineered nanomaterials in the pathogenesis of type 2 diabetes mellitus. *Front Endocrinol* 9:704. doi: 10.3389/fendo.2018.00704.

Qi, W., J. Bi, X. Zhang et al. 2014. Damaging effects of multi-walled carbon nanotubes on pregnant mice with different pregnancy times. *Sci Rep* 4:4352.

Rossi, S., M. Savi, M. Mazzola et al. 2019. Subchronic exposure to titanium dioxide nanoparticles modifies cardiac structure and performance in spontaneously hypertensive rats. *Part Fibre Toxicol* 16(1):25. doi: 10.1186/s12989-019-0311-7.

Ryman-Rasmussen, J. P., J. E. Riviere, and N. A. Monteiro-Riviere. 2006. Penetration of intact skin by quantum dots with diverse physicochemical properties. *Toxicol Sci* 91(1):159–165.
Sahu, Y. S. 2016. Nanomaterials market by type and end-user: Global opportunity analysis and industry forecast, 2014–2022. *Allied Market Research* MA_16296. https://www.alliedmarketresearch.com/nano-materials-market (accessed January 10, 2020).
Sarlo, K., K. L. Blackburn, E. D. Clark et al. 2009. Tissue distribution of 20 nm, 100 nm and 1000 nm fluorescent polystyrene latex nanospheres following acute systemic or acute and repeat airway exposure in the rat. *Toxicology* 263(2–3):117–126. doi: 10.1016/j.tox.2009.07.002.
SCCS [Scientific Committee on Consumer Safety]. 2013. Opinion on titanium dioxide (nano form) https://ec.europa.eu/health/scientific_committees/consumer_safety/docs/sccs_o_136.pdf (accessed March 31, 2020).
SCCS [Scientific Committee on Consumer Safety]. 2015. Opinion on silicium dioxide (nano form). https://ec.europa.eu/health/scientific_committees/consumer_safety/docs/sccs_o_175.pdf (accessed January 10, 2020).
SCENIHR [Scientific Committee on Emerging and Newly Identified Health Risks]. European Commisssion. 2007. Opinion on the appropriateness of the risk assessment methodology in accordance with the Technical Guidance Documents for new and existing substances for assessing the risks of nanomaterials. https://ec.europa.eu/health/archive/ph_risk/committees/04_scenihr/docs/scenihr_o_010.pdf (accessed January 10, 2020).
SCENIHR [Scientific Committee on Emerging and Newly Identified Health Risks]. European Commission. 2009. Risk assessment of products of nanotechnologies. http://ec.europa.eu/health/ph_risk/committees/04_scenihr/docs/scenihr_o_023.pdf (accessed January 10, 2020).
SCENIHR [Scientific Committee on Emerging and Newly Identified Health Risks]. 2014. Opinion on nanosilver: Safety, health and environmental effects and role in antimicrobial resistance. Luxembourg: European Commission. doi: 10.2772/76851.
Schleh, C., U. Holzwarth, S. Hirn et al. 2012. Biodistribution of inhaled gold nanoparticles in mice and the influence of surfactant protein D. *J Aerosol Med Pulm Drug Deliv* 26(1):24–30.
Schulte, P., V. Leso, M. Niang et al. 2018. Biological monitoring of workers exposed to engineered nanomaterials. *Toxicol Lett* 298:112–124. doi: 10.1016/j.toxlet.2018.06.003.
Schulte, P. A., V. Leso, M. Niang et al. 2019. Current state of knowledge on the health effects of engineered nanomaterials in workers: A systematic review of human studies and epidemiological investigations. *Scand J Work Environ Health* 45(3):217–238. doi: 10.5271/sjweh.3800.
Schumacher, C., and D. Pallapies. 2009. Criteria for assessment of the effectiveness of protective measures. IFA DGUV. https://www.dguv.de/ifa/fachinfos/nanopartikel-am-arbeitsplatz/beurteilung-von-schutzmassnahmen/index-2.jsp (accessed January 10, 2020).
Seipenbusch, M., A. Binder, and G. Kasper. 2008. Temporal evolution of nanoparticle aerosols in workplace exposure. *Ann Occup Hyg* 52(8):707–716. doi: 10.1093/annhyg/men067.
Seixas, N. S., L. Sheppard, and R. Neitzel. 2003. Comparison of task-based estimates with full-shift measurements of noise exposure. *AIHA J* 64(6):823–829. doi: 10.1202/524.1.
Sellers, K., N. M. E. Deleebeeck, M. Messiean et al. 2015. Grouping nanomaterials: A strategy towards grouping and read-across. RIVM Report 2015-0061. Bilthoven, The Netherlands: National Institute of Public Health and the Environment (RIVM).
Shah, P. S., and T. Balkhair. 2011. Air pollution and birth outcomes: A systematic review. *Environ Int* 37(2):498–516.
Shirasuna, K., F. Usui, T. Karasawa et al. 2015. Nanosilica-induced placental inflammation and pregnancy complications: Different roles of the inflammasome components NLRP3 and ASC. *Nanotoxicology* 9(5):554–567. doi: 10.3109/17435390.2014.956156.

Singh, S., and H. S. Nalwa. 2007. Nanotechnology and health safety: Toxicity and risk assessments of nanostructured materials on human health. *J Nanosci Nanotechnol* 7(9):3048–3070. doi: 10.1166/jnn.2007.922.

Stapleton, P. A., Q. A. Hathaway, C. E. Nichols et al. 2018. Maternal engineered nanomaterial inhalation during gestation alters the fetal transcriptome. *Part Fibre Toxicol* 15(1):3. doi: 10.1186/s12989-017-0239-8.

Stockmann-Juvala, H., P. Taxel, and T. Santonen. 2014. Formulating occupational exposure limits values (OELs) (inhalation & dermal). http://scaffold.eu-vri.eu/filehandler.ashx?file=13717 (accessed January 10, 2020).

Stone, V., M. R. Miller, M. J. D. Clift et al. 2017. Nanomaterials versus ambient ultrafine particles: An opportunity to exchange toxicology knowledge. *Environ Health Perspect* 125(10):106002. doi: 10.1289/EHP424.

Stone, V., S. Pozzi-Mucelli, L. Tran et al. 2014. ITS-NANO-prioritising nanosafety research to develop a stakeholder driven intelligent testing strategy. *Part Fibretoxicol* 11:9. doi: 10.1186/1743-8977-11-9.

Sun, J., Q. Zhang, Z. Wang et al. 2013. Effects of nanotoxicity on female reproductivity and fetal development in animal models. *Int J Mol Sci* 14(5):9319–9337.

Thorley, A. J., P. Ruenraroengsak, T. E. Potter et al. 2014. Critical determinants of uptake and translocation of nanoparticles by the human pulmonary alveolar epithelium. *ACS Nano* 8(11):11778–11789. doi: 10.1021/nn505399e.

Tian, L., Y. Shang, R. Chen et al. 2019. Correlation of regional deposition dosage for inhaled nanoparticles in human and rat olfactory. *Part Fibre Toxicol* 16(1):6. doi: 10.1186/s12989-019-0290-8.

Umezawa, M., A. Onoda, I. Korshunova et al. 2018. Maternal inhalation of carbon black nanoparticles induces neurodevelopmental changes in mouse offspring. *Part Fibre Toxicol* 15(1):36. doi: 10.1186/s12989-018-0272-2.

Utembe, W., K. Potgieter, A. B. Stefaniak et al. 2015. Dissolution and biodurability: Important parameters needed for risk assessment of nanomaterials. *Part Fibre Toxicol* 12:11. doi: 10.1186/s12989-015-0088-2.

Van Broekhuizen, P., W. Van Veelen, W. H. Streekstra et al. 2012. Exposure limits for nanoparticles: Report of an international workshop on nano reference values. *Ann Occup Hyg* 56(5):515–524. doi: 10.1093/annhyg/mes043.

Van den Brule, S., J. Ambroise, H. Lecloux et al. 2016. Dietary silver nanoparticles can disturb the gut microbiota in mice. *Part Fibre Toxicol* 3(1):38.

Vincent, J. H. 1999. *Particle size-selective sampling for particulate air contaminants.* Cincinnati, OH: American Conference of Governmental Industrial Hygienists (ACGIH).

Wang, M., X. Lai, L. Shao et al. 2018. Evaluation of immunoresponses and cytotoxicity from skin exposure to metallic nanoparticles. *Int J Nanomed* 13:4445–4459.

Wang, Y., L. Xiong, and M. Tang. 2017. Toxicity of inhaled particulate matter on the central nervous system: Neuroinflammation, neuropsychological effects and neurodegenerative disease. *J Appl Toxicol* 37(6):644–667.

Warheit, D. B., R. Kreiling, and L. S. Levy. 2016. Relevance of the rat lung tumor response to particle overload for human risk assessment: Update and interpretation of new data since ILSI 2000. *Toxicology* 374:42–59. doi: 10.1016/j.tox.2016.11.013.

Węsierska, M., K. Dziendzikowska, J. Gromadzka-Ostrowska et al. 2018. Silver ions are responsible for memory impairment induced by oral administration of silver nanoparticles. *Toxicol Lett* 290:133–144. doi: 10.1016/j.toxlet.2018.03.019.

WHO [World Health Organization]. 2017. *WHO guidelines on protecting workers from potential risks of manufactured nanomaterials.* Geneva: World Health Organization.

Wichmann, H. E., and A. Peters. 2000. Epidemiological evidence of the effects of ultrafine particle exposure. *Phil Trans R Soc Lond A Math Phys Eng Sci* 358(1775):2751–2769. doi: 10.1098/rsta.2000.0682.

Wohlleben, W., M. D. Driessen, S. Raesch et al. 2016. Influence of agglomeration and specific lung lining lipid/protein interaction on short-term inhalation toxicity. *Nanotoxicology* 10(7):970–980.

Xie, W., A. Weidenkaff, X. Tang et al. 2012. Recent advances in nanostructured thermoelectric half-heusler compounds. *Nanomaterials* 2(4):379. doi: 10.3390/nano2040379.

Yang, J., S. Hu, M. Rao et al. 2017. Copper nanoparticle: Induced ovarian injury, follicular atresia, apoptosis, and gene expression alternations in female rats. *Int J Nanomed* 12:5959–5971.

Zapór, L. 2016. Strategia grupowania nanomateriałów. *Podstawy i Metody Oceny Środowiska Pracy* 3(89):5–15. doi: 10.5604/1231868X.1217548.

Zapór, L., and W. Zatorski. 2011. *Działanie cytotoksyczne modyfikowanych glinokrzemianów.* V Krajowa Konferencja Nanotechnologii NANO2011. http://docplayer.pl/2157996-V-krajowa-konferencja-nanotechnologii-nano-2011-3-7-lipca-2011-politechnika-gdanska-streszczenia-wystapien.html (accessed January 10, 2020).

Zhang, R., Y. Dai, X. Zhang et al. 2014. Reduced pulmonary function and increased pro-inflammatory cytokines in nanoscale carbon black-exposed workers. *Part Fibre Toxicol* 11:73. doi: 10.1186/s12989-014-0073-1.

Zhao, L., Y. Zhu, Z. Chen et al. 2018. Cardiopulmonary effects induced by occupational exposure to titanium dioxide nanoparticles. *Nanotoxicology* 12(2):169–184. doi: 10.1080/17435390.2018.1425502.

3 Reprotoxic and Endocrine Substances

Katarzyna Miranowicz-Dzierżawska

CONTENTS

3.1 CHARACTERIZATION OF REPROTOXIC AND ENDOCRINE SUBSTANCES

In recent years, an increase in sick absenteeism of working pregnant women can be seen in many countries. More and more potential parents are unable to get pregnant at all. An increasing number of pregnancy complications and offspring developmental disorders are also being noted [Schmitz-Felten et al. 2016].

According to the Centers for Disease Control and Prevention (CDC), in the United States alone 12% of women aged 15–44 have problems getting pregnant (due to male or female infertility) within six months to a year of sexual activity with no contraceptives [Gul et al. 2019].

Occupational health and disease surveillance have led, through several decades, to the identification of workplace chemicals adversely affecting reproductive health. The main adverse effects of reprotoxic substances – identified through animal testing and, to a lesser extent, in epidemiological studies on humans – are either a negative impact on reproductive functions and fertility or on offspring development.

Reprotoxic chemicals (also known as substances toxic for reproduction) are defined as chemical substances which disrupt the reproductive processes and proper offspring development. In the European Union, they are classified according to Regulation (EC) No 1272/2008 of the European Parliament and of the Council on classification, labeling, and packaging of substances and mixtures (the CLP Regulation). The regulation is based on the Globally Harmonized System of Classification and Labeling of Chemicals (GHS), which is an internationally agreed-upon standard, currently used, among others, as a basis for the international and national regulations concerning the transport of hazardous materials.

Among the effects of reprotoxic substances, infertility comes at the top, as it is a growing problem in many developed and highly developed countries. Due to its prevalence, it is considered a disease of society by the World Health Organization (WHO). Based on WHO data, there are 60–80 million couples worldwide permanently or temporarily affected by fertility problems. In highly developed countries, 10–12% of couples at the reproductive age face infertility [Koperwas and Głowacka 2017].

A comprehensive analysis of reprotoxicity problems requires an understanding of the nature of the risk to the health of workers exposed to reprotoxic substances, along with the knowledge of where and how they are used in the workplace. This understanding needs to encompass both the reprotoxic activity of the substances raising concerns as occupational health hazards and the legal restrictions regarding their use, which can influence exposure levels [RPA 2013].

Taking into account that reproduction is a multi-stage process, involving the production of reproductive cells (gametogenesis), fertilization, implantation of the fertilized egg, embryo-fetal development, birth, and development from birth to adolescence, the process can be disrupted at any stage by various endogenous and exogenous factors that are capable of influencing the reproductive and developmental processes in a number of ways. These ways include direct damage to male and female reproductive cells causing infertility or reduced fertility, metabolic disorders in the mother's organism influencing homeostasis and leading to impaired fetal development, as well as abnormal embryogenesis or organogenesis. Additionally, processes leading to direct fetotoxic effects, influencing the progress and course of birth, as well as early and later stages of offspring postnatal development, can also be classified into this category. Such processes may lead to the damage or death of reproductive cells, intrauterine fetal death, and developmental malformations such as abnormal ossification, impaired physical development, functional impairment of organs and systems or enzymatic deficiencies [Schmitz-Felten et al. 2016].

The most common cause of female infertility, occurring in 30–35% of women, is fallopian tube obstruction, which requires surgical intervention. For both sexes, however, immunological and hormonal disorders are the leading causes, constituting 25–30% of cases [Koperwas and Głowacka 2017].

As mentioned before, when risks of occupational exposure to reprotoxic substances were considered, women, especially potential mothers, were initially the main focus. However, since 1993 more and more attention has been paid to the issue of male fertility. Studies have found that there is an increase in incidence of developmental disorders of male genitals and testicles. Jørgensen et al. [2011] published a study which found oligospermia in young Finns as well as a higher incidence of testicular cancer. The authors concluded that young people might have more reproductive health problems than the previous generations.

Reproductive health problems are a growing threat in developed countries throughout the world and attempts to explain them have indicated the considerable impact of hormone-like substances, many of which are prevalent in the workplace [Schmitz-Felten et al. 2016].

Risk factors in the workplace capable of influencing broadly defined reproductive health include both organic and inorganic substances, such as solvents, pesticides, heavy metals, pharmaceuticals, as well as biological, physical, ergonomic, and psychosocial factors.

These factors can cause hormonal disorders in both men and women. They can lead to both anovulatory cycles as well as deterioration of sperm quality biological parameters and the so-called reduced fertility, identified in 7% of males (according to the Recommendation of the Polish Society of Andrology and National Chamber of Laboratory Diagnosticians) [PTA 2016].

Exposure to hormone-disrupting substances can not only influence the functioning of the reproductive system and reproductive processes, but can also lead to disorders in other organs and cause diseases not related to reproduction. In recent years, significant interest has been shown in substances capable of not only causing particular effects in the environment and bioaccumulation in fatty tissues, but also binding to estrogen receptors. These substances are referred to as hormonally active agents, hormone modulators, endocrine disrupting chemicals, or endocrine disruptors (EDCs or ED). Exposure to endocrine disruptors may cause, among others, metabolic disorders, obesity, diabetes, and hormone-related cancers (such as breast, prostate, ovarian, and testicular cancers).

The significance of this threat is constantly growing and its health effects may be felt a long time after the exposure has ended. It is estimated that in the European Union alone, more than 150 billion euros are spent on the treatment of diseases caused by EDCs and to support the livelihoods of people who cannot work due to health loss.

The growing problem of exposure to reprotoxic substances and endocrine disruptors warrants particular attention from all interested parties. Therefore, in this monograph we discuss the effects of substances toxic to particular stages of reproduction, as well as the classification and impact of reprotoxic substances and endocrine disruptors. We also present the legal side of the issue, including acts aimed at the limitation of exposure and exclusion of reprotoxic substances and endocrine disruptors from work and living environments. In this chapter, the effects of exposure (including combined exposure) to such substances will be discussed.

3.1.1 The Influence of Xenobiotics on Male Fertility Disorders and Female Reproductive Functions

Reproductive disorders can stem from functional disorders as well as morphological abnormalities and encompass a range of abnormalities, from infertility due to the lack of ovulation, decrease in sperm quality, amenorrhea, and decreased libido to the inability of a man to have sexual intercourse [Indulski and Sitarek 1998].

Multi-center questionnaire studies conducted by WHO from 1982 to 1985 led to the conclusion that 20% of infertility cases were mainly due to the male partner, 38% due to the female partner, 27% due to both the male and the female partner, and in 15% of cases no underlying cause was diagnosed or identified [Gul et al. 2019].

Chemical substances can not only directly damage male and female reproductive cells, but also induce metabolic disorders, disrupting homeostasis, resulting in reduced fertility.

Gonads (female – ovaries, male – testicles) perform two functions: (1) endocrine, involving the release of sex hormones and (2) non-endocrine, involving the production of reproductive cells (gametes). Gametogenesis, the process in which reproductive cells (male and female gametes) are formed through precursor cell division and differentiation, as well as endocrine functions of both ovaries and testicles, is dependent on glycoprotein molecules – gonadotropins, synthesized and released from the anterior pituitary lobe: follicle-stimulating hormone (follitropin, FSH), and luteinizing hormone (lutropin, LH). LH and FSH bind to receptors in ovaries and testicles, where they regulate the gonadal function.

In males, LH stimulates testosterone secretion by interstitial cells of Leydig (the Leydig cells) and FSH stimulates growth processes of the testicles and induces the production of androgen-binding proteins in Sertoli cells, which are a part of seminiferous tubules, required to maintain sperm maturation processes. The androgen-binding protein ensures a high local testosterone concentration in the testicles, constituting an important factor in the development of proper spermatogenesis (sperm production and maturation). Therefore, sperm cell maturation requires the presence of both LH and FSH.

In females, LH stimulates theca folliculi cells, surrounding the ovarian follicle, stimulating the secretion of sex hormones: progesterone and estrogens. A sudden, significant increase of LH in the middle of the menstrual cycle is responsible for ovulation, and continued LH secretion stimulates the corpus luteum to produce progesterone. The development of the Graafian follicles is controlled by FSH, and estrogen secretion depends on both FSH and LH [Greenspan and Gardner 2004; Klaassen and Watkins III 2014].

Chemicals temporarily blocking the LH influx can delay or prevent ovulation, causing infertility or reduced fertility linked to delayed egg cell fertilization [Klaassen and Watkins III 2014].

LH and FSH secretion are controlled by the hypothalamus through a gonadotropin-releasing hormone (gonadoliberin; GnRH), released in pulses, responsible for gonadotropin secretion during ovulation and determining the onset of puberty. In turn, sex steroids present in the circulatory system influence GnRH secretion – thereby affecting LH and FSH secretion – through both a positive and negative feedback mechanism [Greenspan and Gardner 2004].

Xenobiotics can directly affect the hypothalamus and anterior pituitary lobe, causing changes in hormones released by the hypothalamus.

Substances which affect the biotransformation processes of endogenic sex steroids in the liver and/or kidneys can interfere with the pituitary feedback system in the hypothalamic-pituitary-gonadal axis (HPG), as it is a very delicately modulated hormonal process. An example of such a substance is carbon tetrachloride (CCl_4), which affects proper sex steroid metabolism, leading to changes in their clearance (mainly in hydroxytestosterone coupling with glucuronides and sulfates), indirectly affecting the HPG axis and influencing fertility.

Endocrine neurons have synaptic vesicles containing monoamines (noradrenaline, dopamine, and serotonin) in the nerve endings. Changes in their levels or activity caused by xenobiotics affect the production of gonadotropin.

Xenobiotics may damage oocytes (the cells from which a mature egg cell is formed through DNA replication and a series of cellular divisions), thus speeding up their loss. At birth, women have approx. 400000 primary follicles, many of which atrophy soon after birth. The number of surviving primary follicles gradually decreases during the lifespan and may lead, in consequence, to decreased fertility in women. Ovary weight can be reduced due to egg cell loss or a break in the HPG axis. Toxic substances can also cause other changes in the ovaries, e.g. polyovular follicles, interstitial cell hyperplasia, corpora albicans, and absence of corpora lutea [Klaassen and Watkins III 2014].

An increased frequency of irregular and prolonged, often heavy and painful menstrual periods has been observed in women exposed to benzene and its homologues, styrene, trichloroethylene, mercury vapors, or inorganic mercury compounds. Menstrual disorders have also been observed in women exposed to toxic metal fumes (lead and manganese) and those working in the oil refining and rubber industries. Chronic ethanol abuse is also considered reprotoxic [Starek 2007].

In case of male exposure to xenobiotics, the blood-testis barrier between the capillaries and seminiferous tubules disrupts or even makes it impossible for chemicals to be exchanged between blood and fluid inside the seminiferous tubules, which limits the health risk of exposure to certain reprotoxic substances affecting the testicles. It is important to note that the importance of male infertility is often underestimated or entirely disregarded [Gul et al. 2019].

Some chemicals (e.g. cadmium) can impact testicular structure and their circulatory systems (responsible for maintaining proper temperature of the scrotum), causing testicular ischemia and reduced fertility.

Chemical substances other than cadmium, known to negatively impact male fertility are, among others, lead, carbaryl, carbon disulfide, toluenodiamine, and addictive substances, such as cannabis and cocaine [Starek 2007].

The spermatogenesis process, which involves a number of biochemical processes and cellular changes, is highly susceptible to toxic effects of xenobiotics such as certain phthalates, glycol ethers, and antiandrogens.

Toxic substances can also affect sperm cells, which are released into the lumen of seminiferous tubules and transported to the epididymis, where they can be affected by toxic substances. It has been shown that chlorocarbohydrates (the derivatives of which are, e.g., pesticides and fungicides) and epichlorohydrin inhibit the energy metabolism of sperm cells, prohibiting their proper functioning. Moreover, the process of moving sperm-containing fluid through the epididymal duct can also be disrupted by chemicals through the creation of an environment inappropriate for sperm cell maturation, when the excess fluid is removed by active transport [Klaassen and Watkins III 2014].

Ethylene glycol methyl ether is also a strong gonadotoxic factor, causing a reduction in the number of spermatocytes and testicular atrophy. Similarly, 2,3,7,8-tetrachlorodibenzo-p-dioxin (TCDD) and its congeners cause testicular atrophy and degenerative changes in the seminiferous tubules and impair spermatogenesis.

An example of a xenobiotic gonadotoxic for males (causing oligospermia: 1–20 million of sperm cells per 1 ml of ejaculate, or even irreversible azoospermia: <1 million of sperm cells per 1 ml of ejaculate) is 1,2-dibromo-3-chloropropane (DBCP), once used as a nematocide in agriculture [Starek 2007].

Pesticides, especially organophosphates, can have a detrimental effect on the reproductive functions also through the impairment of neuroendocrine processes involved in the erection and ejaculation. In fact, all the substances which affect the autonomic nervous system can in principle affect male fertility. They might be responsible for either a complete lack of erection or short-lived erection.

Epidemiological studies on the relationship between work in greenhouses and the risk of reproductive disorders indicate a potentially significant threat to fertility and reproductive health of workers employed in agriculture. Most of them indicate the exposure to pesticides capable of influencing pregnancy and its outcome to be a potential cause. Studies on time to pregnancy as well as sperm quality and quantity show that work in greenhouses can increase the risk of infertility. In sperm studies, a higher risk of abnormalities in sperm cell morphology and survival, as well as a decrease in their numbers, was observed. Likewise, data on the time to pregnancy period indicates a risk of delay in pregnancy in women working in agriculture. Results of these studies confirm the necessity of increasing awareness in people using plant protection products as to potential health risks, including fertility disorders, in order to minimize their exposure. During the period of pregnancy planning, it is necessary for both men and women to minimize exposure to pesticides [Jurewicz and Hanke 2007].

3.1.2 Embryotoxic and Teratogenic Effects

Less than half of human fertilization results in the birth of a completely healthy infant. Toxicological studies have shown that a number of chemicals in the environment exhibit teratogenic properties – teratology being the study of abnormalities of physiological development, including congenital malformations.

One of the first twentieth-century attempts at a synthetic approach to the fundamental studies in the field of teratology was a 1959 paper published in the *Journal of Chronic Diseases* by James G. Wilson from Florida Medical School in Gainesville, titled 'Experimental Studies on Congenital Malformations'. Based on the results of studies conducted on birds and mammals during the previous four decades, the author formulated general principles and guidelines, including issues such as species recommended for teratologic studies, rules governing human embryonic and fetal development, and factors influencing these processes. Some of these assumptions are still valid today [Klaassen and Watkins III 2014].

The first principle states that an embryo's susceptibility to an external agent depends on the embryo's developmental stage at the time when that agent interferes with it. Early stages of organogenesis, the phase of embryonic development in which internal organs are formed, taking place from the third to eighth week of pregnancy, are considered particularly dangerous as a period of increased susceptibility to malformations.

The second principle states that each teratogenic agent has particular effects on a specific aspect of cellular metabolism. This principle implies that every agent produces a particular, though not necessarily unique, type or pattern of malformation when applied to a certain species at a certain time.

The third principle states that the genotype of the test animal plays a role in how the animal reacts to the teratogenic agent (a given species or strain might be genetically susceptible to given teratogenic effects).

The fourth principle states that any agent capable of producing malformations also causes an increase in embryonic mortality. The author also states that abnormal development and death are likely different degrees of reaction to teratogenic agents. The incidence and degree of symptom severity increase with the increase in dose, from slight changes to a lethal effect.

In the fifth and final principle, Wilson states that a teratogen might not harm the maternal organism, citing the rubella virus as an example. Pregnant women with the virus may not experience any discomfort or adverse reaction to the virus, but the fetus might be affected. The author notes that there is no standard relationship between the reaction of the pregnant female and the embryo's reaction [Tantibanchachai 2017].

Conversely, although developmental toxicity is the result of damage to the embryo at the cellular level, damage can occur not only as a result of direct action on the embryo or fetus, but also indirectly through the action of the toxic factor on the mother's body and/or placenta, or by a combination of effects direct and indirect.

It is important to distinguish between direct and indirect developmental toxicity to properly assess the safety of pregnant animals in toxicological studies, because the maximum dose is assessed based on the toxic activity on the maternal organism.

If developmental toxicity only occurs when the chemical is toxic to the maternal organism, the influence of the agent may be indirect – caused by the improper developmental conditions in the maternal organism, not the direct interaction with the fetus [Klaassen and Watkins III 2014].

Congenital defects are defined as structural or functional abnormalities, occurring during the intrauterine period, regardless of their time of identification – by USG during the prenatal period, at birth or later in life, as an accidental discovery. Congenital defects can be caused by genetic disorders (5%), chromosome aberrations (7%), epigenetic changes (changes in gene expression) (3%), or environmental factors (10%). In 20% of cases, the underlying cause is multifactorial. Surprisingly, in about 60% of cases, the congenital defect etiology is still unknown [Carachi and Doss 2019].

Birth defects such as cleft lip or palate, anencephaly, spina bifida, certain congenital heart defects, pyloric stenosis, hypospadias, inguinal hernia, clubfoot, or congenital hip dislocation can be categorized both as multifactorial diseases and hereditary diseases. The multifactorial hypothesis assumes that the genetic predispositions are modulated by both internal and external (environmental) factors. The latter are medicinal drugs, chemicals, radiation, and viruses. In extreme cases, they can be the cause of embryo loss of fetal death (a fetus is less vulnerable to morphological changes than an embryo, but functional impairment of mental development, reproduction, or renal function can occur) [Soni and Singh 2018].

It is important to note that psychological changes (mental impairment) are not considered teratogenic effects with no accompanying anatomical deformities. The term 'teratogenic' should be distinguished from the term 'embryotoxic effect', which encompasses all adverse changes, which a chemical compound can cause in an embryo of a fetus. It might be difficult to precisely distinguish between such changes – it should therefore be considered that teratogenic changes are anatomical deformities of a fetus occurring with doses of the substance not yet causing clearly visible toxic effects in the mother's body [Piotrowski 2006].

Damage to a fetus or an embryo at different stages of its development (embryotoxic and teratogenic effects) as well as congenital malformations in the offspring (teratogenic effects) can be caused by polychlorinated biphenyls (PCB), polychlorinated dibenzodioxins (PCDD), and polychlorinated dibenzofurans (PCDF) [Starek 2007].

The following substances and medicinal drugs can also influence human prenatal development: lead, ethylene oxide, carbon monoxide, toluene, organic mercury compounds, iodides, ethanol, methylene blue, lithium, cocaine, aminoglycosides, folic acid antagonists (aminopterin, methotrexate), angiotensin receptor blockers (sartans), large doses of quinine, drugs used for the treatment of cancers (busulfan, chlorambucil, cyclophosphamide, cytarabine), danazol, diethylstilbestrol, ergotamine, fluconazole, androgens, ACE inhibitors (captopril, enalapril), coumarins, anticonvulsive drugs (diphenylhydantoin, trimethadione, valproic acid, carbamazepine), metamizol, mizoprostol, thalidomide, tetracyclines, penicilamine, retinoids (isotretinoin, etretinate, acitretin), or large doses of vitamin A [Klaassen and Watkins III 2014].

In a Spanish study on the correlation between parental occupational exposure to pesticides and the frequency of congenital malformations on their offspring [Garcia et al. 1999 – quoted by Jurewicz and Hanke 2008], the analysis of 261 congenital malformations in infants showed that when mothers worked in agriculture for one month before the fertilization and in the first trimester of the pregnancy, the odds ratio of a congenital malformation was significantly higher. Most of the congenital malformations were defects of the nervous system and the palate, or complex defects. A significant correlation between parental exposure and congenital malformations in the offspring was found for pyridyl derivatives, and no increase in the share of congenital malformations in infants has been found in the case of parental exposure to organochlorine and organophosphorus pesticides, organosulfur compounds, and carbamates.

In Finland, an analysis of a group of 1,306 infants with cleft lips and/or cleft palate has shown a higher incidence of these defects in offspring born to mothers employed in agriculture, who have been exposed to pesticides [Nurminen et al. 1995 – quoted by Jurewicz and Hanke 2008].

Among women exposed to pesticides in the first trimester of pregnancy, the risk of cardiovascular disorders in infants was twice as high. Mothers exposure to herbicides and rodenticides almost tripled the risk of having a child with this type of defect [Loffredo et al. 2001 – quoted by Jurewicz and Hanke 2008].

In farmers exposed to chloro-phenoxy carboxylic herbicides, a higher risk of having a baby with congenital disorders of the cardiovascular, respiratory, and musculoskeletal systems was found [Schreinemachers 2003 – quoted by Jurewicz and Hanke 2008].

In a study carried out in Minnesota, USA, a higher risk of central nervous system and urinary tract disorders was found among the offspring of workers handling and spraying pesticides. An increased incidence of these defects was present in the case of fathers exposed to atrazine, trifluralin, and chlorophenoxyacetic acids [Garry et al. 1996 – quoted by Jurewicz and Hanke 2008].

In a study conducted in the state of Washington, USA, a nearly 4 times greater risk of limb defects was found in the group of infants whose mothers worked in agriculture than in the group in which none of the parents worked in agriculture. The risk of these defects was more than 4.5 times for cases in which only the father worked in agriculture. Exposure of any of the parents to plant protection products, increased the rish of having a dead child due to birth defects was also reported [Jurewicz and Hanke 2008].

While examining the causes of congenital malformations in the offspring, it is also important to note that, besides genetic factors, the mother's exposure, before and during pregnancy, to certain pesticides and other chemicals, medicinal drugs, alcohol, tobacco, radiation, and socioeconomic and demographic factors (e.g. low income), as well as maternal nutrition, also play an important role.

Working or remaining in the vicinity or at landfill disposal sites, smelters, or mines can also be a risk factor, especially if the mother is exposed to other environmental risk factors, suffering from nutritional deficiencies, or afflicted with syphilis or rubella [WHO 2016].

In recent years, reports started to appear, stating that maternal exposure to particulate matter with an aerodynamic diameter no larger than 2.5 μm (PM2.5) can also affect fetal development. A critical 'time window' is the second and fourth week after conception, when maternal exposure to PM2.5 can cause clinically recognized early pregnancy loss (CREPL) [Zhang et al. 2019]. Maternal exposure to PM2.5 is also a risk factor involved in low birth weight (LBW) and pre-term birth (PTB) [Zhang et al. 2019; Hyder et al. 2014]. Oxidative stress, DNA methylation, changes in mitochondrial DNA (mtDNA) content, and endocrine disruption can all play a role in the mechanism of adverse effects induced by PM2.5 particles on pregnant women and fetuses [Li et al. 2019]. In the described cases, it was not specifically said whether such effects had been caused by the particles themselves, or the adsorbed substances.

A number of published literature studies contain the experimental results concerning teratogenic effects of chemical substances on animals. During the analysis of such results, it is important to remember that the extrapolation of toxicological data from animal studies to humans is particularly difficult in this case, as teratogenic effects are dependent on the species to a larger extent than other toxic effects. This can stem from the fact that when a metabolite of a given chemical substance is the active teratogen, the efficiency of its production is genetically predetermined and can vary strongly between species. For example, methylmercury has strong teratogenic effects in rodents, causing cleft palate, where no anatomical deformities due to exposure to this compound have been found in humans and there is no basis to classify methylmercury as a human teratogen [Piotrowski 2006].

3.1.3 Influence on Postnatal Developmental Disorders

When speaking of reprotoxic substances, it is important to also mention xenobiotics, especially lipophilic alkaline substances, which pass into breast milk.

Breastfed children can be exposed in this way to persistent organic pollutants (POPs): polychlorinated dibenzodioxins (PCDDs), polychlorinated dibenzofurans (PCDFs), polychlorinated biphenyls (PCBs), and polychlorinated hydrocarbons, such as organochlorine insecticides [Starek 2007; Mead 2008].

First reports on the possibility of human milk contamination by chemicals present in the environment were published in Laug et al. [1951], who established that the milk of healthy African Americans contained significant amounts of an organochlorine insecticide 1,1,1-trichloro-2,2-bis(p-chlorophenyl)ethane (DDT). Since then, a number of studies have been conducted worldwide, focusing on the issue of human milk containment, with most of them confirming the presence of DDT and other organochlorine pesticides. In the late 1960s, the presence of PCBs was confirmed in a majority of human milk samples. Currently, the following substances are also considered human milk contaminants: hexachlorobenzene, dieldrin, chlordane, heptachlor, and bisphenol A [Mead 2008].

In recent years, researchers have also addressed the problem of human milk contamination with mycotoxins, such as aflatoxins, ochratoxins, patulins, zearalenone, fumonisins, nivalenol, and deoxynivalenol. These compounds can cause acute or chronic toxic symptoms with teratogenic, mutagenic, cancerogenic, immunotoxic, or hepatotoxic effects. Recent studies unambiguously show that constant exposure to mycotoxins is, among others, a cause of growth impairment in children. Aflatoxin B1 is considered to be the most toxic due to the presence of a lactone ring and two furan rings, including those with double bonds at positions 8 and 9, thanks to which the aflatoxin molecule can bind even more effectively to a protein or DNA molecule, impairing cellular activity [Kowalska et al. 2017].

Potential occupational exposure to mycotoxins mainly occurs in the agricultural and food industries, but is also possible in other economic sectors. Workers can be exposed to mycotoxins during cultivation, harvesting, and storage of the crops. Mycotoxin-related health risk is also connected to further crop marketing or processing of crops (trade, grinding, fodder production, etc.). Exposure analysis of the workers handling raw plant materials (storage, grinding, packaging of products such as coffee, pepper, nutmeg, and cocoa beans) has shown that this line of work can carry the risk of exposure to mycotoxins contained in organic dust of a plant material attacked by fungi [Soroka et al. 2008].

Breast milk mycotoxin contamination and subsequent infant exposure may cause problems in breastfeeding women in the aforementioned lines of work; however, the main source of mycotoxins found in breast milk seems to be the intake of contaminated foodstuff. This fact does not, however, diminish the importance of occupational breast milk contamination exposure.

Fakhri et al. [2019], based on the results of over 200 studies, have established that the minimal values of aflatoxins in breast milk have been found in Sierra Leone (0.80 ng/L), and maximum values – in the United Arab Emirates (UAE) (465.76 ng/L). The lowest prevalence of ATM1 in human breast milk was reported in Brazil (2%) and the highest in Gambia, Tanzania, and Jordan (100%). The lowest concentrations were found in America (10.30 ng/L) and the highest in Southeast Asia (358.99 ng/L). Moreover, the lowest and highest prevalence of ATM1 in human breast milk were observed in the Western Pacific region (7%) and in Africa (52%). It has been established that the prevalence of ATM1 in human breast milk significantly increased with the increase in average annual rainfall as well as poverty (as per gross domestic product).

Memiş and Yalçın [2019] have found AFM1 and ochratoxin A in human breast milk at concentration levels of 3.07 pg/mL and 1.38 ng/mL, respectively. Levels of zearalenone and deoxynivalenol (produced by certain species of *Fusarium*) were higher than 0.3 ng/mL in 59% of tested samples and higher than 10 ng/mL in 37.7% of the tested samples.

3.1.4 Hormonal Balance Disorders

A large number of xenobiotics cause fertility disorders through their hormonal or hormone-like activity; examples are: polychlorinated biphenyls (PCB) and polychlorinated aromatic hydrocarbons, particularly isomers of 1-chloro-4-[2,2,2-trichlo ro-1-(4-chlorophenyl)ethyl]benzene/1,1,1-trichloro-2,2-bis(p-chlorophenyl)ethane (DDT) and hexachlorocyclohexane (HCH) as well as methoxychlor [Starek 2007].

Endocrine disrupting chemicals are substances which occur naturally or as contaminants in food and the environment. In recent years, several definitions of these substances have been proposed.

The US Environmental Protection Agency (EPA) defined endocrine disruptors as 'exogenous agents that interfere with the synthesis, secretion, transport, binding, action, or elimination of natural hormones in the body that are responsible for the maintenance of homeostasis, reproduction, development, and/or behavior' [Pałczyński 2018].

According to WHO, an endocrine disruptor is an 'exogenous substance or mixture that alters the function(s) of the endocrine system and consequently causes adverse health effects in an intact organism, or its progeny, or (sub)populations' [Pałczyński 2018].

In the European Union, the definition is as follows: 'An ED is an exogenous substance that causes adverse health effects in an intact organism, or its progeny, secondary to changes in endocrine function'. A potential EDC is a substance with properties, which can lead to endocrine disruption in a healthy organism [Rutkowska et al. 2015].

According to the Polish Bureau for Chemical Substances (www.chemikalia.gov.pl/substancje_zaburzajace_gospodarke_hormonalna.html – accessed January 22, 2020), the commonly accepted definition of substances interrupting the functioning of the endocrine system, called 'endocrine disruptors' or 'endocrine disruptor chemicals', is that given in a report of the International Programme on Chemical Safety from 2002: 'An endocrine disrupter is an exogenous substance or mixture that alters function(s) of the endocrine system and consequently causes adverse health effects in an intact organism, or its progeny, or (sub)populations'.

As the aforementioned Bureau for Chemical Substances reports, 'the most commonly known substances, that have caused or are suspected to cause endocrine disruptions in wildlife, belong to the group of polychlorinated biphenyls. These substances caused fertile and immunological disorders in Baltic seals, leading to a decrease in their population, a significant decrease in egg shell thickness and abnormal gonadal development in birds of prey feeding on animals exposed to these substances, and embryonic development disorders in birds feeding on fish contaminated with chemicals. Developmental disorders, including those of gonadal functioning and structure have also been observed in reptiles, amphibians, fish, and snails environmentally exposed to anthropogenic chemicals. In the majority of cases, as the studies were not carefully executed experiments, but rather observations of wild animals, the evidence on a causal relation between exposure, often quite frequent, and biological effects, is rather weak and insufficient to establish a dose–response characteristic. A number of health changes of unknown etiology, which appear in certain populations, are suspected of a causal relation to environmental exposure to chemicals, which can cause endocrine disruption. Sperm quality decreased in Western Europe after the year 1930, when research enabling the analysis of this phenomenon was started (oligospermia, decrease in sperm vitality, increasing incidence of abnormal sperm morphology). In addition, an increase in the incidence of fertility disorders in humans, including in those who were occupationally exposed to certain substances, an increase in the frequency of spontaneous abortions, a decrease in the number of male newborns in certain human populations, and an increase in the incidence of developmental disorders of the male reproductive system (cryptorchidism, hypospadias) were linked, among others, to environmental exposure to endocrine disrupting chemicals, in particular, regulation through sex hormones. According to WHO, studies have not provided sufficient evidence to confirm hypotheses of a causal link between these health changes and exposure to endocrine disrupting chemicals. Similarly, hypotheses of the leading role of hormonal imbalances have not been confirmed in clinical and epidemiological studies, imbalances which can be tied to environmental exposure to chemicals, in etiology of endometriosis, precocious puberty, neurological and immunological disorders, as well as breast, endometrial, testicular, prostate, and thyroid cancers. However, the results of these studies do not make it possible to refute such a hypothesis, as the test did not have sufficient statistical power. Therefore, it is not possible to exclude hormone regulation disorders, especially the reproductive system in adults and the development in children, caused by environmental exposure to exogenous substances. It is known that some of these substances can bind to endogenous hormone receptors or disrupt hormonal regulation in different ways. One of the fundamental weaknesses of studies explaining the prevalence of adverse health effects in humans is the lack of sufficiently reliable data on the extent of exposure to substances which can cause endocrine disruptions' (www.chemikalia.gov.pl/substancje_zaburzajace_gospodarke_hormonalna.html – accessed January 22, 2020).

Compounds from the EDCs group are commonly used as pesticides in agriculture, and as plasticizers and intermediates in the industry. The best known and dangerous endocrine disruptors, besides polychlorinated biphenyls (PCB), are bisphenol A (BPA), phthalates, and dioxins, the exposure to which may occur in daily life during the consumption of meals, drinking contaminated water (tap or bottled), or breathing contaminated air. These compounds tend to accumulate in animal fatty tissue, from which they can cross into the organisms of humans, who are at the top of the food chain. In humans, the presence of endocrine disruptors was found in fatty tissue as well as body fluids: serum, urine, milk, and even amniotic fluids [Rutkowska et al. 2015].

Currently, there are over 800 compounds which have been shown to have endocrine disrupting properties [Klaassen and Watkins III 2014].

Because the endocrine disruptor structure shows similarities to steroid hormones, EDCs are capable of interacting with receptors, making it impossible for them to bind to endogenous hormones [Rutkowska et al. 2015].

Chemical substances belonging to a range of classes cause developmental toxicity in at least three ways related to the endocrine system: as steroid receptor ligands, by modifying steroid hormone metabolic enzymes, and by disrupting the secretion of tropic hormones regulating the secretion of other hormones by the hypothalamic-pituitary axis [Klaassen and Watkins III 2014].

Apart from chemical agents used in industry and agriculture, pharmaceuticals such as diethylstilbestrol, estradiol, or antiestrogenic drugs, such as tamoxifen and clomiphene citrate, also belong to the category of toxic estrogenic and antiestrogenic substances. It has been established that female offspring is in general more susceptible to such substances than male offspring, with the most often observed abnormalities being changes in the course of puberty, lower fertility rates, and reproductive system anomalies [Klaassen and Watkins III 2014].

Antiandrogens belong to another large class of chemical substances causing endocrine disruption. The most important consequences of developmental exposure to antiandrogens usually concern males and involve defects of the phallus (such as hypospadias), retained nipples, testicular atrophy and additional sex glands, as well as lower sperm production [Klaassen and Watkins III 2014].

3.2 LEGAL ASPECTS OF SUBSTANCES TOXIC TO REPRODUCTION AND ENDOCRINE DISRUPTORS

3.2.1 Substances Included in Annex XIV to Regulation (EC) No 1907/2006

A legal act which is in force from June 1, 2007, in all member states of the European Union and which concerns, among others, substances toxic to reproduction, is the Regulation of the European Parliament and of the Council for the Registration, Evaluation, Authorization and Restriction of Chemicals (REACH).*

It has been specified in the preamble that the REACH Regulation should contribute to the realization of the Strategic Approach for International Chemicals Management (SAICM)† and assure a high level of human health and environmental protection as well as the free circulation of goods while enhancing competitiveness and innovation and promoting the development of alternative methods for the assessment of hazardous substances.

An important objective of the system set by this regulation is to encourage that substances of very high concern are ultimately replaced (set out in the regulation as, among others, substances toxic to reproduction Category 1 and 2) with less hazardous substances or technologies.

Substances evaluated to have adverse health or environmental effects according to the REACH Regulation require permits issued by the European Commission to be placed on the market (including reprotoxic substances). The list of substances subject to authorization is included in Annex IV to the REACH Regulation. The first six substances were included in this list with the Commission Regulation (EU) No 143/2011 of February 17, 2011. Three of them have shown reprotoxic activity: Bis(2-ethylhexyl) phthalate (DEHP), butyl benzyl phthalate (BBP), and dibutyl phthalate (DBP).

Successive commission regulations changed Annex XIV (so far, these have been Regulations No 125/2012 of February 14, 2012, No 348/2013 of April 17, 2013, No 895/2014 of August 14, 2014, and No 2017/999 of June 13, 2017), expanding the list, the current of which can be found on the European Chemicals Agency (ECHA) website: https://echa.europa.eu/en/authorisation-list (accessed January 22, 2020).

---

* Regulation (EC) No 1907/2006 of the European Parliament and of the Council of December 18, 2006, for the Registration, Evaluation, Authorisation and Restriction of Chemicals (REACH), the formation of European Agency of Chemicals, which amends the Directive 1999/45/EC and annuls the Ordinance of the Council No. 793/93 (EEC) and the Ordinance of the Commission No. 1488/94, well as Council Directive 76/769/EEC and Commission Directives 91/155/EEC,93/67/EEC, 93/105/EC and 2000/21/EC (*Journal of Laws of 2009*, item 396, as amended).

† SAICM – Strategic Approach to International Chemicals Management. It includes the Dubai Declaration for Strategic Approach to International Chemicals Management, a global policy framework to foster the sound management of chemicals, adopted on February 6, 2006, in Dubai.

3.2.2 Substances Included in the SVHC Candidate List

According to Article 57 of the REACH Regulation, Substances of Very High Concern (SVHC) may be:

- carcinogenic, mutagenic, or reprotoxic substances (CMR) of Category 1 or 2,
- persistent, bioaccumulative and toxic (PBT), or very persistent and very bioaccumulative (vPvB) in accordance with the criteria set out in Annex XIII to the REACH Regulation,

  and/or
- other substances which give rise to the equivalent level of concern to those of other substances listed above, e.g. endocrine disruptors.

Substances identified to have any of the abovementioned properties are first placed on a candidate list published by the ECHA, which then recommends substances to be included in Annex XIV (list of substances subject to authorization) by the Commission.

The current candidate list is available on the ECHA website: https://echa.europa.eu/candidate-list-table (accessed January 22, 2020).

3.2.3 Substances Classified in Regulation (EC) No 1272/2008 of the European Parliament and of the Council

The REACH Regulation has been changed by the Regulation of the European Parliament and of the Council (EC) No 1272/2008* of December 16, 2008, on the classification, labeling, and packaging of substances and mixtures, which implements on a European level the principles of classification, labeling, and packaging of substances and mixtures in the Globally Harmonized System of Classification and Labeling of Chemicals (GHS). Regulation No 1272/2008, known as the CLP Regulation ('Classification, Labeling and Packaging') came into force on January 20, 2009.

According to the CLP Regulation, reprotoxic activity encompasses adverse effects to reproductive functions and fertility in male and female adults as well as reproductive toxicity in the offspring.

In this classification system, reproductive toxicity is differentiated into two main categories:

a) adverse effects on reproductive functions and fertility;
b) adverse effects on offspring development.

* Regulation of the European Parliament and of the Council (EC) 1272/2008 of December 16, 2008, on classification, labeling and packaging of substances and mixtures, amending and repealing Directives 67/548/EEC and 1999/45/EC, and amending Regulation (CE) No 1907/2006 (*EU Journal of Laws* L 353/2, of 31.12.2008, as amended).

For the purposes of classification, the hazard class 'reproductive toxicity' is divided into:

a) adverse effects on sexual function and fertility;
b) adverse effects on offspring development; and
c) adverse effects on or via lactation.

Reprotoxic substances/mixtures, classified as per the Regulation of the European Parliament and of the Council (EC) No 1272/2008, are labeled as shown in Table 3.1.

3.2.4 Other Legal Aspects

In the United States, the first Toxic Substances Control Act (TSCA) was introduced in 1976. The legislator's intention was to assure that the obtained chemical substances are safe during the entire period of their use, since production through to disposal. In the following 35 years, the US Environmental Protection Agency (EPA) has tested approximately 200 chemical compounds/groups of compounds, five of which (polichlorinated biphenyls, chloro-fluoro alkanes, dioxin, asbestos, and chromium(VI) compounds) were regulated [Sobczak 2012].

On May 24, 2016, the US House of Representatives adopted the Chemical Safety for the 21st-Century Act, which had replaced the TSCA. This act enabled the Environmental Protection Agency (EPA) to re-examine over 85,000 chemical substances, which had been deemed safe based on the TSCA. The new act states that chemical substances must be considered safe for children, pregnant women, workers, and other vulnerable social groups to be applied and used in the production process [Lee 2016].

The Environmental Protection Agency will be able to prohibit distribution, use, and sale of chemical substances it considers to be unsafe. Under the previous act (TSCA), the decision to prohibit the use of particular chemical substances was taken with an allowance for the financial repercussions of such a prohibition; the new act takes into account only the health effects of using a given substance[Sobczak 2012].

The official text of TSCA as amended by the Frank R. Lautenberg Chemical Safety Act of the twenty-first century is available in the US Code on FDSys, from the US Government Printing Office.

In turn, on May 22, 2001, in Stockholm, the Convention on Persistent Organic Pollutants was prepared, which included the decision of the UNEP Governing Council (United Nations Environmental Programme) No 19/13C of February 7, 1997, to take international action for the protection of human health and the environment through ventures aiming at the decrease and/or elimination of the emission and discharges of persistent organic pollutants (POPs). Knowing that POPs are toxic, resist degradation, and bioaccumulate, the parties to the convention were aware of the need for global action, as POPs are transported through air, water, and migratory species across international boundaries and can be deposited far from their place of release where they accumulate in terrestrial and aquatic ecosystems.

The objective of this convention is to eliminate or limit the production or use of all internationally produced POPs, and continuous minimization, and where possible, ultimate elimination of unintentional release of POPs.

TABLE 3.1
Label Elements for Reproductive Toxicity

Classification	Category 1A or Category 1B	Category 2	Additional Category for Effects on or via Lactation
GHS pictogram* (* - pictograms are red-framed)			No pictogram
Signal word	Danger	Caution	No signal word
Hazard statement	H360: May damage fertility or the unborn child.	H361: Suspected of damaging fertility or the unborn child (state specific effect if known) (state route of exposure if it is conclusively proven that no other routes of exposure cause the hazard)	H362: May cause harm to breastfed children.
Precautionary Statement **Prevention**	P201: Obtain special instructions before use. P202: Do not handle until all safety precautions have been read and understood. P281: Use personal protective equipment as required.	P201: Obtain special instructions before use. P202: Do not handle until all safety precautions have been read and understood. P281: Use personal protective equipment as required.	P201: Obtain special instructions before use. P260: Do not breathe dust/fume/ gas/mist/ vapor/spray. P263: Avoid contact during preg nancy/while nursing. P264: Wash … thoroughly after handling. P270: Do not eat, drink or smoke during use.
Precautionary Statement **Response**	P308 + P313: If exposed or concerned, get medical advice/attention.	P308 + P313: If exposed or concerned, get medical advice/attention.	P308 + P313: If exposed or concerned, get medical advice/attention.
Precautionary Statement **Storage**	P405: Store locked up.	P405: Store locked up.	
Precautionary Statement **Disposal**	P501: Dispose of contents/container to …	P501: Dispose of contents/container to …	

Modified from Regulation (EC) No 1272/2008 of 16 December 16, 2008, on classification, labeling and packaging of substances and mixtures, amending and repealing Directives 67/548/EEC and 1999/45/EC, and amending Regulation (CE) No 1907/2006 (EU Journal of Laws L 353/2, of 31.12.2008, as amended), no permission required.

Substances subject to the convention have been listed in three annexes:

1. Substances intended to be completely withdrawn from production, marketing, and use: aldrin, alpha-hexachlorocyclohexane, beta-hexachlorocyclohexane, chlordane, chlordecone, C10-C13 chloroalkanes (short-chain chlorinated paraffins) (SCCPs), decabromodiphenyl ether (commercial mixture, c-decaBDE), dieldrin, endrin, technical endosulfan and its isomers, hexabromodiphenyl ether, heptabromodiphenyl ether, tetrabromodiphenyl ether, pentabromodiphenyl ether, heptachlor, heksabromodiphenyl, hexabromocyclododecane (HBCDD), hexachlorobenzene (HCB), hexachlorobutadiene, lindane, mirex, pentachlorobenzene, pentachlorophenol and its salts and esters, polichlorinated biphenyls, polichlorinated napthalenes, toxaphene.
2. Substances the use of which is limited to specific purposes: DDT (1,1,1-tri chloro-2,2-bis(p-chlorophenyl)ethane) and perfluorooctane sulphonic acid and its derivatives (PFOS).
3. Substances that are formed as unintentional by-products as a result from human activity: polychlorinated dibenzo-p-dioxins and dibenzofurans (PCDD/PCDF), hexachlorobenzene (HCB), hexachlorobutadiene (HCBD), pentachlorobenzene, polychlorinated biphenyls (PCBs), and polychlorinated naphthalenes.

    (quoted by Polish Bureau for Chemical Substances: www.chemikalia.gov.pl/konwencja_sztokholmska.html – accessed January 22, 2020)

Persistent organic pollutants in the EU member states are subject to the Regulation of the European Parliament and of the Council (EC) No 2019/1021 of June 20, 2019, concerning persistent organic pollutants, which implements in Union law the commitments set out in the Stockholm Convention on Persistent Organic Pollutants and in the Protocol to the 1979 Convention on Long-Range Transboundary Air Pollution on Persistent Organic Pollutants.

3.2.5 Occupational Exposure Limit Values

Occupational exposure limit values (OEL) are a part of occupational risk management linked to exposure to chemical substances. OELs are established for specific chemical substances. The main task of establishing OEL values, as put by the Scientific Committee for Occupational Exposure Limits to Chemical Agents (SCOEL), is that 'the recommended levels must be set to ensure that exposure, even when repeated on a regular basis throughout a working life, will not lead to adverse health effects for people who were exposed to the chemicals, and/or their progeny'.

As a rule, OEL values established for particular chemicals should ensure the protection of workers, unborn children, and future generations.

It is important to note, that when establishing occupational exposure limit values, a review of available literature also on reprotoxicity is carried out as a typical step. For many chemical substances, however, no data on their negative influence on reproductive health is available or sufficient. Moreover, occupational exposure limit values developed based on health effects, should be established only in the cases in which the assessment of all available scientific data leads to the conclusion that it is possible to set a distinct threshold dose, below which the exposure to a particular substance should not cause adverse health effects [Bertazzi 2010 – quoted by Schmitz-Felten et al. 2016].

During a workshop ('Workplace risks to reproductivity: from knowledge to action'), organized by the EU-OSHA in Paris, 2014, a discussion took place about whether reprotoxic substances can be considered as substances for which a safe threshold dose is possible to establish. Currently, a number of sources argue that it is impossible to establish dose–response or dose–effect relationships for occupational exposure to such chemicals, similarly as for most of the carcinogenic and mutagenic substances.

PEL (Permissible Exposure Limits), hygienic standards, legally binding in the United States, are also based on threshold effects of chemical substances, with the NOAEL (No Observed Adverse Effect Level) or LOAEL (Lowest Observed Adverse Effect Level) being used for their establishment as the baseline. The PEL values are published by OSHA* (Occupational Safety and Health Administration) in the Federal Register, along with arguments for the change or introduction of a standard [Gromiec and Czerczak 2002].

In the European Union, BOELV (Binding Occupational Exposure Limit Values) are established for particular chemical substances as legally binding values. However, there is no single BOELV value the basis of which would be the reproductive toxicity or endocrine activity of the substance (currently, BOELVs only concern some of carcinogenic and/or mutagenic substances).

Similarly, among the indicative limit values (IOELV), which are implemented under Council Directive, there is no reference to reprotoxicity or endocrine activity of chemical substances (there is only one possible annotation present: 'skin', indicating the possibility of significant dermal absorption).

It is important to note that substances toxic to reproduction and development can be considered as such only after they have been appropriately tested. For chemical substances, there is a significant discrepancy between the number of existing chemicals and the amount of chemicals which have been assessed for their reproductive toxicity [Lawson et al. 2003 – quoted by Schmitz-Felten et al. 2016]. This can explain why e.g. current lists of substances toxic to reproduction in the European Union include only approx. 150 chemicals (including pesticides) classified as reprotoxic substances (category 1A: substances toxic to reproduction, classified based on evidence from studies on humans; category 1B: substances toxic to reproduction, classified based mostly on animal studies) among thousands of chemicals in the lists of substances with harmonized classifications [RPA 2013 – quoted by Schmitz-Felten et al. 2016].

* TLVs (Threshold Limit Values), published by the ACGIH (The American Conference of Governmental Industrial Hygienists) do not have a legal binding character but can act as significant indicators on the recommended occupational exposure levels.

3.3 OCCUPATIONAL EXPOSURE TO REPROTOXIC SUBSTANCES AND ENDOCRINE DISRUPTORS

According to epidemiological data, exposure to reprotoxic chemicals mostly concerns workers:

- employed in the production or use of pesticides
- employed in the cosmetics industry and in hairdressing
- employed in hospitals and other health services
- employed in plastic industry
- employed in jobs involving the production of wood and timber products
- employed in the rubber industry
- employed in jobs involving the refining of copper and copper electrolytes
- employed in the construction sector
- exposed to organic solvents
- exposed to metals [Schmitz-Felten et al. 2016]

Occupational exposure to reprotoxic substances is linked to the production of plastics, inks, textiles, dyes, and glues, as well as intermediate products and solvents. The highest number of hazards has been noted in the occupational groups of ISCO 8 and ISCO 3.

Developmental toxicity is also linked to the exposure to chemicals used in the production of glues, textiles, dyes, insecticides, greases, varnishes, coolants, cements, and cellulose, as well as solvents in the rubber and plastics industries. In this case, the greatest number of hazards also involved groups ISCO 8 and 3 with groups 9, 7, and 6 having a significant part as well [Montano 2014].

According to the International Standard Classification of Occupations ISCO-88, group 8 consists of plant and machine operators and assemblers, group 3 – of technicians and associate professionals, group 9 elementary occupations, group 7 – craft and related trades workers, and group 6 – skilled agricultural and fishery workers. (www.klasyfikacje.gofin.pl/kzis/6,0.html – accessed January 22, 2020).

The total number of EU workers identified to be potentially exposed to specific reprotoxic substances was estimated in the

> Study to collect recent information relevant to modernizing EU Occupational Safety and Health chemicals legislation with a particular emphasis on reprotoxic chemicals with the view to analyze the health, socio-economic and environmental impacts in connection with possible amendments of Directive 2004/37/EC and Directive 98/24/EC. Final Report prepared for DG Employment, Social Affairs & Inclusion, in 18 March 2019

and is given in Table 3.2.

The authors stress that the numbers given include workers who may not be at all exposed or be exposed for very short periods of time or to a substance in concentrations below the observed adverse effect level of exposure, therefore being overestimated in relation to the actually exposed population.

TABLE 3.2
Potentially Exposed Workforce – Conclusions from Substance Evaluations (Rounded)

Substance	Estimate	Total No. of Exposed Workers	Men	Women of Reproductive Age
Lead	Central estimate	17,800	17,000	800
	High estimate	43,300	41,500	1,700 (all women)
Bisphenol A (BPA)	Central estimate	1,000*	690	230
Borates	Central estimate	250,000	190,000	60,000
Imidazolidine-2-thione (ETU)	Central estimate	Not relevant	Not relevant	45,000
4-tert-butylbenzoic acid (pTBBA)	Central estimate	110,000	100,000	10,000
2-ethoxyethanol	Central estimate	1,530	1,360	170
	High estimate	1,650	1,450	200
2-(4-tert-butylbenzyl) propionaldehyde (2,4-TBP)	Central estimate	22,000	13,000	8,000
Dodecyl phenols	Central estimate	337,900	304,100	33,800
Organotins	Low estimate	1,480	1,030	330
	Central estimate	4,430	3,100	990
	High estimate	7,390	5,170	1,660
Retinol	Low estimate	6.23m (6.2m in agriculture)	4.02m (4m in agriculture)	1.24m (1.2m in agriculture)
	Central estimate	6.28m	4.045m	1.225m
	High estimate	6.33m (6.2m in agriculture)	4.07m (4m in agriculture)	1.21m (1.2m in agriculture
Dinoseb	Low estimate	1,600	1,400	300
	High estimate	3,300	2,700	500
Aprotic solvents	DMAC[a] and DMF[b]	245,450	204,250	41,210
	NMP[c]	339.80	253,780	85,910
Maximum total potentially exposed – low to central estimates and no adjustment for multiple exposures		7,557,430 (or 1,357,430 without workers using retinol in agriculture)	5,106,600 (or 1,106,600 without workers using retinol in agriculture)	1,495,740 (or 295.740 without workers using retinol in agriculture)
Maximum total potentially exposed – central to high estimates and no adjustment for multiple exposures		7,690,670 (or 1,490,670 without workers using retinol in agriculture)	5,186,630 (or 1,186,630 without workers using retinol in agriculture)	1,530,940 (or 330,940 without workers using retinol in agriculture)

Source: Vencovsky et al. 2019; reuse is authorized, provided the source is acknowledged.

Note: The vast majority are at extremely low levels of exposure and have therefore been discounted. Numbers shown are workers in BPA manufacturing plants who could be exposed above any of the thresholds.

[a] DMAC: N,N-dimethylacetamide (EC No: 204-826-4; CAS No: 127-19-5)

[b] DMF: N,N-dimethylformamide (EC No: 200-679-5; CAS No: 68-12-2)

[c] NMP: 1-methyl-2-pyrrolidone (EC No: 212-828-1; CAS No: 872-50-4)

Among reprotoxic substances with the highest number of exposed persons are retinol, aprotic solvents (1-methyl-2-pyrrolidone, N,N-dimethylacetamide, and N,N-dimethylformamide), dodecylphenols, borates, 4-tert-butylbenzoic acid.

Exposure to retinol is linked to the production of cosmetics of which it is an ingredient, as well as animal feed and medicinal drugs [Ries and Hess 1999].

Boric acid and borates are used to pest control, including insects, dust mites, algae, fungi, and higher plants. They are typically used for eliminating ants, fleas, termites, cockroaches, as well as insects and fungi in wood. Sodium metaborate is used as an herbicide, and disodium octaborate is used for woodworking [Cox 2004].

Lead is also among reprotoxic substances with the highest number of exposed people. It is also used in over 100 sectors, and many types of work involve the use of lead. Industry sectors with the highest number of workers exposed to lead are production, construction, and mining. The main types of work known to involve occupational exposure to lead include radiator repair, battery production and regeneration, primary smelting, home repair, repair of steel structures, foundry works, production of ceramics, scrap recycling, production and use of ammunition, demolition, and work with lead paints. Lead ore is still being mined all in a number of countries worldwide, with the largest share of production coming from China, Australia, Peru, Canada, Mexico, and Sweden. Because lead is mined and melted, simultaneously undergoing distribution in the environment, but is never destroyed, there is also a secondary market for this toxic metal. In 2016, in the United States, approx. 1.07 million tonnes of secondary lead were produced, which corresponds to 69% of domestic lead consumption. It has been established that in about 95% of cases (when the source of exposure is known), when average blood lead level (BLL) is 25 μg/dL or more, it is an occupational exposure case. In the years 1976–1980 to 2015–2016, average blood lead level in the US population has decreased from 12.8 to 0.82 μg/dL, i.e. by 93.6%. However, despite significant progress in the reduction of lead exposure in the United States, currently (2015–2016), in 0.2% of 1–5-year-olds, there were found BLL values of 10 μg/dL or more, and in 1.3% – BLL values of 5 μg/dL or more. It is known that average blood lead level of less than 10 μg/dL leads to adverse effects on IQ and behavior [Dignam et al. 2019].

Workers employed in the welding and soldering sectors are often exposed to vapors of lead and its compounds. Soldering is one of the main processes in the electronics industry, and lead is one of the metals most commonly used in soldering wires. In Mohammadyan et al. [2019a], the exposure of brazers to lead (in concentrations 0.09 ± 0.01 mg/m^3) was twice as high as the standard exposure limit of 0.05 mg/m^3. In Mohammadyan et al. [2019b], the average blood lead level for brazers was determined to be 10.59 ± 3.25 μg/dL.

In the risk assessment of occupational exposure to lead, it is important to note that the natural environment can also be a significant source of exposure. In high-income countries, accumulated lead residue is a source of constant population exposure, while in lower- and middle-income countries the lack of proper legislation or no possibility of its enforcement additionally influence population exposure. The main sources of lead exposure in the United States are paints, lead-containing industrial residues, and batteries. In Australia, these are paints, dusts, imported toys, and traditional medicines. In China, the main sources of lead exposure are electronic waste, traditional medicines and industry, and in Nigeria – electronic waste, paints, and batteries. As for Mexico, these sources are glazed pottery, utensils, and water contaminated with lead. The sources of lead in India are cosmetics and traditional medicines, and in France – lead paint from old houses, imported pottery and cosmetics, as well as industrial emissions [Obeng-Gyasi 2019].

Occupational exposure to PCDDs and PCDFs exist in sectors of the industry in which processes generating these compounds are used:

- production and further processing of chlorophenols (wood conservation, fungicides, tanning industry, paints and coatings industry),
- production of phenoxyacetic acids and herbicides based on them,
- production of cellulose and paper (cellulose pulp bleaching),
- production of chlorine using graphite electrodes (electrode sludges contain large amounts of PCDDs/PCDFs),
- iron and steel production
- ore sintering and foundry processes, processing of non-ferrous metal scrap,
- copper production,
- fossil fuel combustion in electrical and heating installations,
- lead gasoline combustion in car engines [Szymańska et al, 2020].

Occupational exposure to cytostatics concerns medical and pharmaceutical personnel of cancer hospitals and pharmacies preparing cytostatic solutions for infusions, coming into contact with cytostatic drugs during their preparation and administration to patients.

3.4 RISKS RESULTING FROM EXPOSURE TO REPROTOXIC SUBSTANCES IN THE WORK ENVIRONMENT

3.4.1 Metals

Metals and metalloids are chemicals for which comprehensive epidemiological studies have been carried out to test their reprotoxicity. A number of animal study results are also available, allowing for the classification of some metals and metalloids as substances toxic to reproduction according to Regulation No 1271/2008 (see legal aspects).

One of the best studied and most serious occupational hazards linked to metals is the exposure to lead. According to the statistics, lead is the fourth most often used element worldwide [Mohammadyan et al. 2019b].

Currently, the largest amount of lead is used in the production and recycling of batteries, but exposure is also present, among others, during construction and demolition works, as well as during scrap melting and works with scrap metals. Lead is found in several different forms, including organic lead, and exposure takes place mainly through the inhalation of lead-containing dust. After absorption, lead accumulates in the body, with its half-life in different tissues ranging from several days to several years. During pregnancy, lead is mobilized from the bone and its concentration in the mother's blood increases, which may be toxic to the child during its fetal development (especially since the placenta is not a barrier for lead) as well as during breastfeeding [Schmitz-Felten et al. 2016].

Mohammadyan et al.'s [2019b] studies of occupational exposure to lead and its blood levels in 40 women employed in soldering in two electrical components factories in the city of Neyshabur (Iran) in 2017–2018 have shown that with the average lead air concentration at the workstations of 0.09±0.01 mg/m^3, worker blood levels of this element are in the range of 10.59 ± 3.25 μg/dL.

There is conflicting data on the influence of lead on the duration of pregnancy and on birth weight of newborns. One of the articles has shown that in women exposed to this metal, premature births and low birth weight in children are more prevalent; however, the results of other studies do not confirm these observations [Indulski and Sitarek 1998].

A critical consequence of lead affecting children is the decrease in the intelligence quotient (IQ), though controversy arose as to whether the observed results are caused by lead exposure during pregnancy or at later stages. Studies from the Mexico area, 2006, indicate that mother's exposure to lead plays a significant role during the fetal stage – particularly around the 28th week of pregnancy, which is considered as the critical period for intellectual development of offspring. The mode of action of lead on the nervous system is based on the disruption of calcium homeostasis (lead substitutes calcium), which affects a number of cellular communication processes. Lead can also substitute zinc in some enzymes and proteins important for the functioning of the nervous system. The main lead activity areas in the brain are the endothelial cells of the blood-brain barrier, and the immature endothelial cells still developing in the brain are, with a high probability, less resistant to lead exposure than mature cells. Rossouw et al. [1987] found that lead absorption in rat brains was higher in the prenatal period than in the postnatal period – with the same level of exposure, the concentration of lead was found to be 6 times more during the fetal stage, 3.3 times more in newborns, and 2 times more in children post the breastfeeding period [Jakubowski 2014].

Even a moderately high blood lead level affects sperm quality and fertility in males. SCOEL estimates that the threshold blood lead level, for which male fertility starts to be affected, is approx. 40 μg/dL. However, there are reports of significantly lower concentrations affecting male fertility [Schmitz-Felten et al. 2016]. Male fertility disorders may be caused by adverse effects of lead on sperm cells – in males exposed to high lead concentrations, abnormal sperm morphology, decreased sperm mobility, and azoospermia have been found [Indulski and Sitarek 1998].

The assessment of the influence of mercury on reproduction and offspring development has been the subject of a number of publications since the 1960s, when occupational health effects have been studied. For example, a higher incidence of menstrual cycle disorders has been found in female employees of dental clinics and in other sites, where the level of exposure to metallic mercury vapors was high [Indulski and Sitarek 1998]. Retrospective epidemiological studies of 296 women occupationally exposed to mercury vapors (concentrations from 1 to 200 μg/m^3) and 394 women not exposed to these vapors have shown a significantly higher incidence of abnormal periods in the exposed group [Yang et al. 2002 – quoted by Sapota and Skrzypińska-Gawrysiak 2010].

Mercury is used in the production of alkaline batteries, fluorescent lamps, mercury lamps, control and measuring devices (thermometers, pressure valves, flow meters) [Sapota and Skrzypińska-Gawrysiak 2010]. The amount of mercury in a fluorescent lamp is 5–50 mg, depending on the size of the lamp and the year of its production (newer lamps tend to contain less mercury than older ones), which makes them a significant source of exposure. Similarly, certain batteries contain mercury, though its use has decreased in recent years [Bjørklund et al. 2019].

Occupational exposure to mercury vapors is present mainly during the mining and processing of cinnabar, as well as in the chlor-alkali industry – in the production of chlorine and sodium hydroxide (lye) using electrolytic methods, metal alloys, dyes, and fungicides, as well as in the production and use of instruments filled with mercury, such as flow meters, various measuring equipment, thermometers, barometers, and rectifiers. Workers employed in laboratories, research offices, dental practices, and photographer's shops are also exposed to mercury [Sapota and Skrzypińska-Gawrysiak 2010].

Some researchers on exposure to this metal say that the personnel of dental practices belong to the professional group most exposed to mercury. In a number of countries, dental personnel are daily exposed to a mixture of metallic mercury vapor and its inorganic compounds because of the use of amalgam composed of 50% mercury. Numerous studies indicate the presence of a significantly higher level of mercury in the blood of dental personnel in comparison with the control group, especially older dentists and dental assistants. Levels of mercury higher than 10 ppm have also been found in the hair strands of 8% of dentists, and higher than 5 ppm, in 25% of dentists [Bjørklund et al. 2019].

Mercury has a negative impact on fertility in both men and women. Results of studies indicating a 2 times higher risk of spontaneous miscarriages in wives of males exposed to mercury vapors date back to the 1990s. Data on reproductive toxicity of mercury has also been provided by the results of animal studies [Indulski and Sitarek 1998].

In women, exposure to mercury can be the cause of infertility due to endocrine disruption – the increase in estrogen levels as a result of mercury inhibits the release of luteinizing hormone (LH), consequently causing female infertility through an increase in the release of prolactin, analogous to the dopamine effect at the pituitary and midbrain levels. Exposure to mercury can also result in polycystic ovary syndrome, premenstrual syndrome, menstrual pains, amenorrhea, early menopause, endometriosis, and galactorrhea [Bjørklund et al. 2019].

There is an adverse effect of mercury on the Sertoli cells in the seminal epithelium and on the functioning of the seminal vesicles [Sapota and Skrzypińska-Gawrysiak 2010].

The underlying toxic effect mechanism of mercury has not yet been sufficiently explained. It is known that mercury has a high affinity for sulfhydryl groups, the presence of which is a characteristic of almost all proteins, including the majority of enzymes. Mercury damages cell membranes, which leads to cell death. It binds to amino and carboxyl groups as well, though to a much lesser extent to the sulfhydryl groups [ATSDR 1999 – quoted by Sapota and Skrzypińska-Gawrysiak 2010]. Cytotoxicity mechanisms of mercury compounds are mainly based on the decrease in reduced glutathione levels, the generation of free radicals, and lipid peroxidation (mercury can disrupt the activity of glutathione reductase and superoxide dismutase). Literature data suggests that mercury causes cytogenetic damage (numerical chromosomal aberrations, increased incidence of chromosomal aberrations and micronuclei), as it most probably negatively affects DNA damage repair. This mechanism might be connected to blocking the enzymes repairing DNA damage and associated proteins, having so-called zinc fingers, which are structures rich in sulfhydryl groups [Cebulska-Wasilewska et al. 2005 – quoted by Sapota and Skrzypińska-Gawrysiak 2010].

Cadmium is a heavy metal common in nature. It is used for the production of, among others, batteries, galvanic covers, fluorescent materials, pigments, rubber, plastic stabilizers, nickel-cadmium batteries, nuclear reactor control rods, and other neutron absorption components. Cadmium is also used in the production of fertilizers based on industrial waste. Its compounds are also used in forensics for better fingerprint detection [Krzywy et al. 2011].

Exposure to cadmium in the prenatal period adversely affects the unborn child, as cadmium can partially cross the placental barrier. Studies have shown that cadmium concentration in cord blood can range from 10 to 70% of the concentration of this metal in the mother's blood. A heightened cadmium concentration level in the placenta causes a decrease in zinc transportation to the fetus and disrupts the metabolism of calcium and vitamin D, which affects growth and development of the fetus. Llanos and Ronco [2009] have shown that in the placenta of newborns with lower birth weight, a higher level of cadmium was determined than in placenta from infants with a normal birth weight. Yang et al. [1997] have proven that cadmium causes oxidative damage to human fetal lung fibroblasts. Exposure to cadmium causes increased production of free radicals, and, consequently, increased lipid peroxidation and loss of mitochondrial membrane potential, which can lead to cell damage and death. Wang et al. [2016] suggest that prenatal cadmium exposures can have a negative impact on neurological development [Jastrzębski et al. 2016].

Studies on laboratory rats have shown that intraperitoneal administration of cadmium causes a decrease in sperm mobility and disrupts spermatogenesis and spermiogenesis. Authors suggest that damage to endothelium of testicular blood vessels might play a key role. Oxygen-free radicals formed due to the influence of cadmium can also be toxic to the testicular tissue. It has been found that chronic exposure to cadmium can cause a decrease in the testicular mass due to the reduction of the number of cells as a result of their necrosis, exacerbated apoptosis, and disruption of intercellular connections. Cadmium is also an endocrine disruptor causing decreased testosterone production by the testicles [Krzywy et al. 2011].

Exposure to cadmium during infancy often causes health effects which can only be observed in adolescence or adulthood. These health effects are mostly the impaired metabolic processes or the impaired functioning of the nervous and hematopoietic systems. Scientists more often indicate that exposure to cadmium in the childhood period results in mental development disorders [Jastrzębski et al. 2016].

3.4.2 Organic Solvents

The term 'organic solvents' encompasses a vast variety of chemical compounds belonging to different chemical groups, many of which can adversely affect the reproductive functions of the exposed people and the development of their offspring. For example, 2-ethoxyethanol disrupts male fertility, causing a decrease in sperm cell count; ethylene glycol methyl ether causes a decrease in the number of spermatocytes and testicular atrophy; exposure to methyl ethyl ketone, trichloroethylene, xylene, and toluene can cause intrauterine growth retardation; and exposure to toluene, xylenes, and styrene – menstrual cycle disorders [Schmitz-Felten et al. 2016].

The application and effects on reproductive health of selected organic solvents are shown below.

2-Methoxyethanol is used in the chemical, metallurgical, machine, electronics, furniture, textiles, leather, and cosmetics industries. It is a solvent for cellulose acetate and nitrocellulose, natural and synthetic resins, chlorinated rubber, paints, varnishes, polishes, and inks. It is also used in the production of photographic films and in photolithographic processes (e.g. in the manufacturing of semiconductors). It is also used as a fixative in the production of perfumes, liquid soaps, and other cosmetics. Negative effects of 2-methoxyethanol on reproductive health and fetal development have been shown in epidemiological studies. Exposure of males to 2-methoxyethanol in concentrations of 17–26 mg/m^3 caused a significant decrease in the size of the testicles. In women exposed to 2-methoxyethanol in the first trimester of pregnancy, the risk of spontaneous miscarriages was 2–3 times higher. In newborns, an increase in the incidence of ossification disorders, congenital malformations of the ribs and the cardiovascular system, cleft palate, and multiple birth defects were noted [Szymańska and Bruchajzer 2010].

Technical xylene is a mixture of xylene isomers, containing 44–70% of *m*-xylene and approx. 20% each of *p*-xylene and *o*-xylene. The mixture often contains 6–10% of ethylbenzene. Xylene is mostly used for fuel enhancement, as solvents for paints, varnishes, glues, resins, and as a cleaning and degreasing agent in organic synthesis. In experimental studies on mice administered a dose 2,060 mg/kg of xylene per day between day 6 and day 15 of gestation, a significant increase in the incidence of cleft palate and lower birth weight was found in the fetuses. Delayed jaw ossification was found in fetuses of pregnant Wistar rat females, exposed to xylene vapors with a concentration of 870 mg/m^3 6 h per day between day 4 and day 21 of gestation. The offspring of mothers exposed to xylene had a slightly larger body mass and less developed motor skills (especially the female offspring). A dose of 3,100 mg/kg per day of xylene has caused a 30% mortality rate of the fetuses [Ligocka 2007].

N,N-Dimethylformamide (DMF) is mostly used as a solvent in organic synthesis and petrochemistry. It is used in the production of low and high molecular vinyl and acrylic polymers, films, fibers, and coatings, as a solvent for printer inks, glues, and polyurethane varnishes, and in the manufacture of artificial leather. An analysis of the semen of workers employed in an artificial leather factory and exposed to DMF has shown a significant decrease in sperm motility in comparison to samples from non-exposed males. It is suspected that DMF metabolite, N-methylformamide (NMF), is responsible for sperm function disorders, but according to Chang et al. [2004] this conclusion requires further research to confirm it. Based on experimental animal studies, it has been established that N,N-dimethylformamide has embryotoxic and teratogenic properties after inhalation, dermal, and oral exposure [Jankowska and Czerczak 2010].

N-Methyl-2-pyrrolidone (NMP) is used as a solvent and paint remover, pigment suspension agent, and an intermediate in chemical, electrical, and electronics industries. Fetotoxic and embryotoxic effects of this compound have been confirmed in laboratory animals. Administered to female mice, rats, or rabbits in non-toxic or slightly toxic doses for mothers, it caused an increase in intrauterine morbidity, incidence of stillbirths, and lower birth weight of the offspring. The compound has caused similar effects after intraperitoneal, inhalation, intragastric or dermal administration. It is also probably reprotoxic to humans [Sitarek 2005].

Methyl ethyl ketone (MEK; 2-butanone) is mostly used as a solvent for surface coatings. It has also found applications in the removal of long-chain paraffins from lubricating oils and in the production of synthetic resins, artificial leather, glues, and aluminum foil. The compound is also used in the pharmaceutical and cosmetics industries. In Scandinavia, a higher incidence of damage to the central nervous system was observed among the offspring of women exposed to organic solvents containing among others methyl ethyl ketone during the first trimester of pregnancy. Altenkirch et al. [1978] report that exposure of pregnant rats to methyl ethyl ketone in a concentration of 2,352 mg/m^3 (800 ppm) or 4,600 mg/m^3 (1,500 ppm) has caused an increase in the number of resorptions in comparison to the non-exposed animals. Methyl ethyl ketone has also probably shown weak fetotoxic effects, but data on its reprotoxic activity is difficult to interpret, as this compound typically occurs in combination with other solvents (acetone, ethyl acetate, n-hexane, toluene, and alcohols) and it cannot be unambiguously stated that MEK is the one responsible for reprotoxic effects of the mixtures [Grunt and Czerczak 2007].

Toluene is used in the production of rubber, resins, detergents, dyes, medicinal drugs, trinitrotoluene, benzoic acid, and toluene diisocyanate. It is a component of paints, shellac, corrosion inhibitors, thinners, and cleaning and sanitary agents. Occupational exposure to toluene can take place at the stages of production, manufacturing, packaging, and storage of intermediate products and products containing toluene, as well as during their use, e.g., during painting or cleaning. In women occupationally exposed to toluene in concentrations of 170–550 mg/m^3, spontaneous abortions were reported; whereas toluene exposure did not cause menstrual disorders [Ng et al. 1992]. In males employed in printing works and exposed to toluene in concentrations of 30–416 mg/m^3, a decrease in the concentration of follicle-stimulating hormone (FSH), luteinizing hormone (LH), and free testosterone was found; the concentration values did however still remain in the reference value ranges, and the effects of toluene exposure were reversible [Svensson et al. 1992]. Courtney et al. [1986] observed fetotoxic effects of toluene in a concentration of 1,500 mg/m^3 in exposed mice. Jones and Balster [1998] established that in mothers exposed to toluene in a concentration of 7,500 mg/m^3 a decrease in body mass and a decrease of reflex speed in the offspring were noted, with no signs of toxic effects in mothers [Jakubowski 2007].

No evidence of embryotoxic and teratogenic effects of hexane have been found based on inhalation exposure studies on rats and intragastric administration to rats (Jakubowski 2006).

In the analysis on the influence of chemicals on reproductive health, it is important to take into account the possible influence of interfering factors, as well as the air concentrations of substances at workstations. Moreover, non-occupational factors, which can also influence reproduction, such as stress, age, socioeconomic conditions, general health, nutrition, and addictions, have to be taken into account as well.

In addition, consideration should also be given to the fact that the symptoms of reprotoxic activity can be linked to the exposure to chemical substances mixtures, e.g., organic solvents, as in the case of chemical industry or chemical laboratories. In this case of exposure, an increased incidence of spontaneous abortions in women, as well as miscarriages in wives of exposed men, has been found. An increased incidence of congenital malformations has been found both for offspring of mothers exposed to organic solvents during pregnancy (especially the first trimester) and for fathers exposed to chemical substances. In women exposed to benzene and its homologues, styrene, and trichlorethylene, an increased incidence of menstrual disorders has been found (irregular cycles, prolonged, painful periods). Such problems have also been reported in the case of women working in the oil refining and rubber industries [Schmitz-Felten et al. 2016].

Laboratory workers employed in the pharmaceutical industry are also exposed to a number of chemical substances, which can adversely affect reproduction [Zhu et al. 2006 – quoted by Attarchi et al. 2012]. A significant correlation between spontaneous abortions and occupational exposure to organic solvent mixtures has been found in a study conducted in a pharmaceutical factory in the suburbs of Teheran in 2010. A correlation between occupational exposure to organic solvent mixtures and time to pregnancy has also been observed. Air concentrations of formaldehyde, phenol, n-hexane, and chloroform have been measured at the workstations at the facility laboratories to be 0.01, 0.5, 20.7, and 3.2 ppm (parts per million), respectively [Attarchi et al. 2012].

3.4.3 Pesticides

The first working environment factor considered to be causally linked to disorders of spermatogenesis was dibromochloropropane (DBCP), used in citrus farming [Indulski and Sitarek 1998].

In rodents, exposure to DBCP (1,2-Dibromo-3-chloropropane) affected reproductive cells and androgen-dependent sex differentiation (decrease in testicle mass, androgen levels, and hypothalamus size) [Kaltenecker Retto de Queiroz and Waissmann 2006]. Azoospermia or oligospermia, chronic or lasting for several years was found in males exposed to this compound [Indulski and Sitarek 1998]. A decrease in the number of male births can also be linked to DBCP exposure. The removal of workers from the site of exposure enables the recovery of proper volume of sperm in the ejaculate, but in the case of azoospermia, the condition persists. Results of testicular biopsies indicate that DBCP targets spermatogonia [Kaltenecker Retto de Queiroz and Waissmann 2006].

The following are also considered reprotoxic pesticides: hexachlorocyclohexanes (HCH) – beta-HCH and gamma-HCH (lindane), carbaryl, chlordane, dicofol, dieldrin, DDT (1,1,1-trichloro-2,2-bis(p-chlorophenyl)ethane) and its metabolites, endosulfan, heptachlor and heptachlor epoxide, malathion, methomyl, methoxychlor, mirex, oxychlordane, parathion, synthetic pyrethroids, toxaphene, and trans-nonachlor [Kaltenecker Retto de Queiroz and Waissmann 2006].

Compounds belonging to this group (carbaryl, chlordecone) cause infertility or reduced fertility in males. Some insecticides, such as DDT, methoxychlor, or chlordecone (kepone) may impair development through the inhibition of blastocyst implantation in the uterine wall. In turn, aldrin, DDT, lindane, and toxaphene, which act similarly to estrogens, cause excessive menstrual bleeding, decrease in female fertility, and increase in the incidence of miscarriages. They also adversely affect the fetus, causing growth retardation, even inducing congenital malformations of the central nervous system [Indulski and Sitarek 1998].

It is worth noting that estrogenic activity of DDT isomers is very weak compared to estradiol (10^3–10^6 weaker), but due to the possibility of accumulation and long half-life, it can have estrogenic effects and act as an androgen antagonist. DDE (1,1-dichloro-2,2-bis(p-chlorophenyl)ethylene), a DDT metabolite, has antiandrogenic properties and can disrupt estrogen metabolism at the stage of its synthesis or elimination.

The exposure of male rats to low concentrations of the procymidone fungicide caused hypospadias and smaller size of androgen-dependent tissues: prostate, seminal vesicle, or glans penis [Kaltenecker Retto de Queiroz and Waissmann 2006].

In workers employed in factories producing lindane (γ-hexachlorocyclohexane), a significant increase of the LH level was noted, along with a slight increase in the FSH level and a slight decrease in the testosterone level. Studies have shown that lindane can accumulate in human testicles, damaging seminal epithelium and Sertoli's cells, as well as affecting the number of spermatids (immature male sex cells) [Kaltenecker Retto de Queiroz and Waissmann 2006].

Carbendazim is a substance causing damage to male fertility in rats. Administered *per os* for 10 consecutive days to female rats in the dose of 400 mg/kg, the compound decreases fertility and even leads to permanent sterility due to seminiferous tubule atrophy. In female rats exposed to this compound, an increase in intrauterine morbidity and offspring congenital malformations were found. Embryotoxic effects were found in cases in which the administered doses were not toxic to the pregnant females [Sitarek 2004].

Endocrine activity of pesticides is described in Section 3.5.

3.4.4 Polichlorinated Biphenyls, Polychlorinated Dibenzo-p-dioxins, and Polychlorinated Dibenzofurans

Dioxins and dioxin-like polychlorinated biphenyls are commonly considered to be, among other components, the most hazardous for the human body due to their chemical properties, persistence, and degradation resistance. Such chemicals are known as Persistent Organic Pollutants (POP) [Stec et al. 2012].

Dioxins have never been manufactured on purpose and have no practical applications. Combustion and incineration processes (of waste, sludge from waste treatment plants, and organic substances, including chlorine-containing plastics) are a direct and significant source of these compounds [Kulik-Kupka et al. 2017; Stec et al. 2012].

Dioxins are the by-products of a number of chemical processes, e.g., cellulose pulp bleaching, ore smelting, the production of steel of certain pesticides, such as herbicides based on polychlorophenoxyacetic acid [Stec et al. 2012]. A decrease in dioxin emission was observed after bleaching paper with chlorine and production of organochlorine plant protection products was stopped. These compounds are also the products of natural reactions or phenomena, such as volcanic eruptions, burning fire, or forest fires [Kulik-Kupka et al. 2017].

The dioxin group includes polychlorinated and polybrominated dibenzofurans (PCDFs, PBDFs), and polychlorinated and polybrominated dibenzodioxins (PCDDs, PBDDs). In their molecules there are 8 positions which can be substituted by atoms or radicals – consequently, there are 75 dioxin congeners and 135 dibenzofuran congeners. These compounds exhibit varying degrees of toxicity, which depends on both the number of atoms of chlorine or bromine (or other elements) in the molecule, as well as the substitution location. 2,3,7,8-Tetra-Chloro-Dibenzo-p-Dioxin (TCDDs) is considered to be the most hazardous to living organisms [Słowińska et al. 2011].

The toxicity mechanism of dioxins is not fully known. These compounds form complexes with cytosolic aryl hydrocarbon receptors (AhR) and induce the expression of genes which drive microsomal monooxygenase systems – with the products of the expression including various forms of cytochrome P450. In the case of dioxin exposure, their increased production impairs liver functioning. Dioxins have a toxic effect on the human body, e.g., through hormonal activity (changes in thyroid hormone concentration, which impairs psychomotor development and increases the concentration of thyroid-stimulating hormone – TSH) and carcinogenicity (increased risk of death due to gastrointestinal and lung cancers, as well as lymphomas). Exposure to dioxins also causes the so-called chloracne [Kulik-Kupka et al. 2017]. Retrospective studies conducted in Seveso 20 years after a catastrophe, involving 472 women who attempted to become pregnant, allowed to establish that the incidence of infertility or a longer time to pregnancy was linked to TCDDs serum levels [Słowińska et al. 2011].

Cells subjected to toxic compounds develop defense mechanisms based on transportation through ATP-dependent membrane pumps MDR1, MRP1, and BCRP. The phenomenon is associated with increased expression of genes coding these proteins [Panczyk et al. 2007]. Wu et al. [2004] noticed that TCDD increased hypermethylation of *H19* and *Sp1* genes (specificity protein 1, involved in MDR1 expression control – multi-drug resistance gene, in conditions of exposure to environmental factors) in mouse embryos which are not nested. Moreover, dioxin administered to pregnant rats caused 50 significant changes to methylation of differential DNA methylation regions (DMR) in sperm cells. These changes are likely connected to puberty disorders and kidney diseases [Rzeszutek et al. 2014].

Studies aimed at the assessment of the influence of dioxins on the development of endometriosis in women are being conducted for years. Igarashi et al. [2005], based on animal tests, suggest that these compounds may have a significant influence on the development of the disease. In turn, Guo et al. [2009] expressed skepticism toward such reports, as they are not based on epidemiological data, which could unambiguously confirm the hypothesis that exposure to dioxins leads to an increased risk of endometriosis in women.

Increased exposure to dioxins might also be linked to a decrease in sperm quality in young males living in certain industrialized areas. Mocarelli et al. [2011] studied

39 sons of mothers who were exposed to dioxins after the Seveso incident and compared them to 58 sons of mothers from the control group (not exposed). Their analysis indicates that exposure to even relatively small doses of dioxins (around 19 ppt) during the prenatal period and breastfeeding can cause a permanent decrease in sperm quality. It can also increase the release of the follicle-stimulating hormone, which is responsible for seminal vesicle enlargement, stimulation of spermatogenesis, as well as increased production of androgen-binding protein, responsible for the proper functioning of testosterone. The researchers have also observed a decrease in inhibin B concentration, which might indicate damage to seminiferous tubules. According to Mocarelli et al. [2011] dioxins can contribute to male infertility. A number of developmental deficits were found in children exposed to high levels of PCDDs and PCDFs both before and after birth.

Moreover, concerns arose as to PCDDs being capable of disrupting the homeostasis of thyroid hormones, which play a key role in proper brain development and cell growth, which could negatively affect neurobehavioral development of prenatally exposed newborns and children. These reports have not, however, been confirmed in epidemiological studies. Wilhelm et al. [2008] have shown that prenatal exposure to dioxins and their congeners does not impair thyroid function and neurological development of newborns [Słowińska et al. 2011].

The presence of polychlorinated biphenyls (PCBs) in the environment is linked to human activity [Stec et al. 2012]. PCBs are released during the incineration of industrial waste and can occur as contaminants in technical industrial, communal, and hospital products containing plastics. They can also contaminate certain plant protection products [Kulik-Kupka et al. 2017]. In the 1970s and 1980, they were used as dielectrics and hydraulic fluids in capacitors and transformers due to their electroinsulating properties, stability, and non-combustibility. They were also used as lubricating oils, paint plasticizers, inks, paper, glue, and sealants [Stec et al. 2012].

Exposure of pregnant women to polychlorinated biphenyls caused thyroid dysfunction and retardation of psychomotor development in their children. Prenatal exposure also causes decreased IQ in the offspring and increases susceptibility to stress. PCBs can also modify the immune system, causing increased susceptibility to infections. Additionally, exposure to PCBs is linked to increased risk of cancer and dysfunction of blood vessel epithelium (a risk factor in the development of atherosclerosis, hypertension, and other cardiovascular diseases) [Kulik-Kupka et al. 2017].

In 2011, the Norwegian Institute of Public Health research had published its objective of determining whether PCBs and PCDDs in the mother's diet could negatively affect the development of the immune system of the fetus. The results indicate that prenatal exposure to PCBs and dioxins contributes to infections of the upper respiratory tract and increases the risk of wheezing, as well as the incidence of communicable diseases in newborns. Authors suggest that decreased consumption of products contaminated with dioxins and PCBs during pregnancy might be beneficial to the child's health [Słowińska et al. 2011].

Gregoraszczuk et al. [2004] established that dioxins and polychlorinated biphenyls negatively affect the course of pregnancy. A correlation has been observed between the concentration of these xenobiotics in blood serum and the risk of spontaneous abortions. The concentration of PCBs in the blood of women who experienced chronic miscarriages was statistically higher than in the control group. Prenatal exposure to PCBs negatively affects fetal development. Low birth weight, decreased head circumference, and body length, as well as reduction in muscular tension have been found in newborns [Stec et al. 2012].

3.4.5 Pharmaceuticals

Drug-induced infertility can result from impairment of potency and disorders of the reproductive system (erectile dysfunction and/or ejaculation disorders, decreased libido, etc.). It can also be caused by the influence of xenobiotics on spermatogenesis, leading to sperm quality disorders, such as azoospermia, asthenozoospermia, or teratozoospermia.

The following pharmaceuticals are known to impair the function of male reproductive organs: cytostatics – used in cancer therapy; lithium salts, which due to their effectiveness remain a common medication in the treatment of bipolar disorder; methyldopa – an antihypertensive medicine; and levodopa – used in the treatment of Parkinson's disease. Moreover, impotence, ejaculation disorders, and decreased libido can occur after exposure to the following pharmaceuticals: spironolactone, clonidine, rilmenidine, guanethidine, prazosin, perhexiline, reserpine – antihypertensive medicine; cimetidine – a histamine receptor antagonist; as well as in therapies using corticosteroids, neuroleptics (thioridazine) and disopyramide, used for the treatment of arrhythmia. Sulfasalazine, used in the treatment of ulcerative colitis and Crohn's disease, can also cause fertility disorders in males [Skowron et al. 2011]. Warfarin (a drug used to reduce blood clotting) or retinoids (used in dermatology) can also have a teratogenic effect.

3.4.6 Others

2,2-Bis(4-hydroxyphenyl)propane (bisphenol A; BPA) is a compound used, among others, as a flame retardant and as the starting substance for the production of flame retardants. It is also used as a fungicide, a component of glues for electronic components, a stabilizer used in the production of PCV, and a brake fluid component, in the production of polyester, polyacrylic, and polysulfone resins used to coat the inside of metal cans and drinking water tanks. Bisphenol A is also present in everyday objects, such as plastic food storage containers, bottles, plastics (in electronic parts, DVDs, CDs), perfumes, deodorants, shampoos, drugs used in dental medicines, and toys, as well as bottles, plates, and pacifiers for children. Such widespread use of BPA causes its large emissions to the environment. Bisphenol A contaminates the air and aquatic environment and recontaminates the aquatic livestock and plants. Small amounts of this compound can migrate into food as a result of damage to plastics and to coatings produced using epoxy resins, and as a result of high temperature affecting other BPA-containing products. The migration of BPA to food intensifies during the contact of containers containing the compound with alkaline or acidic liquids (juices), under mechanical stress, e.g. crushing, or when they were frequently used, as the compound can be washed out. The presence of bisphenol A was found in the placenta and cord blood, human milk, urine, blood of infants, and amniotic fluid [Kulik-Kupka et al. 2017].

Apart from respiratory tract irritation and sensitization, BPA is an endocrine-active compound, as it binds to estrogen receptors [Szymańska and Frydrych 2006].

Endocrine activity of bisphenol A is described in Section 3.5.

Bisphenol A has also been found to cause uterine and ovarian cancers, diabetes, obesity, metabolic syndrome, precocious puberty, impaired genital development and fertility disorders, and difficulties with conception and maintaining pregnancy. It is assumed that the compound can also induce polycystic ovary syndrome. Studies also confirm its effects on sperm quality. Prolonged exposures to BPA can also contribute to an increased risk of prostate cancer [Kulik-Kupka et al. 2017].

There is a lot of literature data on embryotoxic and teratogenic effects of BPA and on the influence of the substance on the reproductive system. However, this data is not sufficient to draw a conclusion on the dose of BPA which can be considered reprotoxic. Changes after the administration of low doses of BPA (such as a decrease in body mass, spermatogenesis, and testes mass) were only observed in the studies conducted by Vom Saal et al. [1998]. Studies conducted by Cagen et al. [1999] and Ema et al. [2001] did not confirm these observations. However, significant changes, such as fetal growth retardation, developmental disorders, and increase in fetal resorption, were reported after large doses of BPA were administered (300–1,250 mg/kg), which indicates the possibility of BPA having embryotoxic and fetotoxic effects [Kim et al. 2001; CHEMINFO 2002 – quoted by Szymańska and Frydrych 2006].

Based on the results of *in vitro* and *in vivo* studies, various hypotheses as to the toxic mechanisms of BPA exposure on the reproductive system were made. Specifically, BPA is considered to have estrogenic and antiandrogenic effects, disrupt the hypothalamic-pituitary-adrenal axis, and cause epigenetic changes (which can be transferred during cell division despite not being dependent on the DNA sequence), affecting the reproductive system. BPA can affect the gonadotropin-releasing hormone (GnRH) levels, gonadotropin release, and signals inducing the proliferation of spermatogonia in Sertoli cells [Cariati et al. 2019].

Due to adverse health effects of BPA, in the recent years, it has been widely substituted in the industry – in the production of epoxy resins, coatings for various applications, such as the production of varnishes, water pipes, and dental sealants – with bisphenol F (BPF, 4,4'-dihydroxydiphenylmethane) (CAS 620-92-8, EC number 210-658-2), which is structurally similar to bisphenol A and occurs naturally in roots, rhizomes, and seeds of *Coeloglossum viride var. Bracteatum*, *Galeola faberi*, *Gastrodia elata*, and *Xanthium strumarium* or *Tropidia curculioides*. *C. viride var. Bracteatum* (rhizomes) is a traditional Tibetan cough and asthma medicine, often called Wang La. *G.elata* is known as Tian Ma in traditional Chinese medicine. However, a systematic review of *in vitro* and *in vivo* studies suggests that hormonal activity of BPF is of a magnitude similar to that of BPA [Rochester and Bolden 2015]. Recent research confirms the similarity in biological effects of BPF and BPA [Goldinger et al. 2015; Kim et al. 2017; Lee et al. 2017; Mesnage et al. 2017; Rosenfeld 2017]. Therefore, herbal products used in traditional Chinese medicine should be considered a potential source of exposure to hormonally active bisphenol compounds [Huang et al. 2019].

Phthalates are salts and esters of the phthalic acid. Examples of these substances include di(2-ethylhexyl) phthalate (DEHP), dioctyl phthalate (DOP), diisononyl phthalate (DINP), diisodecyl phthalate (DIDP), diisobutyl phthalate (DIBP), dibutyl phthalate (DBP), and benzyl butyl phthalate (BBP) [Kulik-Kupka et al. 2017].

The main sources of exposure to phthalates are contaminated food (*per os* exposure), cosmetics (dermal exposure), and air and dust, to which these compounds migrate (inhalation exposure). Occupational exposure to phthalates occurs during the production of glyceryl phthalate resins, glues (synthetic gum arabic), air fresheners, detergents, and cleaning agents. Phthalates are mostly used as plasticizers to improve elasticity and hardness of plastic materials, e.g., waterproof clothing, vinyl flooring, car plastics, degreasers, and cosmetics such as soap, hairspray, or nail polish. They are also the components of polyvinyl products (plastic bags, blood containers, catheters, toys) [Kulik-Kupka et al. 2017].

Phthalates can cross the placental barrier. In males, they can cause hypospadias, underdeveloped sex organs at the time of birth, lower testosterone blood concentration levels, and benign tumors in the testicles [Kulik-Kupka et al. 2017]. Certain phthalate esters inhibit steroidogenesis in the Leydig cells [Kaltenecker Retto de Queiroz and Waissmann 2006]. They can contribute to a decrease in cognitive abilities, e.g., depth perception, memory, perception, overactivity, behavioral problems (aggression), or decrease in social skills. In women, exposure to phthalates may lead to precocious puberty, damage to the liver, kidneys, and heart [Kulik-Kupka et al. 2017].

3.5 HEALTH EFFECTS OF EXPOSURE TO ENDOCRINE DISRUPTORS

Ongoing urbanization, industrialization, and consumerism has lead to increased environmental pollution, which negatively affects living organisms including humans. Increased incidence of endocrine disorders, including genital development disorders and metabolic disorders, draws particular attention to the potential role of environmental factors in their emergence. Endocrine-active compounds occur both in the external environment and within enclosed spaces, which causes constant contact of humans and animals with these substances. Even though they occur in very low concentrations, they have the capacity to change hormonal system activity, bioaccumulate in the environment, and contribute to the development of reproductive anomalies [Szewczyńska and Dobrzyńska 2018].

Definitions and the influence of endocrine disruptors on human reproductivity are discussed in Section 3.1.4.

Substances disrupting hormonal balance can have adverse effects on the endocrine system and hormones, which work in very small amounts and in strictly defined moments in times of growth regulation, reproduction, metabolism, immunity, and behavior. Health effects of endocrine disruption may be felt for a long time after exposure. The activity of endocrine disruptors, apart from fertility disorders, genital development disorders, and fetal damage (including damage to the nervous system), can also cause hormone-dependent cancers (breast, prostate, ovarian, and testicular cancers) or metabolic disorders, obesity, and diabetes. Currently, there are over 560 compounds which have been shown to have endocrine disrupting properties [Szewczyńska and Dobrzyńska 2018].

Phytoestrogens are included in the group of endocrine-active substances. They are compounds structurally similar to 17β-estradiol, which includes isoflavones (genistein, daidzein), coumestan (coumestrol), or stilbenes (resveratrol).These effects can also be ascribed to certain detergents and other chemicals used in the industry, such as nonylphenol, bisphenol A, octylphenol, and their derivatives. Certain medicinal drugs, such as diethylstilbestrol (DES), used in hormonal therapy and contraception in humans and as a growth stimulant in cattle, are xenoestrogens. Urban waste, containing metabolites of oral contraceptives can also pose a threat due to the possibility of them reaching ground and drinking waters. Estrogenic effects can also be ascribed to metals such as zinc, copper, cadmium, cobalt, nickel, lead, mercury, tin, chromium, vanadium anion, and arsenates. It was established that cadmium affects prolactin secretion in a number of species, including humans. In rats, it accumulates in the pituitary gland, disrupting lactotropic cell activity. High doses of cadmium inhibit prolactin secretion both *in vivo* and *in vitro*. Depending on the route of administration, nickel can cause a decreased iodine uptake by the thyroid in rats, fetal growth retardation in mice, and hyperinsulinemia in dogs. In experimental studies, dyshormonogenesis due to exposure to zinc, lead, and mercury was observed and manifested in fertility impairment and disturbances in sex differentiation. Epidemiological studies of women exposed to lead and mercury, in whom an increased incidence of miscarriages, premature births, and menstrual disorders were found, might indicate the probability of dyshormonogenesis caused by metalloestrogens [Langauer-Lewowicka and Pawlas 2015].

Xenoestrogens are also used in agriculture. They are mostly pesticides. Endosulfan (organochlorine insecticide) is an example of a xenoestrogen, which is capable of increasing prolactin expression, and competes with estrogens for the nuclear estrogen receptor. It has been shown that fungicides affect a number of aspects of hormonal signaling. Fungicides such as prochloraz and procymidone exhibit antiandrogenic properties [Demeneix and Slama 2019], binding to androgen receptors and acting as their antagonist and as a dihydrotestosterone-induced transcription inhibitor in laboratory conditions, as established in *in vitro* studies. The herbicide linuron also exhibits antiandrogenic activity, causing an increased incidence of testicular tumors, which can lead to stimulation of the pituitary gland and to an increased level of LH secretion. The fungicide vinclozolin can also exhibit antiandrogenic activities. Its metabolites competitively inhibit the binding of androgens to the androgen receptors in mammals (the substance itself does not have such effects) and can disrupt the functioning of the hypothalamic-pituitary axis. In workers employed in factories producing lindane (γ-hexachlorocyclohexane), a significant increase in the LH levels and a slight decrease in testosterone levels have been found [Kaltenecker Retto de Queiroz and Waissmann 2006].

The mode of action of endocrine disruptors is based on their influence on the formation of active hormone-receptor complexes or inhibition of the signal passing to the effector. Such external factors are known as hormone antagonists. Conversely, substances which increase hormone activity or act similarly to a hormone (hormone mimicking), despite its absence, are called agonists. Apart from the abovementioned activity, endocrine disruptors can disrupt the synthesis of endogenous hormones, modify their metabolism, or modify receptor levels [Biernacki et al. 2015].

The best documented mode of action of endocrine disruptors is their ability to bind to nuclear receptors as a total agonist (causing a positive effect after binding to a receptor), partial agonist, or an inverse agonist (binding to the same part of the receptor as the agonist, but causing an opposite effect), or as an antagonist (replacing another agonist at the receptor binding location, not affecting the receptor activation state) [Lauretta et al. 2019].

Nuclear receptors create a group of intracellular, structurally homologous proteins, which regulate the transcription of genes responsible for proper cell functioning. A number of these proteins require activation to bind small-molecule lipophilic ligands (e.g. steroid, retinoic acid, thyroxine), freely diffusing into the cell. Nuclear receptors after ligand binding move from the cytoplasm to the cell nucleus, where they act as transcription factors for a given receptor. Binding to a particular nucleotide sequence, they can activate or inhibit gene transcription processes. Active nuclear receptors can also indirectly inhibit gene transcription through interaction with other transcription factors [Kopij and Rapak 2008].

Endocrine disruptors in particular might bind to and activate different hormonal receptors (androgen receptor – AR, estrogen receptor – ER, aryl hydrocarbon receptor – AhR, pregnane X receptor – PXR, constitutive androstane receptor – CAR, glucocorticoid receptor – GR, thyroid hormone receptor – TR, retinoid X receptor – RXR) and mimic the activity of the natural hormone [Lauretta et al. 2019].

3.5.1 Chemical Hazards Causing Adrenal Dysfunction

Adrenals are characterized by excellent vascularization, high lipophilicity caused by high content of polyunsaturated fatty acids in the cell membrane, and a large biotransformation potential in the metabolic activation of xenobiotics due to the presence of CYP 450 enzymes, which are responsible for the formation of toxic metabolites and free radicals. All these cause the adrenal glands to be particularly susceptible to toxic effects of xenobiotics [Lauretta et al. 2019].

There are, however, few studies on the influence of EDCs on adrenal glands – especially on substances linked to chemical exposure risks. This might seem surprising, as the proper functioning of the hypothalamic-pituitary-adrenal (HPA) axis is vital for human health and life, and the axis is a common target of a number of medicinal drugs and chemical compounds.

The best researched influence of endocrine disruptors on the HPA axis is linked to their ability to disrupt biosynthesis and the metabolism of steroid hormones through affecting enzymes involved in steroidogenesis. Aromatase, 5α-reductases, and hydroxysteroid dehydrogenases 3-beta, 11-beta, 17-beta play a key role in metabolic pathways of adrenal steroidogenesis, and their proper functioning is disrupted by EDCs, particularly xenoestrogens. EDCs also target the steroidogenic acute regulatory protein (StAR), which regulates the first stage of adrenal steroidogenesis. A number of medicinal drugs and chemicals can interact with the HPA axis and affect every stage of steroidogenesis. Moreover, various chemicals can disrupt different stages of steroidogenesis [Lauretta et al. 2019].

It has been established that spironolactone and its metabolite are mineralocorticoid antagonists, competitively block the binding of aldosterone to cytoplasmic receptors of endothelial cells of the distal renal tubule and collecting tubule, which prevents the synthesis of the so-called aldosterone inducing proteins in potassium channels with the involvement of Na^+/K^+-ATPase. This leads to decreased sodium absorption and potassium secretion in the cells. The effect has not been observed in animals with removed adrenal glands. Repeated parenteral administration of thioguanine (purine antimetabolite with cytostatic effects) to laboratory animals caused hemorrhagic necrosis of the adrenal cortex. Hexadimethrine bromide, a heparin antagonist, causes damage to zona glomerulosa and zona reticularis of the adrenal cortex. Their cells produce mineralocorticoids as well as androgens and estrogens, respectively. Carbon tetrachloride induces necrosis of the adrenal cortex due to damage to the zona reticularis. Acrylonitrile, pyrazole, and cystamine, a natural derivative of cysteine, causes hemorrhagic necrosis of the adrenal glands. Repeated administration of 7,12-dimethylbenz[a]anthracene to laboratory animals induces a slow-onset adrenal gland necrosis. Xenobiotics may directly or indirectly influence the biosynthesis of adrenal hormones and the secretory function of the adrenal cortex. Aminoglutethimide, an aromatase inhibitor, used in the treatment of breast cancer in postmenopausal women inhibits the biosynthesis of progesterone and glucocorticoids when administered in large doses. Tri-o-cresyl-phosphate and tri-tert-butyl phenyl phosphate inhibit the activity of cholesterol esterase, prohibiting cholesterol from being used in the biosynthesis of adrenal gland hormones. DDD, 1,1′-(2,2-dichloroethane-1,1-diyl)bis(4-chlorobenzene), a metabolite of DDT, an organochloride insecticide, caused the degeneration of the adrenal gland cell and athropy of the gland in dogs. Such changes were not observed in other species, including humans. In dogs, DDD administered intravenously in large doses (60 mg/kg body weight) inhibited ACTH-induced steroid production, establishing a correlation between steroidogenesis inhibition and the degree of damage to zona fasciculata and zona reticularis of the adrenal cortex by the compound. DDD, under the name mitotane, was used in the treatment of adrenal gland tumors, linked to abnormal adrenal cortex hormone secretion [Starek 2007].

It has also been established that hexachlorobenzene can disrupt the functioning of corticoid hormone in Wistar rats [Lauretta et al. 2019]. Cortisol deficiency resulting from adrenal insufficiency causes weakness, fatigue, lack of appetite, nausea and vomiting, hypotension, hyponatremia, and hypoglycemia. Mineralocorticoid deficiency causes loss of sodium by the kidneys and potassium retention, which can lead to severe dehydration, hypotension, hyponatremia, hypokalemia, and acidosis [Greenspan and Gardner 2004].

In the assessment of the influence of EDCs on the HPA axis, it is important to consider that even partial disruption in the proper functioning of adrenal glands can cause serious health effects. It is particularly important due to the possibility of bioaccumulation of chemical substances disturbing the hormonal balance in the fatty tissue, which might be the cause of exposure to a ‘cocktail of disruptors’. Moreover, clinical effects may only be observed after several years of constant exposure to low doses [Lauretta et al. 2019].

3.5.2 Chemical Hazards Causing Thyroid Dysfunction

Thyrotoxic effects of xenobiotics can show, in the disruption of thyroid hormone biosynthesis, their storage, secretion, transportation to distant target tissues, and morphological changes to the thyroid itself [Starek 2007].

Chlorates(I) (hypochlorites), chlorates(V), iodates(V), chlorates(VII) (perchlorates), and thiocyanates inhibit the uptake of iodine ions in the thyroid gland. The mechanism is based on the inhibition of the Sodium-Iodide symporter channel (NIS protein), which takes part in the active transport of iodine ions from the blood into thyrocytes [Lauretta et al. 2019].

Carbimazole and its metabolite, methimazole, as well as propylthiouracil, medications used in hyperthyroidism, para-aminosalicylic acid, phenylbutazone, resorcin, and cobalt(II) ions, inhibit the oxidation of iodine ions by thyroid peroxidase and block the formation of monoiodotyrosine. In turn, anabolic steroids, corticosteroids, and phenothiazines lower the levels of thyroxine-producing globulin, leading to an increased blood concentration of free hormones in blood [Starek 2007].

The functioning of the thyroid gland and the hypothalamic-pituitary-thyroid axis is essential for the pathophysiology, clinical course and treatment of bipolar disorder. Meta-analysis of potential toxicity of long-term lithium use in this disease has shown that it causes a 5 times increase in the risk of hypothyroidism [Kraszewska et al. 2014].

Lithium is also used in a number of industrial sectors: lithium oxide is used in the melting of silicon dioxide, a material for the manufacture of glass and ceramic products; lithium is used in lithium-ion batteries; lithium soap is used for oil densification and the production of universal greases; lithium is also used in the production of tritium [Wasilonek 2019].

Lithium decreases iodine uptake in the thyroid and disrupts the synthesis of iodotyrosine through the inhibition of tyrosine iodization, changing the thyreoglobulin structure and inhibiting the formation of colloid in the upper part of the thyroid cells. The administration of lithium decreases deiodination in the liver and clearance of free thyroxine (fT4) in blood serum, which leads to lower activity of type I 5'deiodinase. It has also been established that lithium decreases the activity of type II deiodinase and changes the structure of thyreoglobulin, affecting its functioning, which leads to defective coupling of iodotyrosine. As a result of lithium use, the production and release of thyroid hormones decreases, which leads to increased concentration of TSH – a factor causing thyroid growth and excessive TSH response to stimulation with thyrotropin-releasing hormone. Lithium also influences cellular and humoral immunological response, which can result in the disruption of the production of antithyroid antibodies. In rat studies, lithium was found to accumulate in the hypothalamus and the pituitary gland. It has also been shown that lithium affects the activity and binding of thyroid hormones to brain receptors and regulates gene expression for thyroid hormones [Kraszewska et al. 2014].

Herbicides, phenylurea derivatives, dicarbamates, pyridazinones, and diphenyl ethers can disrupt thyroid function. Impaired iodine uptake by the thyroid, decreased thyroxine levels in plasma, as well as its clearance have been found in workers employed in the packaging of the 2,4-D herbicide (2,4-dichlorophenoxyacetic acid). Another herbicide, nitrofen (2,4-Dichlorophenyl 4-nitrophenyl ether) causes a decrease in vesicle size and colloid density in the thyroid gland. Amitrol (3-Amino-1H-1,2,4-triazole) inhibits thyroid peroxidase, increases TSH level, and contributes to goiter formation [Starek 2007].

A number of fungicides, including bifonazole, imazalil (enilconazole, chloramizole), and flusilazole, which have been shown to inhibit the activity of human placental aromatase, act as endocrine disruptors. Other modes of action of these compounds include the inhibition of thyroid peroxidase, affecting thyroid hormone production [Demeneix and Slama 2019].

Chlorpyrifos (an organophosphate insecticide) has also been described to affect thyroid function. The compound exhibits a lower level of toxicity for mammals, but a significant level for fish [Demeneix and Slama 2019].

Polychlorinated biphenyls can also contribute to a decrease in thyroid hormone levels: thyroxine (T4) and triiodothyronine (T3), directly affecting the thyroid and leading to hyperplasia or hypertrophy [Kulik-Kupka et al. 2017]. PCBs can also increase thyroid hormone metabolism. Disruption of thyroid hormone homeostasis by the PCBs results from their structural similarity to hormones. Moreover, certain PCBs and their hydroxylated metabolites, in particular, exhibit 4–8 times more affinity for thyroid hormone carrier protein [Stec et al. 2012]. PCBs can also cause hypothyroidism, disrupting the release of pituitary hormones and hypothalamus neurohormones [Kulik-Kupka et al. 2017].

Polychlorinated dibenzodioxins and dibenzofurans decrease the blood serum concentration of T4 in rats, which leads to a compensative increase in the TSH level [Starek 2007]. As dioxins are structurally similar to steroid hormones, the main locations of their negative effects are, apart from the thyroid, male and female gonads and other organs in which steroid hormones are produced [Stec et al. 2012].

Phthalates disrupt the functioning of both the thyroid and the pituitary gland, as well as, among others, impair perception, cognitive abilities, and memory, cause behavioral disorders and excessive motor activity [Pałczyński 2018].

Disruption of thyroid hormone synthesis causes thyroid enlargement (non-toxic goiters), hypothyroidism ranging from low to severe, low T3 and T4 concentrations, and increased TSH concentration in the blood serum. Hypothyroidism, which begins in adulthood, causes a general slowdown of metabolic processes in the body, with the accumulation of glycosaminoglycans in the intercellular space, particularly in the skin and muscles – presenting a clinical picture of myxedema. A result of hypothyroidism in newborns and children is a significant decrease in the rate of development, with serious and permanent consequences, including mental retardation [Greenspan and Gardner 2004].

Conversely, hyperthyroidism may lead to cardiovascular system disorders (atrial fibrillation, congestive heart failure, angina pectoris, and acute myocardial infarction) or nervous system disorders (apathy, depression, confusion, and fatigue). A state of increased catabolism also leads to muscular atrophy, especially of the quadriceps.

Hyperthyroidism is also linked to bone mass loss and a tendency for bone fracture. One of the most common forms of thyreotoxicosis is the Graves' disease, currently seen as an autoimmune disease, characterized, among others, by ophthalmopathy and pretibial myxedema. Hyperthyroidism is also caused by hormonally active adenomas, secreting excessive amounts of T3 and T4. A thyroid crisis (thyrotoxic crisis), an acute exacerbation of hyperthyroidism symptoms, is life-threatening due to its severe course [Greenspan and Gardner 2004].

3.5.3 Chemical Hazards Causing Pancreatic Dysfunction

The pancreas is an organ composed of two functionally different organs: exocrine pancreas, the largest digestive gland of the human body, and endocrine pancreas, the source of insulin, glucagon, somatostatin, and pancreatic polypeptide. While the exocrine pancreas plays the largest role in the digestive process (digestive enzymes), as it transforms food into substances which can be absorbed by the body, the endocrine pancreas influences every aspect of nutritional and intracellular changes. This function is applicable both to the amount of absorbed food and the storage of the products of the digestive process in the cells. The impairment of intrapancreatic secretion or abnormal response of target tissues to hormone activity leads to severe abnormalities in digestive homeostasis, including clinically relevant syndromes, collectively known as diabetes [Greenspan and Gardner 2004].

A number of medicines and xenobiotics are toxic to the pancreatic islets β cells [Starek 2007], including their functioning and viability, insulin release, and glucose-level regulation. Linking metabolic changes to exposure to dioxins, especially TCDDs, was among the first reported experimental study results. It has been shown that TCDDs decrease glucose uptake in the pancreas and disrupt insulin secretion. Exposure to TCDDs led to the depletion of cellular insulin reserves and 'exhaustion' of beta cells due to the constant promotion of insulin secretion. This suggests that prolonged exposure to this compound might cause insulin deficiency [Papalou et al. 2019].

Dioxin molecules are structurally similar to steroid hormones, which results in them chiefly targeting the thyroid, male and female gonads, endometrium, and other organs producing steroid hormones. Therefore, dioxins can disrupt the hormonal balance, e.g. through the induction of aryl hydrocarbon receptor (AhR) in the cytosol. They affect the synthesis, transport, and elimination of hormones. Dioxins also disrupt the binding of hormones to their proper receptors and affect feedback processes in the hypothalamic-pituitary-peripheral gland axis. Całkosiński et al. [2003; 2004] have shown that intrauterine effects of dioxins change gonadal development during fetal life and lead to the disruption of fetal endocrine processes. In the study, the researchers have discovered significant disorders of rat fertility linked to changes of estrogen levels [Słowińska et al. 2011].

The negative effect of TCDDs on the metabolism was also observed in humans. In long-term studies of veterans exposed to TCDDs during the Vietnam War, TCDD serum levels were clearly linked to the incidence of type 2 diabetes and insulin resistance [Papalou et al. 2019]. This hypothesis can also be confirmed by the results of studies conducted by Taiwanese researchers, who tested 1,449 people living in the vicinity of an abandoned plant which used to produce pentachlorophenol and alkali metals. Chang et al. [2011] studied the concentration of dioxins and mercury in blood serum and its relationship to insulin resistance and liver damage. Their results unambiguously showed a significant correlation between PCDDs concentration, mercury blood level, and insulin resistance, even in people not suffering from diabetes. This proves that exposure to dioxins and mercury can be a significant risk for human health, as it contributes to an increased risk of incidence of insulin resistance and type 2 diabetes [Słowińska et al. 2011].

Eslami et al. [2016 – quoted by Kulik-Kupka et al. 2017] determined the concentrations of 10 congeners of polichlorinated biphenyls and 8 congeners of polybrominated diphenyl ethers in the plasma of pregnant women with identified gestational diabetes. The authors reported a positive correlation between the incidence of gestational diabetes and the total concentration of polychlorinated biphenyls and polybrominated diphenyl ethers. A similar relationship has also been observed for the incidence of gestational diabetes and the concentration of polychlorinated biphenyls and polybrominated diphenyl ethers.

Oral administration of tributyltin to test animals inhibits proliferation and induces apoptosis of pancreatic islets cells, causing a decrease in relative area of pancreatic islets and disruption of glucose homeostasis [Papalou et al. 2019].

In turn, cobalt(II) parenterally administered to guinea pigs, rabbits and dogs caused degranulation and vacuolization of pancreatic α cells [Starek 2007].

Animal studies and cell culture studies indicate that arsenic can influence the incidence of diabetes, as it exhibits biochemical properties similar to phosphorus, enabling it to substitute the latter in the reaction in which adenosine triphosphate (ATP) is formed. Marchewka and Grzebinoga [2009] report that, as a result of this reaction, adenosine diphosphate arsenate is produced, which causes abnormal energy generation and the inhibition of metabolic pathways which require ATP as a source of energy. Arsenic may also participate in the reaction which produces glucose 6-phosphate, leading to the formation of glucose 6-arsenate, which can disrupt the metabolism of glucose. The first reported causes of correlation between exposure to arsenic were found in the Taiwanese population. It has been observed that the Taiwanese who came in contact with large doses of arsenic through contaminated drinking water were 2 times at a higher risk of diabetes than the general population [Grotowska et al. 2018].

Arsenic can also disrupt insulin secretion through downregulation of the insulin gene expression and interference in proteolysis carried out in the presence of calpain-10 and the activation of a synaptosomal-associated protein 25, playing a key role in the release of insulin as a result of exocytosis [Papalou et al. 2019].

It has also been established that exposure to organophosphate pesticides causes damage to the pancreas and strongly contributes to the incidence of diabetes. A review of the literature data indicates that people exposed to such pesticides experience temporary increases in blood glucose levels and glycosuria, hyperlipidaemia, and hyperinsulinaemia. In turn, a heightened risk of gestational diabetes was found in the wives of farmers using organophosphate pesticides in Australia. Łukaszewicz-Hussain [2011] quotes a study on rats, in which dimethoate was administered in doses of 20 mg/kg body weight and 40 mg/kg body weight for 30 days. An increase in lipase and amylase activity in the serum with a simultaneous decrease of activity of those enzymes in the pancreas has been observed. In rats receiving the larger dose, a decrease in body weight and an enlargement of the organ have been noted [Grotowska et al 2018].

Pesticide influence on reproductivity was discussed in Section 3.4.

Vinclozolin (RS)-3-(3,5-Dichlorophenyl)-5-methyl-5-vinyloxazolidine-2,4-dione) is a fungicide causing antiandrogenic effects and negatively affecting the endocrine system. [Rzeszutek et al. 2014].

Beta cell functioning can also be disrupted by 2,2-bis(4-hydroxyphenyl)propane (bisphenol A; BPA), even though older publications [Ashby 2001] on the estrogenic activity of BPA, as tested on animals, showed inconsistencies and a lack of properly documented control values for specific parameters. It has also been established that the BPA influence on the endocrine system depends on a number of factors, including the breed of the animals, e.g., F-344 rats are susceptible to BPA, while Sprague-Dawley rats are not. Based on *in vitro* studies, it has been established that BPA binds to estrogen receptors (ERs) [Heinrich-Hirsch et al. 2001 – quoted by Szymańska and Frydrych 2006]. BPA is however qualified as a relatively weak xenoestrogen, because its affinity for ERα and ERβ receptors is 1,000–10,000 times lower than 17β-estradiol. Despite that, low doses of BPA still induce a response similar to that present in the case of estrogen stimulation [Kulik-Kupka et al. 2017].

Alonso-Magdalena et al. [2015] have shown that after exposing pregnant mice to BPA, glucose intolerance and changes in insulin sensitivity were found in their offspring several months after birth, including an increased body mass, which has been linked mainly to beta cell function impairment and the effects of intrauterine exposure on the mass of the pancreas.

In vivo studies indicate that exposure to BPA increases insulin secretion and glucose-stimulated insulin secretion dependent on estrogen receptor alpha (ERα). Apart from their primary reproductive role, sex steroids play a key role in affecting metabolic target tissues, controlling insulin secretion form beta cells, both in the cGMP-dependent and -independent pathways. Therefore, BPA might have a partial metabolic influence on the pancreas through estrogen-dependent pathways. Accordingly, it has been demonstrated that BPA urine concentration in adults is positively correlated with hyperinsulinemia and insulin resistance, especially in males [Papalou et al. 2019].

Tai and Chen [2016 – quoted by Kulik-Kupka et al. 2017] have analyzed the correlation between bisphenol A concentration and the incidence of diabetes. Patients aged 3–79 were divided into two groups: the first one, of 1,915 people had their HbA1c and BPA urine concentration measured; the second one of 2,405 had their glucose levels, HbA1c, and BPA urine concentration measured. Based on these analyses, a positive correlation between the concentration of bisphenol A in urine in males and HbA1c has been found. Similar conclusions have not been drawn for groups of women and children [Kulik-Kupka et al. 2017].

A correlation between exposure to BPA and phthalates and diabetes in women was found in the study by Sun et al. [2014 – quoted by Kulik-Kupka et al. 2017]. Based on surveys conducted on 971 couples, it has been established that exposure to bisphenol A and phthalates can be linked to the incidence of type 2 diabetes in middle-aged women. This correlation has not been observed in older women.

It has been recently established that oxidative stress of the endoplasmic reticulum (ER) might be a key pathogenetic mechanism of diabetes. It is known that beta cells have lower levels of antioxidative enzymes (superoxide dismutase, catalase, and glutathione peroxidase) and are more susceptible to adverse effects of reactive oxygen species (ROS). Certain endocrine disruptors, including BPA, arsenic, and DEHP, can disturb beta cell functioning through the promotion of oxidative stress. In rats exposed to phenol compounds, such as octylphenol, nonylphenol, and BPA, abnormal morphology of pancreatic islets and beta cells was found, mainly due to changes in the mitochondrial structure and gene expression. It is also suggested that BPA may accelerate the depletion of beta cell reserve through the immunomodulation of pancreatic islets, which indicates that EDCs can also contribute to a higher incidence of type 1 diabetes [Papalou et al. 2019].

It is also important to note that a number of xenobiotics cause pancreatic inflammation. Substances causing acute pancreatic inflammation rarely cause chronic inflammation of the organ. Ethanol and medicines such as tetracyclines, salicylates, and vinca alkaloids (vincristine and vinblastine) are capable of causing acute pancreatic inflammation, characterized by an increase in pancreatic enzyme activity in the blood, particularly amylase. Complex mechanisms are responsible for these changes, including the following: pancreatic duct dysfunction (indometacin, salicylates, opiates), immunosuppression (steroids, azathioprine), cytotoxicity (azathioprine, L-asparaginase), arterial thrombosis (estrogens), metabolic effects, including ion distortions (thiazides), direct influence on pancreatic cells (sulphonamides, furosemide, chlorothiazide), and free radical mechanism involving the liver (paracetamol, tetracyclines, and ethanol) [Starek 2007].

3.5.4 Combined Effects of Endocrine Disruptors

Particular EDCs can even be harmless on their own, but their combined effects with other EDCs might be hazardous to human health (the so-called cocktail effect). Taking into account the possibility of interactions between particular components of a mixture, risk assessment of these effects becomes extremely complex. Each endocrine disruptor might target every hormonal axis. Their activity is not typically limited to a single axis or an organ, but the most common targets of EDCs are the following axes: hypothalamic-pituitary-thyroid (HPT), hypothalamic-pituitary-gonadal (HPG), and hypothalamic-pituitary-adrenal (HPA).

Exposure to multiple compounds can cause their effects to accumulate, and the effect may be additive or synergic [Lauretta et al. 2019].

When the body is exposed to several substances at once, the following scenarios might take place:

a) every chemical is present in a dose not causing detectable adverse effects in case of exposure to this chemical substance only, and their mixture does not cause adverse effects either (no observable effect);
b) every chemical is present in a dose which might cause or not cause adverse effects in case of exposure to this chemical substance in itself, but their mixture results in adverse effects, foreseeable based on the dose–response

relationship of every single substance constituting the mixture. The effect of the mixture may be equivalent to the sum of responses induced by every chemical in itself (response addition, 'independent action') or the effect of a single chemical in a dose equivalent to the sum (or weighted sum) of the doses of all the chemicals (dose addition, 'cumulative effect');

c) every chemical is present in a dose which might or might not cause adverse effects in case of exposure to this chemical substance in itself and their mixture results in adverse effects not foreseeable based on the dose–response relationship of every single substance constituting the mixture – the effect of the mixture might be stronger (synergism) or weaker (antagonism) than the sum of singular effects of all the chemicals [Demeneix and Slama 2019].

As for endocrine disruptors, it has been proven that bisphenol A causes changes in gut permeability, which might suggest its influence on the effects of other hazardous chemicals absorbed in the gastrointestinal tract.

Synergy specific to endocrine disruptors was characterized at the molecular level by Delfosse et al. 2015, who tested 40 chemicals individually or in two-component mixtures for antagonistic effects on the pregnane X receptor. The authors observed additive effects for most of the mixtures, but in one case an over additive effect was found – a mixture of trans-nonachlor (organochlorine pesticide) and 17α-ethinyl estradiol (synthetic estrogen, a component of the most currently used combined contraceptives). A similar greater than additive (synergistic) effect was observed for other combinations of steroid and organochlorine compounds. Molecular studies allowed for further understanding of the over additive effect, as both chemicals can bind to PXR, with a 10–30 times increase in avidity (affinity-dependent bond) in comparison to the individual compounds. Moreover, it has been proven that the trans-nonachlor and 17α-ethinyl estradiol form a supramolecular compound on interacting with PXR [Demeneix and Slama 2019].

Apart from the pregnane X receptor (PXR), the gamma receptor activated by peroxisome proliferator-activated receptor (PPARγ) can also take part in endocrine disruptor interactions. The major role of the PPARγ is to control the metabolism of fatty acids and to maintain glucose homeostasis – PPARγ transcription factors belong to the superfamily of nuclear hormone receptors, which also includes steroid receptors as well as thyroid hormone, vitamin D, and retinoic acid receptors. Both the receptors undergo heterodimerization (the formation of a macromolecular complex by two protein monomers) with retinoid X receptor (RXR), which induces changes in spatial conformation and the activation of various pathways in target cells. It is important to note that both the receptors are capable of binding two ligands at once, which has been established based on X-ray crystallography.

In *in vitro* studies, a synergic effect of mono(2-ethylhexyl) phthalate and perfluorooctanoic acid was found for the PPARγ transcription-dependent activation [Synergistic and toxic cocktail effects of low dose endocrine disruptors – TOXSYN. https://anr.fr/Project-ANR-13-CESA-0017 – accessed January 22, 2020].

An example of a synergy between EDCs is a study where rats were exposed to both bis(2-ethylhexyl) phthalate, two fungicides present in food, and the drug finasteride (an organic steroid compound, type II steroid 5-α-reductase inhibitor). All four compounds exhibit antiandrogenic activity, but have different modes of action. In the case of a change in the anogenital distance and influence on the mass of sex organs, the mixtures exhibited a cumulative effect, but as for congenital malformations of external sex organs, the effects of the mixture were synergistic [Demeneix and Slama 2019].

In a study by Lukowicz et al. [2018] the administration of commonly used pesticides to rats caused an increased incidence of metabolic disorders: obesity and diabetes, in comparison to their individual administration, in doses not causing adverse effects.

The influence of mixtures of antiandrogenic chemicals on the production of testosterone in an organotypic model of human testicle explants was larger in the case of exposure to several compounds than in the case of exposure to a single one – the effect of the mixture was basically consistent with the cumulative effect consistent with dose addition [Demeneix and Slama 2019].

Yu et al. [2019] assessed the share of individual chemical substances in the total hormone activity of mixtures and established that compared to 17β-estradiol as a standard, estrone, estriol, ethinyl estradiol, bisphenol A, and genistein have shown estrogenic activity, while dibutyl phthalate, butyl benzyl phthalate, bis(2-ethylhexyl) phthalate, nonylphenol, and 4-tert-octylphenol have shown estrogenic activity. The EDC 11 mixture has also exhibited estrogenic activity. Additive effects have been observed only in the case of mixtures with 6 estrogenic compounds. Moreover, in the EDC 11 mixture, the presence of ethinyl estradiol in environmentally significant concentrations did not cause an increase in estrogenic activity in comparison with mixture 10 of EDCs without ethinylestradiol.

3.6 SUMMARY

Due to the fact that reproductive health and endocrine system disorders are an increasing problem in the modern society, measures are taken in a number of countries to ensure that workers exposed to chemicals causing such effects are protected from their adverse health effects.

Efforts are being made to include all types of occupational risks in the legislation – physical, chemical, biological, and organizational. They are regulated both in general and in specific legislations, though framework legislation on occupational safety and health usually does not specifically refer to reprotoxic substances and endocrine disruptors, despite the fact that they lead to increasingly significant risks, the health effects of which may be felt a long time after exposure.

Therefore, the intensification of legislative and organizational activities, e.g. the currently considered expansion of the scope of the Carcinogens and Mutagens Directive (CMD*), is an area which needs further efforts to be made in the light of risks discussed in this chapter. The scope of the Directive could be expanded to include substances toxic to reproduction, or the implementation of screening programs, similar to the US Environmental Protection Agency in relation to endocrine

* Directive 2004/37/EC of the European Parliament and of the Council of April 29, 2004, on the protection of workers from the risks related to exposure to carcinogens or mutagens at work (sixth individual Directive within the meaning of Article 16(1) of Council Directive 89/391/EEC) (the *Journal of Laws of 2004*, No.158 item 50, as amended).

disruptors (Endocrine Disruptor Screening Program – EDSP), which uses a two-tiered approach to screen pesticides, chemicals, and environmental contaminants for their potential effects on estrogen, androgen, and thyroid hormone systems (//www.epa.gov/endocrine-disruption/endocrine-disruptor-screening-program-edsp-overview – accessed January 22, 2020).

Risks for the reproductive health and maintenance of hormonal balance of workers, as discussed in this chapter, indicate that it is necessary to take measures to intensify and further detail epidemiological and toxicological studies to focus on particular problems.

There is a particular need for studies to be conducted and scientifically proven data to be gathered on the combined effects of chemical substances due to a significant lack of results in the available literature. It is especially important due to the fact that workers are occupationally exposed to mixtures, not individual substances, and the combined effects might be significantly different from the effects of exposure to a single chemical.

The actions taken in this respect should guarantee the safety not only of people occupationally exposed to reprotoxic substances and endocrine disruptors, but also of the future generations.

REFERENCES

Alonso-Magdalena, P., M. Garcia-Arevalo, I. Quesada et al. 2015. Bisphenol-A treatment during pregnancy in mice: A new window of susceptibility for the development of diabetes in mothers later in life. *Endocrinol* 156(5):1659–1670. doi: 10.1210/en.2014-1952.

Altenkirch, H., G. Stoltenburg, and H. M. Wagner. 1978. Experimental studies on hydrocarbon neuropathies induced by methyl ethyl ketone. *J Neurol* 219(3):159–170. doi: 10.1007/BF00314531.

Ashby, J. 2001. Testing for endocrine disruption post-EDSTAC: Extrapolation of low dose rodent effects to humans. *Toxicol Lett* 120(1–3):233–242. doi: 10.1016/s0378-4274(01)00299-5.

ATSDR [Agency for Toxic Substances and Disease Registry]. 1999. Toxicological profile for mercury (update). US Department of Health & Human Services. https://www.atsdr.cdc.gov/toxprofiles/tp.asp?id=115&tid=24 (accessed January 22, 2020).

Attarchi, M. S., M. Ashouri, Y. Labbafinejad et al. 2012. Assessment of time to pregnancy and spontaneous abortion status following occupational exposure to organic solvents mixture. *Int Arch Occup Environ Health* 85(3):295–303. doi: 10.1007/s00420-011-0666-z.

Bertazzi, P. A. 2010. Health-based occupational exposure limits: An European experience in perspective. *Ital J Occup Environ Hyg* 1(2):87–95.

Biernacki, B., K. Bulenger, A. Woźniak et al. 2015. Czynniki antropogeniczne a układ endokrynny. *Życie Weterynaryjne* 90(11):720–724.

Bjørklund, G., S. Chirumbolo, M. Dadar et al. 2019. Mercury exposure and its effects on fertility and pregnancy outcomes. *Basic Clin Pharmacol Toxicol* 125(4):317–327. doi: 10.1111/bcpt.13264.

Cagen, S. Z., J. M. Waechter Jr., S. S. Dimond et al. 1999. Normal reproductive organ development in CF-1 mice following prenatal exposure to bisphenol A. *Toxicol Sci* 50(1):36–44. doi: 10.1093/toxsci/50.1.36.

Całkosiński, I., L. Borodulin-Nadzieja, M. Stańda et al. 2003. Influence of a single dose of TCDD on estrogen levels and reproduction in female rats. *Med Wet* 59(6):536–538.

Całkosiński, I., L. Borodulin-Nadzieja, U. Wasilewska et al. 2004. Wpływ dioksyn na procesy rozrodcze u szczurów w badaniach in vivo. *Adv Clin Exp Med* 13:885–890.

Carachi, R., and S. H. E. Doss. 2019. *Clinical embryology: An atlas of congenital malformations*. Cham, Switzerland: Springer. doi: 10.1007/978-3-319-26158-4.

Cariati, F., N. D'Uonno, F. Borrillo et al. 2019. Bisphenol A: An emerging threat to male fertility. *Reprod Biol Endocrinol* 17(6):1–8. doi: 10.1186/s12958-018-0447-6.

Cebulska-Wasilewska, A., A. Panek, Z. Żabiński et al. 2005. Wpływ par rtęci podczas narażenia zawodowego na limfocyty in vivo, na ich podatność na promieniowanie UV-C lub X oraz wydajność naprawy in vitro. *Med Pr* 56(4):303–310.

Chang, H. Y., T. S. Shih, Y. L. Guo et al. 2004. Sperm function in workers exposed to N,N-dimethylformamide in the synthetic leather industry. *Fertil Steril* 81(6):1589–1594. doi: 10.1016/j.fertnstert.2003.10.033.

Chang, J. W., H. L. Chen, H. J. Su et al. 2011. Simultaneous exposure of non-diabetics to high levels of dioxins and mercury increases their risk of insulin resistance. *J Hazard Mater* 185(2–3):749–755. doi: 10.1016/j.jhazmat.2010.09.084.

CHEMINFO. 2002. Canadian Centre for Occupational Health and Safety: Database. http://ccinfoweb2.ccohs.ca/cheminfo/records/753E.html (a paid subscription – accessed January 22, 2020).

Courtney, K. D., J. E. Andrews, J. Springer et al. 1986. A perinatal study of toluene in CD-1 mice. *Fundam Appl Toxicol* 6(1):145–154. doi: 10.1016/0272-0590(86)90270-8.

Cox, C. 2004. Boric acid and borates. *J Pest Reform* 24(2):10–15.

Delfosse, V., B. Dendele, T. Huet et al. 2015. Synergistic activation of human pregnane X receptor by binary cocktails of pharmaceutical and environmental compounds. *Nat Commun* 6(8089):1–10. doi: 10.1038/ncomms9089.

Demeneix, B., and R. Slama. 2019. Endocrine disruptors: From scientific evidence to human health protection. Policy Department for Citizens' Rights and Constitutional Affairs. Directorate General for Internal Policies of the Union. PE 608.866, European Union, 2019. https://www.europarl.europa.eu/RegData/etudes/STUD/2019/608866/IPOL_STU(2019)608866_EN.pdf (accessed January 13, 2020).

Dignam, T., R. B. Kaufmann, L. LeStourgeon et al. 2019. Control of lead sources in the United States, 1970–2017: Public health progress and current challenges to eliminating lead exposure. *J Public Health Manag Pract* 25(1):13–22. doi: 10.1097/PHH.0000000000000889.

Ema, M., S. Fujii, M. Furukawa et al. 2001. Rat two-generation reproductive toxicity study of bisphenol A. *Reprod Toxicol* 15(5):505–523. doi: 10.1016/S0890-6238(01)00160-5.

Eslami, B., K. Naddafi, N. Rastkari et al. 2016. Association between serum concentrations of persistent organic pollutants and gestational diabetes mellitus in primiparous women. *Environ Res* 151:706–712. doi: 10.1016/j.envres.2016.09.002.

Fakhri, Y., J. Rahmani, C. A. Fernandes Oliveira et al. 2019. Aflatoxin M1 in human breast milk: A global systematic review, metaanalysis, and risk assessment study (Monte Carlo simulation). *Trends Food Sci Technol* 88:333–342. doi: 10.1016/j.tifs.2019.03.013.

Garcia, A. M., T. Fletcher, and F. G. Benavides. 1999. Parental agricultural work and selected congenital malformations. *Am J Epidemiol* 149(1):64–74. doi: 10.1093/oxfordjournals.aje.a009729.

Garry, V. F., D. Schreinemachers, M. E. Harkins et al. 1996. Pesticide appliers, biocides and birth defects in rural Minnesota. *Environ Health Perspect* 104(4):394–399. doi: 10.1289/ehp.96104394.

Goldinger, D. M., A. L. Demierre, O. Zoller et al. 2015. Endocrine activity of alternatives to BPA found in thermal paper in Switzerland. *Regul Toxicol Pharmacol* 71(3):453–462. doi: 10.1016/j.yrtph.2015.01.002.

Greenspan, F. S., and D. G. Gardner. 2004. *Endokrynologia ogólna i kliniczna*. Lublin: Wydawnictwo CZELEJ.

Gregoraszczuk, E., K. Augustowska, and A. Ptak. 2004. Ksenoestrogeny środowiskowe jako jedna z przyczyn zaburzeń endokrynnych będących powodem poronień i przedwczesnych porodów. *Pol J Endo* 6(55):819–824.

Gromiec, J. P., and S. Czerczak. 2002. Kryteria oceny narażenia na substancje chemiczne w Polsce i na świecie: Procedury ustalania i stosowania. *Med Pr* 53(1):53–59.

Grotowska, M., K. Janda, and K. P. Jakubczyk. 2018. Wpływ pestycydów na zdrowie człowieka. *Pomeranian J Life Sci* 64(2):42–50. doi: 10.21164/pomjlifesci.403.

Grunt, H., and S. Czerczak. 2007. Butan-2-on: Dokumentacja proponowanych wartości dopuszczalnych wielkości narażenia zawodowego. *Podstawy i Metody Oceny Środowiska Pracy* 1(51):5–27.

Gul, S., H. Ashraf, O. Khawar et al. 2019. Prevalence and preventive measures of infertility in male by Kruger's criteria, a randomized study in private and government health care hospitals. *Bangladesh J Med Sci* 18(1):94–99. doi: 10.3329/bjms.v18i1.39557.

Guo, S. W., P. Simsa, C. M. Kyama et al. 2009. Reassessing the evidence for the link between dioxin and endometriosis: From molecular biology to clinical epidemiology. *Mol Hum Reprod* 15(10):609–624. doi: 10.1093/molehr/gap075.

Heinrich-Hirsch, B., S. Madle, A. Oberemm et al. 2001. The use of toxicodynamics in risk assessment. *Toxicol Lett* 120(1–2):131–141. doi: 10.1016/S0378-4274(01)00291-0.

Huang, T., L. A. Danaher, B. J. Brüschweiler et al. 2019. Naturally occurring bisphenol F in plants used in traditional medicine. *Arch Toxicol* 93(6):1485–1490. doi: 10.1007/s00204-019-02442-5.

Hyder, A., H. J. Lee, K. Ebisu et al. 2014. PM2.5 exposure and birth outcomes: Use of satellite- and monitor-based data. *Epidemiol* 25(1):58–67. doi: 10.1097/EDE.0000000000000027.

Igarashi, T. M., K. L. Bruner-Tran, G. R. Yeaman et al. 2005. Reduced expression of progesterone receptor-B in the endometrium of women with endometriosis and in cocultures of endometrial cells exposed to 2,3,7,8-tetrachlorodibenzop- dioxin. *Fertil Steril* 84(1):67–74. doi: 10.1016/j.fertnstert.2005.01.113.

Indulski, J., and K. Sitarek. 1998. *Czynniki środowiska pracy upośledzające płodność*. Łódź: Instytut Medycyny Pracy im. prof. dra Jerzego Nofera w Łodzi.

Jakubowski, M. 2006. Heksan: Dokumentacja proponowanych dopuszczalnych wielkości narażenia zawodowego. *Podstawy i Metody Oceny Środowiska Pracy* 1(47):109–129.

Jakubowski, M. 2007. Toluen: Dokumentacja proponowanych dopuszczalnych wielkości narażenia zawodowego. *Podstawy i Metody Oceny Środowiska Pracy* 3(53):131–158.

Jakubowski, M. 2014. Ołów i jego związki nieorganiczne, z wyjątkiem arsenianu(V), ołowiu(II) i chromianu(VI) ołowiu(II) – w przeliczeniu na ołów, frakcja wdychalna: Dokumentacja proponowanych dopuszczalnych wielkości narażenia zawodowego. *Podstawy i Metody Oceny Środowiska Pracy* 2(80):111–144. doi: 10.5604/1231868X.1111932.

Jankowska, A., and S. Czerczak. 2010. N,N-Dimetyloformamid: Dokumentacja proponowanych dopuszczalnych wielkości narażenia zawodowego. *Podstawy i Metody Oceny Środowiska Pracy* 4(66):55–92.

Jastrzębski, T., A. Kowalska, I. Szymala et al. 2016. Narażenie na kadm w okresie pre- i postnatalnym: Jego wpływ na płodność i na zdrowie dzieci. *Med Środow* 19(3):58–64. doi: 10.19243/2016307.

Jones, H. E., and R. L. Balster. 1998. Inhalant abuse in pregnancy. *Obstet Gynecol Clin North Am* 25(1):153–167. doi: 10.1016/s0889-8545(05)70363-6.

Jørgensen, N., M. Vierula, R. Jacobsen et al. 2011. Recent adverse trends in semen quality and testis cancer incidence among Finnish men. *Int J Androl* 34:37–48. doi: 10.1111/j.1365-2605.2010.01133.x.

Jurewicz, J., and W. Hanke. 2007. Ryzyko zaburzeń reprodukcji wśród osób pracujących w gospodarstwach ogrodniczych. *Med Pr* 58(5):433–438.

Jurewicz, J., and W. Hanke. 2008. Zawodowa i środowiskowa ekspozycja na pestycydy a ryzyko wystąpienia wad wrodzonych: Przegląd badań epidemiologicznych. *Probl Hig Epidemiol* 89(3):302–309.

Kaltenecker Retto de Queiroz, E., and W. Waissmann. 2006. Occupational exposure and effects on the male reproductive system. *Cad Saúde Pública* 22(3):485–493. doi: 10.1590/s0102-311x2006000300003.

Kim, J. C., H. C. Shin, S. W. Cha et al. 2001. Evaluation of developmental toxicity in rats exposed to the environmental estrogen bisphenol A during pregnancy. *Life Sci* 69(22):2611–2625. doi: 10.1016/s0024-3205(01)01341-8.

Kim, J. Y., H. G. Choi, H. M. Lee et al. 2017. Effects of bisphenol compounds on the growth and epithelial mesenchymal transition of MCF-7 CV human breast cancer cells. *J Biomed Res* 31(4):358–369. doi: 10.7555/jbr.31.20160162.

Klaassen, C. D. and J. D. Watkins III. 2014. Casarett & Doull: *Podstawy toksykologii*. Wrocław: MedPharm Polska.

Koperwas, M., and M. Głowacka. 2017. Problem niepłodności wśród kobiet i mężczyzn: Epidemiologia, czynniki ryzyka i świadomość społeczna. *Aspekty Zdrowia i Choroby* 2(3):31–49.

Kopij, M., and A. Rapak. 2008. Rola receptorów jądrowych w procesie śmierci komórek. *Postepy Hig Med Dosw* 62:571–581.

Kowalska, A., K. Walkiewicz, P. Kozieł et al. 2017. Aflatoksyny: Charakterystyka i wpływ na zdrowie człowieka. *Postepy Hig Med Dosw* 71:315–327. doi: 10.5604/01.3001.0010.3816.

Kraszewska, A., M. Abramowicz, M. Chłopocka-Woźniak et al. 2014. Wpływ stosowania litu na czynność gruczołu tarczowego u pacjentów z chorobą afektywną dwubiegunową. *Psychiatr Pol* 48(3):417–428. doi: 10.12740/PP/21684.

Krzywy, I., E. Krzywy, J. Peregud-Pogorzelski et al. 2011. Kadm: Czy jest się czego obawiać? *Ann Acad Med Stetin* 57(3):49–63.

Kulik-Kupka, K., J. Nowak, I. Korzonek-Szlacheta et al. 2017. Wpływ dysruptorów endokrynnych na funkcje organizmu. *Postepy Hig Med Dosw* 71:1231–1238. doi: 10.5604/01.3001.0010.7748.

Langauer-Lewowicka, H., and K. Pawlas. 2015. Związki endokrynnie czynne: Prawdopodobieństwo niepożądanego działania środowiskowego. *Med Środ* 18(1):7–11. doi: 10.15199/2.2015.5.4.

Laug, E. P., F. M. Kunze, and C. S. Prickett. 1951. Occurrence of DDT in human fat and milk. *AMA Arch Ind Hyg Occup Med* 3(3):245–246.

Lauretta, R., A. Sansone, M. Sansone et al. 2019. Endocrine disrupting chemicals: Effects on endocrine glands. *Front Endocrinol* 10(178):1–7. doi: 10.3389/fendo.2019.00178.

Lawson, C. C., T. M. Schnorr, G. P. Daston et al. 2003. An occupational reproductive research agenda for the third millennium. *Environ Health Perspect* 111(4):584–592. doi: 10.1289/ehp.5548.

Lee, S., C. Kim, H. Younet et al. 2017. Thyroid hormone disrupting potentials of bisphenol A and its analogues: In vitro comparison study employing rat pituitary (GH3) and thyroid follicular (FRTL-5) cells. *Toxicol In Vitro* 40:297–304. doi: 10.1016/j.tiv.2017.02.004.

Lee, V. 2016. Wpływ reformy TSCA na produkcję artykułów powszechnego użytku: Paszport do Wall Street. https://www.paszport.ws/wp%C5%82yw-reformy-tsca-na-produkcj%C4%99-artyku%C5%82%C3%B3w-powszechnego-u%C5%BCytku (accessed January 13, 2020).

Li, Z., Y. Tang, X. Song et al. 2019. Impact of ambient $PM_{2.5}$ on adverse birth outcome and potential molecular mechanism. *Ecotoxicol Environ Saf* 169:248–254. doi: 10.1016/j.ecoenv.2018.10.109.

Ligocka, D. 2007. Dokumentacja proponowanych wartości dopuszczalnych wielkości narażenia zawodowego: Ksylen – mieszanina izomerów. *Podstawy i Metody Oceny Środowiska Pracy* 4(45):139–165.

Llanos, M. N., and A. M. Ronco. 2009. Fetal growth restriction is related to placental levels of cadmium, lead and arsenic but not with antioxidant activities. *Reprod Toxicol* 27:88–92. doi: 10.1016/j.reprotox.2008.11.057.

Loffredo, C. A., E. K. Silbergeld, C. Ferencz et al. 2001. Association of transposition of the great arteries in infants with maternal exposures to herbicides and rodenticides. *Am J Epidemiol* 153(6):529–536. doi: 10.1093/aje/153.6.529.

Łukaszewicz-Hussain, A. 2011. Wpływ pestycydów fosforoorganicznych na trzustkę. *Med Pr* 62(5):543–550.

Lukowicz, C., S. Ellero-Simatos, M. Régnier et al. 2018. Metabolic effects of a chronic dietary exposure to a low-dose pesticide cocktail in mice: Sexual dimorphism and role of the constitutive androstane receptor. *Environ Health Perspect* 126(6):067007. doi: 10.1289/EHP2877.

Marchewka, Z., and A. Grzebinoga. 2009. Substancje chemiczne – czynnikami ryzyka nefropatii cukrzycowej. *Post Hig Med Dosw* 63:592–597.

Mead, M. N. 2008. Contaminants in human milk: Weighing the risks against the benefits of breastfeeding. *Environ Health Perspect* 116(10):426–434. doi: 10.1289/ehp.116-a426.

Memiş, E. Y., and S. S. Yalçın. 2019. Human milk mycotoxin contamination: Smoking exposure and breastfeeding problems. *J Matern Fetal Neonatal Med* 8:1–10. doi: 10.1080/14767058.2019.1586879.

Mesnage, R., A. Phedonos, M. Arno et al. 2017. Transcriptome profiling reveals bisphenol A alternatives activate estrogen receptor alpha in human breast cancer cells. *Toxicol Sci* 158(2):431–443. doi: 10.1093/toxsci/kfx101.

Mocarelli, P., P. M. Gerthoux, L. L. Needham et al. 2011. Perinatal exposure to low doses of dioxin can permanently impair human semen quality. *Environ Health Perspect* 119(5):713–718. doi: 10.1289/ehp.1002134.

Mohammadyan, M., M. Moosazadeh, A. Borji et al. 2019a. Exposure to lead and its effect on sleep quality and digestive problems in soldering workers. *Environ Monit Assess* 191(184):1–9. doi: 10.1007/s10661-019-7298-2.

Mohammadyan, M., M. Moosazadeh, A. Borji et al. 2019b. Investigation of occupational exposure to lead and its relation with blood lead levels in electrical solderers. *Environ Monit Assess* 191(126):1–9. doi: 10.1007/s10661-019-7258-x.

Montano, D. 2014. Chemical and biological work-related risks across occupations in Europe: A review. *J Occup Med Toxicol* 9(28):1–13. doi: 10.1186/1745-6673-9-28.

Ng, T. P., S. C. Foo, and T. Yoong. 1992. Risk of spontaneous abortion in workers exposed to toluene. *Brit J Ind Med* 49:804–808. doi: 10.1136/oem.49.11.804.

Nurminen, T., K. Rantala, K. Kurppa et al. 1995. Agricultural work during pregnancy and selected structural malformations in Finland. *Epidemiol* 6(1):23–30. doi: 10.1097/00001648-199501000-00006.

Obeng-Gyasi, E. 2019. Sources of lead exposure in various countries. *Rev Environ Health* 34(1):25–34. doi: 10.1515/reveh-2018-0037.

Pałczyński, C. 2018. Związki hormonalnie czynne: Źródła narażenia i skutki zdrowotne. *Alergia* 3:47–50.

Panczyk, M., A. Sałagacka, and M. Mirowski. 2007. Gen MDR1 (ABCB1) kodujący glikoproteinę P (P-gp) z rodziny transporterów błonowych ABC: Znaczenie dla terapii i rozwoju nowotworów. *Postepy Biochem* 53(4):361–373.

Papalou, O., E. A. Kandaraki, G. Papadakis et al. 2019. Endocrine disrupting chemicals: An occult mediator of metabolic disease. *Front Endocrinol* 10(112):1–14. doi: 10.3389/fendo.2019.00112.

Piotrowski, J. 2006. *Podstawy toksykologii*. Warszawa: Wydawnictwa Naukowo-Techniczne.

PTA [Polskie Towarzystwo Andrologiczne]. 2016. Rekomendacje Polskiego Towarzystwa Andrologicznego i Krajowej Izby Diagnostów Laboratoryjnych. Podstawowe badanie nasienia według standardów Światowej Organizacji Zdrowia z roku 2010. Warszawa: Wydawnictwo THORG.

Ries, G., and R. Hess. 1999. Retinol: Safety considerations for its use in cosmetic products. *J Toxicol Cutaneous Ocul Toxicol* 18(3):169–185. doi: 10.3109/15569529909044238.

Rochester, J. R., and A. L. Bolden. 2015. Bisphenol S and F: A systematic review and comparison of the hormonal activity of bisphenol A substitutes. *Environ Health Perspect* 123(7):643–650. doi: 10.1289/ehp.1408989.

Rosenfeld, C. S. 2017. Neuroendocrine disruption in animal models due to exposure to bisphenol A analogues. *Front Neuroendocrinol* 47:123–133. doi: 10.1016/j.yfrne.2017.08.001.

Rossouw, J., J. Offenheimer, and J. Van Rooyen. 1987. Apparent central neurotransmitter receptor changes induced by low-level lead exposure during different developmental phases in the rat. *Toxicol Appl Pharmacol* 91(1):132–139. doi: 10.1016/0041-008X(87)90200-6.

RPA [Risk and Policy Analysts]. 2013. Final report for the analysis at EU-level of health, socioeconomic and environmental impacts in connection with possible amendment to Directive 2004/37/EC of the European Parliament and of the Council of April 29, 2004, on the protection of workers from the risks related to exposure to carcinogens and mutagens at work to extend the scope to include category 1A and 1B reprotoxic substances. Brussels: European Commission. https://www.labourline.org/Record.htm?idlist=1&record=19126248124919444209 (accessed January 13, 2020).

Rutkowska, A., D. Rachoń, A. Milewicz et al. 2015. Polish Society of Endocrinology position statement on endocrine disrupting chemicals (EDCs). (Stanowisko Polskiego Towarzystwa Endokrynologicznego dotyczące związków endokrynnie czynnych (EDC)). *Endokrynol Pol* 66(3):276–285. doi: 10.5603/EP.2015.0035.

Rzeszutek, J., S. Popek, M. Matysiak et al. 2014. Zmiany epigenetyczne spowodowane ekspozycją na pestycydy. *Probl Hig Epidemiol* 95(3):561–567.

Sapota, A., and M. Skrzypińska-Gawrysiak. 2010. Pary rtęci i jej związki nieorganiczne: Dokumentacja proponowanych wartości dopuszczalnych wielkości narażenia zawodowego. *Podstawy i Metody Oceny Środowiska Pracy* 3(65):85–149.

Schmitz-Felten, E., K. Kuhl, K. Sørig Hougaard et al. 2016. State of the art report on reproductive toxicants. European Agency for Safety and Health at Work (EU-OSHA). doi: 10.2802/87916.

Schreinemachers, D. M. 2003. Birth malformations and other adverse perinatal outcomes in U.S. Wheat-producing states. *Environ Health Perspect* 111(9):1259–1264. doi: 10.1289/ehp.5830.

Sitarek, K. 2004. Karbendazym: Dokumentacja proponowanych wartości dopuszczalnych wielkości narażenia zawodowego. *Podstawy i Metody Oceny Środowiska Pracy* 1(39):45–63.

Sitarek, K. 2005. 1-Metylo-2-pirolidon: Dokumentacja proponowanych wartości dopuszczalnych wielkości narażenia zawodowego. *Podstawy i Metody Oceny Środowiska Pracy* 1(43):103–115.

Skowron, B., K. Juszczak, and P. J. Thor. 2011. Niepłodność męska indukowana lekami. *Folia Med Cracov* LI(1–4):99–106.

Słowińska, M., M. Koter-Michalak, and B. Bukowska. 2011. Wpływ dioksyn na organizm człowieka. *Med Pr* 62(6):643–652.

Sobczak, A. 2012. Czynniki chemiczne w środowisku zagrażające zdrowiu ludzi. *Med Środ* 15(1):7–17.

Soni, K. K., and S. Singh. 2018. Risk of congenital malformations: A systematic review. *Int J Med Sci Clin Invent* 6(1):4278–4282. doi: 10.18535/ijmsci/v6i01.08.

Soroka, P. P., M. Cyprowski, and I. Szadkowska-Stańczyk. 2008. Narażenie zawodowe na mykotoksyny. *Med Pr* 59(4):333–345.

Starek, A. 2007. *Toksykologia narządowa.* Warszawa: PZWL.

Stec, M., E. Kurzeja, A. Kościołek et al. 2012. Zagrożenia wynikające z narażenia na dioksyny i dioksynopodobne polichlorowane bifenyle. *Probl Hig Epidemiol* 93(4):639–646.

Sun, Q., M. C. Cornelis, M. K. Townsend et al. 2014. Association of urinary concentrations of bisphenol A and phthalate metabolites with risk of type 2 diabetes: A prospective investigation in the Nurses' Health Study (NHS) and NHSII cohorts. *Environ Health Perspect* 122(6):616–623. doi: 10.1289/ehp.1307201.

Svensson, B. G., G. Nise, E. M. Erfurth et al. 1992. Neuroendocrine effects in printing workers exposed to toluene. *Br J Ind Med* 49(6):402–408. doi: 10.1136/oem.49.6.402.

Szewczyńska, M., and E. Dobrzyńska. 2018. Substancje endokrynnie aktywne: Występowanie, zagrożenia i metody ich oznaczania. *Przemysł Chemiczny* 97(2):230–237. doi: 10.15199/62.2018.2.9.

Szymańska, J., B. Frydrych, P. Struciński, W. Szymczak, A. Hernik, and E. Bruchajzer. 2020. Mieszanina polichlorowanych dibenzo-p-dioksyn i polichlorowanych dibenzofuranów: Dokumentacja proponowanych wartości dopuszczalnych wielkości narażenia zawodowego. *Podstawy i Metody Oceny Środowiska Pracy* 1(103):71–142. doi: 10.5604/01.3001.0013.7815

Szymańska, J., and E. Bruchajzer. 2010. 2-Metoksyetanol: Dokumentacja proponowanych wartości dopuszczalnych wielkości narażenia zawodowego. *Podstawy i Metody Oceny Środowiska Pracy* 4(66):93–139.

Szymańska, J., and B. Frydrych. 2006. 2,2-Bis(4-hydroksyfenylo)-propan – pyły: Dokumentacja proponowanych wartości dopuszczalnych wielkości narażenia zawodowego. *Podstawy i Metody Oceny Środowiska Pracy* 3(49):101–117.

Tai, X., and Y. Chen. 2016. Urinary bisphenol A concentrations positively associated with glycated hemoglobin and other indicators of diabetes in Canadian men. *Environ Res* 147:172–178. doi: 10.1016/j.envres.2016.02.006.

Tantibanchachai, C. 2017. Experimental studies on congenital malformations (1959), by James G. Wilson. Embryo Project Encyclopedia. http://embryo.asu.edu/handle/10776/11696 (accessed January 13, 2020).

Vencovsky, D., M. Postle, D. Fleet et al. 2019. Study to collect recent information relevant to modernising EU Occupational Safety and Health chemicals legislation with a particular emphasis on reprotoxic chemicals with the view to analyse the health, socioeconomic and environmental impacts in connection with possible amendments of Directive 2004/37/EC and Directive 98/24/EC: Final report, Report 1: Baseline assessment. Luxembourg: Publications Office of the European Union. doi: 10.2767/964906.

Vom Saal, F. S., P. S. Cooke, D. L. Buchanan et al. 1998. A physiologically based approach to the study of bisphenol A and other estrogenic chemicals on the size of reproductive organs, daily sperm production, and behavior. *Toxicol Ind Health* 14(1–2):239–260. doi: 10.1177/074823379801400115.

Wang, Y., L. Chen, Y. Gao et al. 2016. Effects of prenatal exposure to cadmium on neurodevelopment of infants in Shandong, China. *Environ Pollut* 211:67–73. doi: 10.1016/j.envpol.2015.12.038.

Wasilonek, M. 2019. Lit: Występowanie, właściwości i zastosowanie najlżejszego z metali. https://www.medonet.pl/zdrowie,lit--wystepowanie--wlasciwosci-i-zastosowanie,artykul,1732970.html (accessed January 13, 2020).

WHO [World Health Organization]. 2016. Fact sheets: Congenital anomalies. https://www.who.int/news-room/fact-sheets/detail/congenital-anomalies (accessed January 13, 2020).

Wilhelm, M., J. Wittsiepe, F. Lemm et al. 2008. The Duisburg birth cohort study: Influence of the prenatal exposure to PCDD/Fs and dioxin like PCBs on thyroid hormone status in newborns and neurodevelopment of infants until the age of 24 months. *Mut Res* 659:83–92. doi: 10.1016/j.mrrev.2007.11.002.

Wu, Q., S. Ohsako, R. Ishimura et al. 2004. Exposure of mouse preimplantation embryos to 2,3,7,8-tetrachlorodibenzo-p-dioxin (TCDD) alters the methylation status of imprinted genes H19 and Igf2. *Biol Reprod* 70(6):1790–1797. doi: 10.1095/biolreprod.103.025387.

Yang, C. F., H. M. Shen, Y. Shen et al. 1997. Cadmium-induced oxidative cellular damage in human fetal lung fibroblasts (MRC-5 cells). *Environ Health Perspect* 105(7):712–716. doi: 10.1289/ehp.97105712.

Yang, J. M., Q. Y. Chen, and X. Z. Jiang. 2002. Effects of metallic mercury on the perimenstrual symptoms and menstrual outcomes of exposed workers. *Am J Ind Med* 42(5):403–409. doi: 10.1002/ajim.10130.

Yu, H., D. J. Caldwell, and R. P. Suri. 2019. In vitro estrogenic activity of representative endocrine disrupting chemicals mixtures at environmentally relevant concentrations. *Chemosphere* 215:396–403. doi: 10.1016/j.chemosphere.2018.10.067.

Zhang, Y., J. Wang, L. Chen et al. 2019. Ambient $PM_{2.5}$ and clinically recognized early pregnancy loss: A case-control study with spatiotemporal exposure predictions. *Environ Int* 126:422–429. doi: 10.1016/j.envint.2019.02.062.

Zhu, J. L., L. E. Knudsen, A. M. Andersen et al. 2006. Laboratory work and pregnancy outcomes: A study within the National Birth Cohort in Denmark. *Occup Environ Med* 63:53–58. doi: 10.1136/oem.2005.021204.

4 Carcinogenic and Mutagenic Substances

Jolanta Skowroń

CONTENTS

4.1 INTRODUCTION

According to the World Health Organization (WHO) and the International Agency for Research on Cancer (IARC), cancer kills 8.2 million people worldwide each year, and 14 million new cancer incidences are reported each year. It is estimated that before the year 2035, cancer mortality will increase by 78%, and cancer incidence by 70% [Takala 2015].

All over the world, the most common and fatal kind of cancer is lung cancer for males and breast cancer for females. Mortality and cancer rates differ, depending on the geographical region. In the United States, the first cancer incidence "map" indicated excess oral cancer rates in the south-west states, which was probably connected to the habit of chewing tobacco. Similarly, the high mortality rate due to lung cancer, seen along the American coasts during the Second World War, can be ascribed to a drastic increase in employment in the shipbuilding industry, where exposure to asbestos was particularly high. Mortality rates for lung cancer in Spanish men were the highest in Extremadura, Asturias, and south-west Andalusia. In Andalusia, they were 20% higher than the national average and twice as high in Navarra. This part of the Spanish region also had the highest share of manual workers in Spain, up to 80% of the active population. The same distribution was observed in Catalonia. The highest indicators were found in the Barcelona region and along the coast of Catalonia. In the city itself, they were concentrated in old working-class districts and in outer suburbs, populated by immigrants. Those geographical inequalities in sickness and deaths also reflected the inequalities in social status [Mengeot et al. 2014].

Cancer is the prominent cause of death due to occupational exposure to carcinogenic/mutagenic substances in highly industrialized countries, including the EU member states. Cancers have a multifactorial etiology. Lung cancer constitutes 54–75% of all occupation-related cancers. Epidemiological studies have shown that occupational exposure to carcinogenic/mutagenic substances causes 5.3–8.4% of all cancers, and, among men, 17–29% of all deaths due to lung cancer.

Of all lung cancer cases, 55–85% were caused by exposure to asbestos, which currently causes other cancers and diseases, and which could have been prevented in the past. In 28 EU member states, in 2015, exposure to asbestos has caused 10,368 deaths out of the total 30,208 deaths worldwide due to occupational cancer. Mortality due to cancers, including occupational cancers, is increasing because of the increasing average life expectancy and because of a gradual decrease in other causes of death, such as infectious diseases and injuries [Takala 2015].

According to the latest data, exposure to asbestos causes 255,000 deaths yearly, with 233,000 deaths being linked to occupational exposure [Furuya et al. 2018]. In the European Union in 2017, 106,307 workers died due to occupational exposure to carcinogenic substances, including 7,874 in Poland alone [Takala 2018]. Occupational cancers can be prevented through the minimization or elimination of exposure to carcinogenic/mutagenic substances.

4.2 CHARACTERISTICS OF THE CARCINOGENIC/MUTAGENIC SUBSTANCES

A carcinogenic substance, according to the Regulation (EC) No 1272/2008 (CLP), is a substance or a mixture of substances causing cancer or increasing its incidence. Substances which have been known to cause benign and malignant tumors in animal models in vivo are also substances assumed to be or suspected of being carcinogenic to human, unless there is strong evidence that the mechanism of tumor formation is not relevant for humans [EC 2008].

Carcinogenic properties of chemical substances can be assessed based on the results of:

- epidemiological studies,
- long-term animal studies,
- short-term tests, allowing for genotoxicity assessment (mutation and non-permanent DNA damage).

Currently known substances with proven carcinogenic properties belong to a number of chemical groups, and no shared characteristic in their structure or other properties can be clearly shown as the one causing carcinogenic effects. Compounds with known carcinogenic properties include:

- inorganic compounds: arsenic, chromium(VI), and nickel salts,
- organic compounds: benzene, 2-naphthylamine, vinyl chloride, polycyclic aromatic hydrocarbons,

- complex substances: soot, tar, mineral oils,
- natural substances: aflatoxins, mitomycin C, safrole, phorbol esters, nitrosamines.

A mutagenic substance, according to the Regulation (EC) No 1272/2008 (CLP), is a substance or a mixture of substances known to cause inherited genetic mutations or suspected of causing genetic mutations in human reproductive cells [EC 2008].

Mutations are random changes in the genetic material, which may be inherited, and are not a result of recombination processes. Mutations can occur spontaneously or be induced as a result of mutagenic effects of e.g. chemicals reacting with the deoxyribonucleic acid (DNA) [Mol and Stolarek 2011].

The terms "mutagenic" and "mutagen" are used for factors contributing to an increased incidence of mutations in cell or organism populations. The more general terms "genotoxic" and "genotoxicity" refer to factors or processes that alter the DNA structure or its segments, including those factors which cause DNA damage through the disruption of replication processes or those which non-physiologically (temporarily) cause changes in its replication. Results of genotoxicity studies are commonly considered to be indicators of mutagenic effects. The hazard class of mutagenic effects on reproductive cells is mainly determined for substances which can cause mutations in human reproductive cells and which can be inherited by the offspring. When classifying substances and mixtures into mutagens, one must also consider the results of in vitro studies on mutagenicity or genotoxicity and those of in vivo studies on mammalian somatic and reproductive cells [EC 2008].

In workers exposed to chemical substances with carcinogenic/mutagenic effects, tumorous lesions can reveal themselves many years after the first exposure. This delay period is called the period of latency and it can last for 4–40 years. Cancers occurring after occupational exposure can be located in various locations in the organism, not necessarily in those that have direct contact with the chemical substance.

It was estimated that in 28 EU member states, in 2012, on average, 122,600 people (range 91,500 to 150,500) were diagnosed with cancer, caused by occupational exposure to carcinogenic substances. The number of deaths due to these cancers was estimated to be 79,700 (range 57,700 to 106,500), which has caused the loss of almost 1.2 (0.8 to 1.6) million years of life due to premature death in the EU population [Jongeneel et al. 2016].

4.3 MECHANISMS OF CARCINOGENICITY/MUTAGENICITY

The carcinogenic process is a multistage process, in which benign tumors can be an intermediate step. Studies on tumor induction have shown that two stages can be distinguished in the carcinogenic process: initiation and promotion. Tumor initiation causes an interaction between a genotoxic compound and the cell's DNA and is an irreversible process (the cell undergoes permanent mutation). Such initiated cells can remain in the body during most of its life, with no tumorous growth ever appearing. Tumor promotion causes the multiplication of the initiated cells under the influence of non-carcinogenic factors and is a reversible process.

Tumor promoter compounds do not usually form reactive metabolites, but modulate the growth or death of cells ("apoptosis") via receptors or other mechanisms. The available initiation–promotion schemes are, however, simplified models, which cannot reflect all the aspects of carcinogenicity [Schrenk 2018].

The first stage, initiation, takes place quickly and seems to be irreversible. The available data indicates that initiation usually results from one or more mutations of cellular DNA. The main reason for these mutations is covalent reactions of electrophilic derivatives of carcinogenic factors with the DNA. Factors causing DNA damage have the capacity for causing permanent genetic disruption and contribute to an increase in the frequency of changes in the genome. DNA damage can cause genotoxic and cytotoxic effects in the cells. Genotoxic activity is the ability to cause changes in the DNA – directly or through a particular compound or its active metabolite. Genotoxic activity can encompass changes at the level of a single gene, a chromosome, or the entire genome. The disruption or inhibition of the replication process and DNA transcription results from the cytotoxicity of a given compound and can lead to apoptosis (death) of a cell. The most common kinds of DNA damage, caused by mutagenic factors, are the loss of a nitrogen base, intercalation of the agent between base pairs, modification of nitrogen bases through hydrolysis, alkylation, and oxidation. Mutagens can cause the formation of crosslinks inside and between DNA strands, as well as causing photodamage and single- or double-strand DNA breakage [Mol and Stolarek 2011].

The second stage, promotion, takes a longer time, as it is a complex process in which the early stages are largely reversible. Critical events seem to be epigenetic. Complete carcinogenic agents have both initiating and promoting properties, but their ratios can vary significantly for different chemical substances. Knowledge of the mode of action of carcinogens constitutes a useful basis to prevent cancers in humans [Miller and Miller 1981].

Carcinogenicity can be caused by several mechanisms, including metabolic activation causing the formation of active metabolites, covalently bonding to DNA and proteins, mutagenic and genotoxic activity, cytotoxicity, and the formation of adducts with proteins, for example with hemoglobin [IARC 2019a].

Most of the carcinogens, by themselves or in combination with others, cause cancer by interacting with the cell's DNA and disrupting their normal functioning. This ultimately leads to the formation of a tumor (abnormal tissue growth), which has the ability to spread (metastasize) from the location of origin as well as the ability to invade and cause dysfunction of other tissues, which in turn results in organ failure and eventually death. Two basic mechanisms by which carcinogens initiate the formation of such cancers are based on changing the DNA to incite cell division and on taking away the cell's ability to auto-destruct when stimulated by the usual triggers, such as DNA damage or cell damage (a process known as apoptosis). There are also carcinogens which induce cancer through non-genotoxic mechanisms, such as immunosuppression and the induction of tissue-specific inflammation.

In 2019, the International Agency for Research on Cancer (IARC), based on a mechanistic data review of substances included in group 1, i.e. substances carcinogenic to humans (Volume 100: A Review of Human Carcinogens, covers all agents previously classified by IARC as "carcinogenic to humans (Group 1)" and was developed by six separate Working Groups), distinguished ten "key properties" of these substances, which collectively represent a number of mechanisms of carcinogenicity. These included:

1) Electrophilic structure of the parent compounds or metabolites, which form adducts with nucleic acids or proteins. An example of a direct electrophilic carcinogen is ethylene oxide. Classic examples of relatively inert chemical agents which require metabolic activation to become carcinogenic are polycyclic aromatic hydrocarbons, aromatic amines, *N*-nitrosamines, aflatoxins, and benzene. A number of enzymes, including cytochrome P450, flavine-containing monooxygenase, prostaglandin synthase, and certain peroxidases can convert these relatively inert chemical substances into strongly toxic and carcinogenic metabolites or reactive intermediates through biotransformation.
2) Genotoxic activity, which refers to factors inducing DNA damage, mutation, or both. DNA damage can be spontaneously caused by errors in nucleic acid metabolism and can be induced by endogenous or exogenous factors. In some cases, exogenous factors can produce endogenous factors, such as formaldehyde or acetaldehyde, creating a background level DNA damage. Examples of DNA damage include DNA adducts (molecules covalently bonded to DNA), DNA strand breakage (phosphodiester bond breakage), DNA crosslinking, and DNA alkylation. DNA damage alone is not a mutation and does not usually change the linear sequence of nucleotides (or bases) in DNA, while mutation is a change in the DNA sequence and typically occurs when a cell attempts to repair DNA damage. Genome mutations encompass duplication or deletion of nucleotide sequences of the entire chromosome, e.g. aneuploidy, or the formation of micronuclei containing a centromere. A number of factors from group 1 according to IARC is genotoxic.
3) The change in DNA damage repair or gene instability. Carcinogenic substances can work not only through direct DNA damage, but also through changing the processes controlling normal DNA replication or DNA damage repair. Examples include the inhibition of DNA repair by cadmium or formaldehyde. Genome instability is a well-recognized characteristic of a number of cancers and is considered to be one of the causes of cancer. Incidents indicating genome instability include chromosome aberrations, gene mutations, microsatellite instability, and apoptosis, which were observed e.g. after exposure to cadmium.
4) Induction of epigenetic changes. The term "epigenetic" refers to stable changes in gene expression and chromatin organization, which are not caused by changes to the DNA sequence itself. Epigenetic phenomena, including changes in the state of methylome and chromatin density along with histone modification, can influence carcinogenesis, affecting gene expression and DNA repair.
5) Causing oxidative stress. A number of carcinogens can affect the redox balance in target cells. An imbalance, if present, is conducive to the formation

of reactive oxygen or nitrogen species at the cost of cell detoxification, which is known as oxidative stress. Reactive oxygen species and other free radicals, produced as a result of tissue inflammation, xenobiotic metabolism, interruption of mitochondrial oxidative phosphorylation or turnover of oxidized cellular components, can play a key role in a number of processes necessary for the conversion of normal cells into cancer cells. Oxidative stress is linked not only to cancer, but to a number of chronic diseases and pathologies. It is also common in cancer tissues and can be a part of the cancer environment. Oxidative damage to DNA can lead to point mutations, deletion, insertion, or chromosomal translocation, which can cause the activation of an oncogene and the inactivation of cancer suppressor genes, as well as potentially initiating or promoting carcinogenicity. Induction of cell damage by oxygen radicals is a characteristic of a number of different carcinogens, including radiation, asbestos, and carcinogenic infectious agents.

6) Causing chronic inflammation. Chronic inflammation, caused by persistent infections, such as those caused by *Helicobacter pylori*, as well as chemical agents, e.g. respirable fraction of crystalline silica or asbestos fibers, is linked to several forms of cancer. Inflammation contributes to a number of aspects of cancer development and progress, and enables cancer initiation. Inflammation acts via both internal and external pathways. Persistent infection and chronic inflammation disrupt local tissue homeostasis and change cell signaling, leading to the recruitment and activation of inflammatory cells. They constitute external pathways, linking inflammation to cancer. There are strong connections between inflammation and induction of oxidative stress and genome instability – it may be therefore difficult to distinguish the role of each of these mechanisms.
7) Immunosuppressive activity. Immunosuppression is the lowering of the capacity of the immune system to respond effectively to foreign antigens, including cancer cell antigens. Chronic immunosuppression creates the risk of cancer, especially an excessive risk of lymphoma. For example, immunosuppression constitutes a significant risk of cancer when accompanied by chronic exposure to foreign antigens, as in patients with organ transplants or in people infected with a carcinogenic virus.
8) Indirect effect modulation through receptors. A number of carcinogens act as ligands for receptors, such as 2,3,7,8-tetrachlorodibenzo-*p*-dioxin or PCBs. Receptor-mediated activation is essentially divided into two categories: (a) intracellular activation through nuclear receptors, which move into the nucleus as transcription factors and affect DNA, and (b) activation of cell surface receptors, which induce signaling pathways, causing biological responses, and involving different protein kinases. Most of the exogenous factors act as agonists by competing for the bond with an endogenous ligand; there are, however, also receptors for which few or no endogenous ligands have been identified, such as the aryl hydrocarbon receptor. Receptor-mediated activation most often causes changes in gene transcription. Molecular pathways regulated through ligand-receptor interaction, and which are the most significant in the context of carcinogenicity, are cell

proliferation (e.g. stimulation of the normal proliferation pathways, as in the case of estrogen-dependent tissues and hormonal therapy), xenobiotic metabolism, apoptosis, and the modulation of endogenous ligand bioavailability through influencing biosynthesis, bioactivation, and degradation.

9) Cell immortalization. Several human DNA and RNA viruses, including different human papillomaviruses, Epstein-Barr virus, Kaposi's sarcoma-associated herpesvirus, hepatitis B virus, hepatitis C virus, HIV, Merkel cell polyomavirus (MCPyV), and human T-lymphotropic virus type 1 (HTLV-1), are carcinogenic to humans. These viruses have developed a number of molecular mechanisms to disrupt particular cellular pathways to facilitate incorrect replication.
10) Affecting cell proliferation, cell cycle, or nutrient supply. There are at least three different carcinogenicity scenarios in which changes in cellular replication or cell cycle control have been described. One of carcinogenicity scenario invokes the predisposition for unrepaired DNA damage leading to cancer-causing mutations in replicating cells; the other tries to identify constant replication as a key mechanistic event, and the third describes the ability of a transformed cell to avoid normal cell cycle control and to continue replication. The common element of all three scenarios is the avoidance of apoptosis or other forms of programmed cell death, such as autophagy (the intracellular system of degradation of macromolecular cytoplasm components, particularly long half-life proteins and entire organelles), by at least a part of the cell population. Necrotic cell death releases proinflammatory signals to the surrounding tissue microenvironment, causing cell inflammation, which can intensify cancer cell proliferation and promote metastasis. However, different kinds of apoptosis and autophagy have an inverse effect, removing potential cancer cells from the population before changes enabling the formation of a malignant tumor happen. A number of agents affect necrosis, apoptosis, or autophagy and can have a varied influence on cancer induction in different tissues. Besides cellular death caused directly by agent toxicity, cells can also die in the tumor due to disrupted nutrient supply. The number of cancer cells can rise rapidly, quickly exceeding the supply capacity of the existing vascular system of the tissue. Neoangiogenesis, in which new blood vessels become tumorous, is the key to assuring the supply of nutrients. Agents which promote or inhibit angiogenesis will contribute to or delay tumor growth [Smith et al. 2016; IARC 2019a].

4.4 CLASSIFICATION OF CARCINOGENIC/ MUTAGENIC SUBSTANCES

The most authoritative organ in relation to carcinogenic substances, agents, or processes is the International Agency for Research on Cancer (IARC) of Lyon, France, a World Health Organization agency. IARC organizes scientific panels to discuss and assess the available data on the carcinogenicity of substances, agents, or mixtures.

These carcinogens are classified under one of the following groups: carcinogenic to humans – 120 agents, Group 1; probably carcinogenic to humans – 83 agents, Group 2A; possibly carcinogenic to humans – 314 agents, Group 2B; not classifiable as to its carcinogenicity to humans – 500 agents, Group 3; and probably not carcinogenic to humans, Group 4 . IARC does not classify mutagenic agents, but as shown in the monograph, the fact that an agent may induce gene and chromosome mutations in mammals in vivo indicates that it may be carcinogenic as well [IARC 2019b].

Among 120 agents classified by IARC under Group 1, there is data on occupational exposure for 70 of them in IARC monographs. From these 70 agents, 63 had significant proof of being carcinogenic to humans. In the case of the remaining 7, occupational exposure data were present, but they were classified under Group 1, based on mechanistic evidence, because evidence of carcinogenic/mutagenic activity in humans was "insufficient". From these 63 Group 1 agents with "sufficient evidence of carcinogenicity" in humans, 59 of the assessments were at least partially based on studies on the exposed workers. The remaining four agents were excluded from the group, because occupational exposure was noted, but no epidemiological data was submitted. Among these 59 agents, 47 were individual substances, mixtures, or types of radiation, and 12 were professions, industries, or processes. Even though IARC monographs aim to identify and assess particular agents, some processes, industries, and professions have been classified under Group 1, with "sufficient evidence of carcinogenicity" in humans. Basing IARC monograph data from 1971 to 2017, and using epidemiological data on workers exposed to carcinogenic/mutagenic substances as a criterion, it has been established that the number of occupational carcinogenic substances has increased from 28 agents in 2004 to 47 in 2017 [Loomis et al. 2018].

The IARC classification is not legally binding, in contrast to the EU definition and carcinogenic/mutagenic substance classification, included in the European Union CLP Regulation for classification, labeling, and packaging of substances and mixtures (Regulation (EC) No 1272/2008), according to the UN Globally Harmonized System of Classification and Labeling of Chemicals (GHS). A substance is classified under Category 1 of carcinogenicity based on the results from epidemiological studies or animal studies. Category 1 of carcinogenicity is further divided into two categories: 1A and 1B. Category 1A (409 substances) encompasses substances known to have carcinogenic potential for humans, with evidence for the categorization being largely based on humans. Category 1B (735 substances) encompasses substances which are presumed human carcinogens, with evidence for the categorization being largely based on well performed animal studies. Category 2 (30 substances) encompasses substances which are suspected human carcinogens, and a chemical is placed under this category on the basis of evidence obtained from human and/or animal studies, but which is not sufficiently convincing to be placed under Category 1A or Category 1B [ECHA 2019a].

Substances known to induce heritable mutations or considered to induce heritable mutations in germ cells in humans are classified under mutagenicity Category 1 according to the CLP Regulation. Mutagenicity Category 1 is further divided into categories 1A and 1B. The mutagenicity Category 1A classification is based on evidence from human epidemiological studies. As per the harmonized classification of the CLP Regulation, no substance has been classified under Category A1 of mutagenicity. Classification of substances under Category 1B is based on (a) positive results from *in vivo* heritable germ cell mutagenicity tests in mammals, or (b) positive results from *in vivo* somatic cell mutagenicity tests in mammals, in combination with some evidence suggesting that the substance has the potential to cause mutations to germ cells. Evidence to support the classification of substances under the mutagenicity category can be derived from mutagenicity/genotoxicity tests in germ cells *in vivo*; or from demonstrating the ability of the substance or its metabolites to interact with the genetic material of germ cells; or from positive results from tests showing mutagenic effects in the germ cells of humans, without transmission to progeny, for example, an increase in the frequency of aneuploidy in sperm cells of exposed people. So far, the mutagenicity Category 1B according to the harmonized classification in Annex VI to the CLP Regulation includes 477 substances/mixtures, and the Category 2, 301 factors [the CLP Regulation]. If a substance is classified as mutagenic to germ cells, under Category 1A or 1B, it is assumed that a genotoxic carcinogenicity mechanism is probably likely [ECHA 2019a].

Directive 2004/37/EC (on the protection of workers from the risks related to exposure to carcinogens or mutagens at work) defines a carcinogen/mutagen as "a substance or mixture which meets the criteria for classification as a Category 1A or 1B carcinogen/mutagen set out in Annex I to Regulation (EC) No 1272/2008 of the European Parliament and of the Council" [EC 2004].

Although IARC and EU classifications generally coincide, they are not identical in every single aspect. For example, diesel exhaust is classified by the IARC under Group 2A, but is not included in the EU list. Crystalline silica dusts (quartz, cristobalite) were classified by the IARC as a Group 1 carcinogen in 1996, but still have not been classified by the European Union as a hazardous substance. Therefore, one should refer to both classifications, as applying only the EU classification would significantly decrease the share of workers exposed to carcinogenic/mutagenic chemicals [Mengeot et al. 2014].

In individual member states of the European Union, and in the United States, regardless of the IARC assessment, there have been developed and published lists of agents, which are considered carcinogenic.

In the Federal Republic of Germany, the Commission for the Investigation of Health Hazards of Chemical Compounds in the Work Area of the German Research Foundation (DFG) has distinguished five categories of agents with different levels of risk for their carcinogenicity [BAuA 2013].

In the United States, the recommendations of the following institutions are essential for the working environment: American Conference of Governmental Industrial Hygienists (ACGIH), National Institute for Occupational Safety and Health (NIOSH), and legal acts of US Federal Government agenda – Occupational Safety and Health Administration (OSHA). These organizations and institutions have published recognized lists of carcinogens.

The ACGIH experts divide carcinogenic agents into five groups:

Group A1 – confirmed human carcinogens (19 substances),
Group A2 – suspected human carcinogens (29 substances),
Group A3 – confirmed animal carcinogens with unknown relevance to humans (127 substances),

Group A4 – regroups agents non-classifiable as to their carcinogenicity to humans (205 substances),

Group A5 – regroups agents suspected not to be carcinogenic to humans (2 substances).

ACGIH recommends that occupational exposure to carcinogens/mutagens be kept at the lowest level possible. Workers exposed to Group A1 agents, for whom no hygienic standards have been established, should be equipped with proper personal protective equipment in order to eliminate this exposure. For A1 carcinogens, for which hygienic standards have been set (TLV-TWA®) and for agents with Group 2A, workers exposed by all routes should be kept at the possible lowest level and for substances with exposure limit should be below these limits [ACGIH 2019].

The National Toxicology Program (NTP) has created several US government agencies, including the National Institutes of Health (NIH), Centers for Disease Control and Prevention (CDC), and Food and Drug Administration (FDA). NTP issues and updates Reports on Carcinogens (RoC) every few years.

NTP divides carcinogens into two groups:

- "Known to be human carcinogens" – "There is sufficient evidence of carcinogenicity from studies in humans, which indicates a causal relationship between exposure to the agent, substance, or mixture, and human cancer". and
- "Reasonably Anticipated to Be Human Carcinogen" – there is limited evidence of carcinogenicity from human studies; or there is sufficient evidence of a carcinogenic effect from studies in experimental animals, indicating an increased incidence of tumors; or there is less than sufficient evidence of carcinogenicity in humans or experimental animals – however, the agent, substance, or mixture belongs to a structurally well-defined class of substances that was mentioned in a previous report on carcinogens that are known to be carcinogenic to humans, or are expected to be carcinogenic to humans; or there is information that the factor operates through appropriate mechanisms and is likely to cause cancer in humans [NTP 2016].

In the 14th Report RoC, NTP has listed 248 agents, substances, mixtures, and exposure conditions which are known to be or are suspected of being carcinogenic to humans. Among those 248 agents, 62 are known to be human carcinogens and 186 agents are suspected of being carcinogenic to humans. In the assessment of the carcinogenicity of an agent, a substance, or a mixture, the following are taken into account: properties (e.g. chemical, physical, or biological), manufacture and use, human exposure, toxicokinetics, results of carcinogenicity studies in humans and experimental animals, carcinogenicity mechanisms, and other effects. Information on exposure, properties of the substance assessed for carcinogenicity, and the results of scientific studies have to come from publicly available sources [NTP 2016].

The US Environmental Protection Agency (EPA) maintains Integrated Risk Information System (IRIS) – an electronic database which includes information on health effects of human exposure to certain substances in work and living environment.

For the assessment of carcinogens, EPA uses a system similar to IARC:

Group A: Human Carcinogen;
Group B: Probable Human Carcinogen;
Group C: Possible Human Carcinogen;
Group D: Not Classifiable as to Human Carcinogenicity;
Group E: Evidence of Non-Carcinogenicity to Humans [EPA 2019].

Scientific Committee for Occupational Exposure Limits to Chemical Agents (SCOEL) was an EU body until 2018, composed of independent scientific experts, advising the European Commission on limits of occupational exposure to chemicals at the workplace. Determining the occupational limit value (OEL) for a carcinogenic or mutagenic substance at the workplace in SCOEL depended on the type and mechanism of its carcinogenic activity – a distinction between substances having and not having genotoxic activity. Taking this into account, carcinogens have been divided into the following groups:

- Group A: genotoxic carcinogens with no limit value – risk assessment based on the linear non-threshold model (LNT) extrapolation of animal studies to humans (small doses), e.g. 1,3-butadiene, vinyl chloride, dimethyl sulfate.
- Group B: genotoxic carcinogens for which there is not enough available data to apply the linear no-threshold model (LNT), e.g. acrylonitrile, benzene, naphthalene, wood dusts.
- Group C: Genotoxic carcinogens, for which a limit value can be established based on existing data, e.g. formaldehyde, vinyl acetate, nitrobenzene, pyridine, crystalline silica, lead.
- Group D: Non-genotoxic carcinogens not affecting DNA, for which a limit value can be established based on No Observed Adverse Effect Level (NOAEL) value, e.g. carbon tetrachloride, chloroform [Bolt and Huici-Montagud 2008; EC 2018]

In SCOEL, OEL values were established in relation to compounds from groups C and D. In the case of compounds from group A or group B, carcinogenicity assessment was based on an LNT model. In the documentation, concentrations of carcinogenic/mutagenic agents, as well as risks linked to them, were presented, but no specific recommendations as to specific permissible risk levels were introduced [EC 2018; van Kesteren et al. 2012].

4.5 EXPOSURE TO CARCINOGEN/MUTAGENIC SUBSTANCES AT WORK

Substances carcinogenic to humans, first identified in the work environment, include arsenic, benzene, benzidine, chromium, dichlorodiethyl sulfide (mustard gas), nickel, radon, and vinyl chloride. There is a very long list of chemical substances in the work environment, which are probably/possibly carcinogenic to humans, and that require further assessment. Substances that are carcinogenic/mutagenic to humans in the workplace also have such effects outside of work, because they are popular in non-occupational environments [Blair et al. 2011].

Carcinogens, similarly to other chemical substances, can pass into the body through the airway (through the inhalation route), from which they can cross to the circulatory system and to various internal organs, including the brain. Other substances can also be absorbed through the skin. Carcinogenic/mutagenic substances do not cause cancers in every case of exposure, as they can have different levels of carcinogenic potential. Some of them can only cause cancer after long-term exposure to high concentrations. For each worker, the risk of developing cancer as a result of occupational exposure depends on many factors, including how the person is exposed to the carcinogenic/mutagenic substance, duration of exposure, concentration, and the person's genetic "map".

A number of carcinogenic substances cause changes in more than one location in the body. For example, occupational exposure to asbestos is a cause of mesothelioma and cancers of lungs, larynx, and the gastrointestinal tract. Exposure to chromium (VI) can cause tumors of the nasal cavity and the lungs; to arsenic, cancers of the liver, lungs, and skin.

2,2′-Dichlorodiethyl sulfide can cause cancer of the pharynx, larynx, and lungs. Lung, skin, and bladder cancers may occur in those working in coking plants. Occupational exposure to ionizing radiation can cause leukemia and liver, lungs, bone, breast, and thyroid cancers. Occupational exposure to carbon black can cause esophageal, lung, and skin cancers [Blair et al. 2011; Loomis et al. 2018].

In 2010, 10% of French workers – almost 2.2 million – were exposed to at least one chemical carcinogen (French database Médicale des Expositions aux Risques Professionnels, SUMER 2010). Manual workers constituted more than two-thirds of workers exposed to at least one carcinogen, although they represented only 29% of the workforce. This involved workers employed in large sectors of the industry, such as maintenance (43%), public and construction works (32%), and metal working (31%). Young people were more exposed to carcinogens than other age groups: 16% of workers under 25 years of age were exposed to at least one chemical carcinogen, in comparison with 7% of workers over the age of 50. In the machinery/metal industry, 70% of trainees were exposed to carcinogenic substances, in comparison with 35% of all manual workers. Similarly, the 2010 study, focused on carcinogenic substances classified under IARC Groups 1 and 2A, found substances most often appearing to be the same as those found in a 2003 study, i.e. diesel exhaust, mineral oils, wood dust, crystalline silica [Mengeot et al. 2014].

Women were generally less exposed to carcinogenic substances than men, but in certain sectors more women than men were exposed to certain substances, e.g. cytostatic agents (used in chemotherapy), formaldehyde (a disinfectant), and aromatic amines. Workers employed in very small companies (less than 10 workers) were more exposed to carcinogenic substances than those employed in companies with over 500 workers. Exposure to carcinogens was considered low in 70% of cases, high or very high in 10% of the cases (2% above the occupational exposure limit value), and unknown in 18% of the cases. Thirty-eight percent of workers were significantly exposed to the products of degradation and decomposition generated by manufacturing processes: smoke, dust, diesel exhaust, crystalline silica derivatives, etc. The most commonly mentioned collective protective measures were extraction systems and general ventilation, but the latter would not be able to effectively protect the workers from carcinogens. Closed systems were listed only in 1% of instances of exposure. In 35% of exposure no collective protection measures were used. In 57% of exposed workers in the construction industry and 37% in the maintenance industry, no collective protective measures were used [Mengeot et al. 2014].

The comparison between the SUMER database 2003 and 2010 data has shown that the percentage of workers exposed to carcinogens in France decreased from 13 to 10%, which was partially due to strengthening provisions (raising awareness of the hazards and preventing the risks). When substances have been identified as carcinogenic at the point of purchase, most commonly, substitution was made. The solution was considerably more difficult in places where the risk stemmed from the manufacturing process itself. It has also been shown that a decrease in the number of workers exposed to carcinogens in companies employing more than 500 was 6% in relation to 2003, in comparison with a very low percentage (less than 1%) in very small companies (less than 10 workers). This indicated poor risk control and the need for developing a more systematic risk prevention method for micro-enterprises [Mengeot et al. 2014].

In the European Union, the Carex database (exposure to carcinogens) is a comprehensive register that contains figures on the percentage of workers exposed to carcinogens [Kauppinen et al. 2000]. Carex is an initiative started at the end of the 1980s as part of the Europe Against Cancer programme. Between 1990 and 1993, the studies were conducted in fifteen EU member states. The share of workers exposed to carcinogens ranged from 17% in the Netherlands to 27% in Greece, which amounts to 32 million workers in total. After 1995, the Carex database was extended to include the Baltic republics and the Czech Republic, where 28% of workers were found to be exposed to occupational carcinogens. The database has never been extended to other EU countries.

The most common occupational carcinogens were crystalline silica (3.2 million workers), diesel exhaust (3.1 million), radon (2.7 million), wood dust (2.6 million), lead and its inorganic compounds (1.5 million), and benzene (1.4 million). The remaining agents were asbestos, ethylene dibromide, formaldehyde, polycyclic aromatic hydrocarbons, glass wool, tetrachloroethylene, chromium (VI) and its compounds, sulphuric acid mists, nickel, styrene, chloromethyl ether, and trichloroethylene. Sectors of the economy in which the exposure to carcinogens was the highest were raw material extraction (silica, diesel exhaust), woodworking and furniture industries (wood dust and formaldehyde), construction (silica, diesel exhaust), and aviation (exhaust fumes). Exposure to benzene was the highest in the vehicle repair sector. The estimates of the Carex database involved all workers and the exposure to all occupational carcinogens, such as solar radiation, radon, and passive smoking, which occurred during 75% of the working time. And although the percentage of workers exposed to passive smoking and asbestos has decreased due to more stringent regulations, the number of known carcinogens in Europe has increased [Mengeot et al. 2014].

In the EU (28), the carcinogens to which the most workers were exposed, were: benzo[*a*]pyrene, diesel exhaust, hard wood dusts, hydrazine, mineral oils (such as waste oil), 4,4′-methylenedianiline, chromium (VI), respirable fraction of crystalline silica, formaldehyde, and asbestos. There is often a significant delay between the start of exposure to carcinogens and the development of cancer; i.e. long-term exposure does not lead to cancer immediately after exposure [Jongeneel et al. 2016].

In Poland, between 2013 and 2017, workers were most commonly exposed to the following carcinogens: formaldehyde, polycyclic aromatic hydrocarbons (PAHs), benzene, and chromium (VI) compounds. Among the technological processes in which carcinogen/mutagen chemicals, their mixtures or agents are released, most plants reported works involving exposure to hardwood dust (about 800 plants every year and from over 10,000 to almost 15,000 exposed workers). On average, between 2013 and 2017, a worker in Poland was exposed to approximately three different carcinogenic/mutagenic substances [Niepsuj et al. 2020].

The first study that identified work-related chemical hazards was conducted in the United States in the early 1970s. Over 9,000 potentially hazardous situations at work were identified, and groups of workers exposed to carcinogens were classified. Another study took place a decade later, and based on it the US National Institute for Occupational Safety and Health (NIOSH) developed a database that allows estimating the number of exposed workers and individual sectors of the economy in terms of carcinogens [Mengeot et al. 2014].

The Carex project, currently run in Canada, includes 229 carcinogens or suspected carcinogens used in various sectors of the economy. The decreasing share of employment in industry and agriculture should contribute to reducing the percentage of workers exposed to carcinogens. However, some service sectors (cleaning, healthcare, transport) may pose an increased risk of cancer, which until now has been ignored [EC 2019; Mengeot et al. 2014].

4.6 CONCEPTS OF OCCUPATIONAL EXPOSURE LIMIT VALUES ESTABLISHED FOR CARCINOGENIC/ MUTAGENIC SUBSTANCES

Brandys and Brandys [2008] published a list of occupational exposure limit (OEL) values collected from around the world, which included over 5,000 various chemicals. Eighteen organizations/institutions established OEL values for 1,341 unions. Although OEL values have been established for the most common substances in the work environment, due to the huge number of chemicals they have still not been established. The Chemical Abstracts Service registered recently 75 million substances, and 5 million new substances were added last year. The rate of innovation in the field of chemicals is very fast, although most of them are not commercially produced. Even taking into account that 84,000 chemicals are commercially available in the United States, that 23,000 domestic substances are included in the Canadian list, and that more than 107,000 different substances are registered under REACH, most of the chemicals used do not have OEL values. In addition, many existing OEL lists include substances that were added many years ago and are no longer commercially valid. Therefore, the number of relevant OEL values is even smaller than the total number included in these lists. Even though traditional OEL values exist for a particular compound, not all bodies setting these values have this chemical on their list. In a comparative study of OEL values set by 18 organizations/institutions, most of the organs dealt with less than half of the 1,341 substances, which constituted the full list of studied compounds. More than one third (460) of substances were listed by just one organization (with Finland being exceptional for having 189 unique OEL values). Eighteen organizations/institutions listed less than 2% of substances. The reason why the choice of substance was not more harmonized can be partly explained by differences in the database between countries. Different organizations obtained different values for the same chemical [Deveau et al. 2015].

Most carcinogens are non-threshold substances, meaning that no safe levels of exposure can be set for them. Health risk assessment in the case of carcinogenic/mutagenic substances is based on the identification of the risk of disease or death due to cancer as a result of occupational exposure to these substances.

Various government agencies, national, or international organizations which establish or propose exposure limit values for carcinogens/mutagens use the term "acceptable risk". The level of acceptable risk depends mostly on the accepted social and economic priorities. In developed countries, three main groups of interested parties decide on the matter: representatives of employees, employers, and state administration, whose task is to supervise compliance with existing regulations.

In the case of carcinogens, different levels of occupational risks have been adopted in the EU countries, ranging from 10^{-6} to 10^{-3}, which means that Member States' administrations have accepted the possibility of an increase in the number of cases of 1 cancer per 1,000,000 workers exposed or 1 cancer per 1,000 workers exposed to a carcinogen at a given concentration.

In Germany, three areas of risk have been defined for carcinogens/mutagens: high, medium, and low. The area between high and medium risk is defined as "tolerable risk". The tolerable risk defines the additional cancer risk of 4:1,000 that is tolerated, meaning that, statistically, 4 out of 1,000 persons exposed to the substance throughout their working life will develop cancer. Workers should not be exposed to carcinogenic substances above the tolerable risk level. The area between the medium and low risk is known as acceptable risk. The acceptable risk defines the additional cancer risk of 4:10,000, meaning that, statistically, 4 out of 10,000 persons exposed to the substance throughout their working life will develop cancer. From 2013, the risk level for carcinogenic substances has been expected to consequently reduce until it reaches a statistical value of 4 out of 100,000, which corresponds to the average risk of cancer incidence in the general population [BAuA 2013].

A comparison between the exposure level at the workplace and the derived substance-specific "acceptable" and "tolerable" concentrations determines the necessity and urgency of protective measures according to a graduated concept, falling into five categories: administrative, technological, organizational, occupational medicine, and substitution. The rule is: the higher the risk, the more stringent the requirements as to the measures to be put in place. Thus, the substitution of a hazardous substance is obligatory in the case of high-risk sectors (if alternatives are available), whereas in the area of low- and medium-risk sectors, technical feasibility and proportionality may also be taken into consideration [BAuA 2013].

In Holland, binding concentration limit values for carcinogenic or mutagenic substances are established using two risk levels: upper, which limits the increase in cancer incidence to 10^{-4}/year (risk level 4×10^{-3} during the 40 years of professional activity of the workers), and target, which limits the additional increase in cancer incidence to 10^{-6}/year (risk level 4×10^{-5} during the 40 years of professional activity of the workers), with exposure below concentration limit values not requiring the use of additional protective measures. The first step toward the establishment of binding values for carcinogens or mutagens is the determination of the risk level by the Dutch Health Council. During the second stage, the Social and Economic Council of the Netherlands (SER) assessed the possibility of introducing the value into national law and refers the matter to the minister of labor and social policy. Representatives of employees, employers, and trade unions are included in the subcommittees. The results of the assessment of the possibility of compliance with the proposed value for carcinogenic or mutagenic substances sometimes lead to establishing a higher concentration limit value. Such a value is verified by the subcommittee every 4 years in order to assess whether it is possible to lower it, with the aim of reaching the target risk level of 10^{-6}/year [van Kesteren et al 2012].

In France, for non-threshold carcinogens, the limit values do not protect the worker from the occurrence of cancers, but are set at a low risk level (e.g. additional risk of cancer at the 10^{-4}, 10^{-5}, or 10^{-6} level) or at a level technically possible to reach, where quantitative risk assessment is not possible (pragmatic value). In a French list, published by the INRS Institute, it was not noted whether the value for a particular carcinogen was established based on a calculated health effect or on risk assessment [van Kesteren et al. 2012].

In 2015, the European Union decided to step up prevention efforts to reduce occupational exposure to carcinogens. European Commission and The Advisory Committee on Safety and Health at Work (ACSH) decided that the basis for these actions will be the introduction of binding occupational exposure limit values (BOELVs) in relation to substances classified as carcinogenic/mutagenic. These values are determined based on the latest scientific data, socioeconomic conditions, and the technical possibilities of achieving such value in industry.

By 2017, binding values at EU level were established for: asbestos (actinolite, anthophyllite, chrysotile, grunerite (amosite), crocidolite, tremolite), benzene, hardwood dusts, lead and its inorganic compounds, vinyl chloride monomer (10 substances; Directive 98/24/WE [EC 1998], 2004/37/WE, 2009/148/WE). Directive of the European Parliament and of the Council (EU) 2019/130 of January 16, 2019, amending Directive 2004/37WE on the protection of workers from the risk of exposure to carcinogens and mutagens at work, sets binding occupational exposure limits in relation to 19 chemical agents: hardwood dusts, chromium (VI) compounds, refractory ceramic fibers, respirable crystalline silica dust, chloroethene (vinyl chloride), ethylene oxide, 1,2-epoxypropane, trichloroethylene, acrylamide, 2-nitropropane, *o*-toluidine, 4,4′-methylenedianiline, epichlorohydrin, ethylene dibromide, 1,3-butadiene, ethylene dichloride, hydrazine, bromoethylene, and diesel engine exhaust emission. Twelve substances also have a "skin" annotation (substantial contribution to the total body burden via dermal exposure possible), with no change in the binding value of benzene. In Annex I "List of Substances, Preparations and Processes", the following paragraphs have been added: "6. Work involving exposure to respirable crystalline silica dust generated by a work process" [EC 2017], "7. Work involving dermal exposure to mineral oils that have been used before in internal combustion engines to lubricate and cool the moving parts within the engine" and "8. Work involving exposure to diesel engine exhaust emissions" [EC 2019a].

Directive (EU) 2019/983 of the European Parliament, and of the Council of June 5, 2019, amending Directive 2004/37/EC on the protection of workers from the risks related to exposure to carcinogens or mutagens at work [EC 2019b], has introduced occupational exposure limits for another five carcinogens or mutagens: cadmium and its inorganic compounds, beryllium and its inorganic compounds, arsenic acid and its salts, and inorganic compounds of arsenic, formaldehyde, and 4,4′-methylene-bis(2-chloroaniline) (MOCA). Among 913 carcinogenic (Category 1A or 1B) or mutagenic substances (Category 1A or 1B) in harmonized classification [EC 2008], binding values were established for 26 chemicals.

In 2016, NIOSH has published a document titled "Current Intelligence Bulletin 68: NIOSH Chemical Carcinogen Policy". It was decided that there is no safe exposure level to carcinogenic agents; therefore it should be limited as much as possible by the elimination or substitution of carcinogens in the work environment and by proper engineering control to prevent occupational cancer. Thus, the term "recommended exposure limit" (REL) with respect to carcinogens has been replaced by the term Risk Management Limit for Carcinogens (RML-CA). For each chemical identified as a carcinogen, this level corresponds to 95% of the confidence limit for estimating the risk of one additional cancer in 10,000 workers exposed to it for 45 years of working life. Maintaining exposure at a risk level of 1 in 10,000 is the minimum level of protection, and it is recommended that lower exposure levels be aimed. Although measuring the concentration of the carcinogen at RML-CA is not analytically feasible when estimating a risk of 1 in 10,000, NIOSH will calculate RML-CA at the limit of quantification (LOQ) of the analytical method. Moreover, NIOSH will still assess data available on existing technical controls and on used respiratory protective equipment, as well as publishing the information along with the RML-CA values for carcinogenic substances. In addition, NIOSH will continue to assess available information on existing technical controls and respiratory protection equipment, and will make this information available when publishing RML-CA values for carcinogens. The RML-CA value is the daily maximum average 8-hour concentration of the carcinogen above which the worker should not be exposed. NIOSH uses a four-step process of occupational risk assessment: NIOSH uses a 4-stage process of occupational risk assessment: hazard identification, workplace exposure assessment, exposure–effect relationship assessment, and workplace risk characterization. NIOSH classifies the chemical as professional carcinogen, taking into account: carcinogenicity and risk assessment by the United States Department of Health and Welfare (HHS), National Toxicology Program (NTP), Integrated Risk Information System (IRIS), US Environmental Protection Agency (EPA) or the International Agency for Research on Cancer (IARC) of the World Health Organization, and a nomination issued by NIOSH for classification by NTP, or classification according to NIOSH criteria [NIOSH 2017].

If a chemical is of particular concern, NIOSH may develop its own chemical risk assessment to determine whether the substance should be classified as an occupational carcinogen. When NIOSH determines that workers may be exposed to it in the work environment, and that the chemical is not classified by NTP, EPA, or IARC for carcinogenicity, this will be a reason to reconsider the evidence underlying the chemical carcinogen assessment. To develop a new classification of carcinogenic substances, NIOSH uses carcinogenicity criteria included in the Globally Harmonized System of Classification and Labeling of Chemicals (GHS), which have been included in the OSHA US assessment. When a substance has been classified under GHS 1A, 1B, or 2 carcinogenic category, then NIOSH recognizes the substance as a "carcinogen at work". After establishing that the chemical is an occupational carcinogen, NIOSH assesses whether there is adequate data to perform a quantitative risk assessment (QRA) and determine RML-CA [NIOSH 2017]. In the publication "NIOSH Pocket Guide to Chemical Hazard, 2007" the most important information and data have been collected in a short tabular form in relation to 677 chemicals or groups of substances commonly found in the work environment, including 131 carcinogens. No mutagenic factors have been identified [NIOSH 2007].

Most of the permissible exposure limits (PEL), established in the United States by OSHA, were released soon after the adoption of the Work Environment Act in 1970, and have not been updated since then. These values are legally binding. In 1989, OSHA US has changed the list of PEL values for chemical air contaminants, including tables Z-1, Z-2, and Z-3. The verification concerned 212 permissible exposure limits (PEL), listed in tables annexed to the Act. New PEL values were established for 164 substances not included in the register. The changes included verification of the PEL value, inclusion of short-term exposure limits (STEL) to supplement the 8-hour weighted average limits (PEL), the introduction of the "skin" designation and the addition of ceiling values. For 22 carcinogens, out of 457 substances listed in Table Z-1, no PEL value was determined. The changed PEL values are contained in Table Z-1-A. Industrial experience, new technological achievements, and scientific data have shown that in many cases the previously established PEL limits did not provide sufficient protection for employees' health. To provide employers, employees and other interested parties with a list of alternative occupational exposure limits, which may serve better protection of the worker's health, OSHA has added to Table Z-1-A, in addition to PEL values set by OSHA, the values of the California Department of Occupational Safety and Health (Cal/OSHA), the recommended exposure limits for NIOSH (REL) and ACGIH® TLV-TWAs®. In Table 1-Z-A, 78 chemical substances are marked with the letters "Ca" – a potential occupational carcinogen according to the NIOSH list [OSHA US 2014].

Canadian provinces set their own occupational exposure limits, although most of them adhere to the recommended values (TLV-TWA®) by ACGIH. There are significant differences in jurisdiction between the United States and Canada. In the United States, OSHA sets safety standards for the whole country, including permissible exposure limits for chemicals at the workplace. Most of the agencies responsible for occupational safety observe federal limits, though some, e.g. in California,

Washington, and Hawaii, set their own PEL values. They are often the same as national ones, but sometimes they are more restrictive [Unifor Canada 2019].

In Australia, workplace exposure standards cover about 700 chemicals. The workplace exposure standards for a chemical have been defined as a working exposure standard (WES) which must not be exceeded [Safe Work Australia 2018].

Exposure to hazardous chemicals in the workplace may pose a serious risk to workers' health. The risk of exposure depends on the air concentration of the chemical, time of exposure, and effectiveness of the control. Occupational exposure standards do not represent acceptable levels of worker exposure, but the maximum, upper limits set in the legislation. In 2018, consultations were started on a regulation to establish the influence and the best way to implement an update to Australian occupational exposure standards. It has been decided for them to result in a change of the phrase "working exposure standards" to "workplace exposure limits". Chemical substances classified as carcinogens, presumed carcinogens, or suspected carcinogens have been labeled as per the Globally Harmonized System of Classification and Labeling of Chemicals (GHS) criteria. The classification of carcinogenic substances is described in Annex VI to Regulation (EC) No 1272/20085 (cat. 1A, 1B, or 2). The list from 2018 includes 11 carcinogens of Category 1A, 32 substances of Category 1B, and 49 substances of Category 2. Mutagenic labeling is not included [Safe Work Australia 2018].

A summary of the number of carcinogen/mutagen substances for which limit values have been established in comparison with the total number of substances included in the appropriate lists in the European Union, Germany [DFG 2018], Poland [Dz. U. poz. 1286], ACGIH [2019], and Australia [Safe Work Australia], is shown in Figure 4.1.

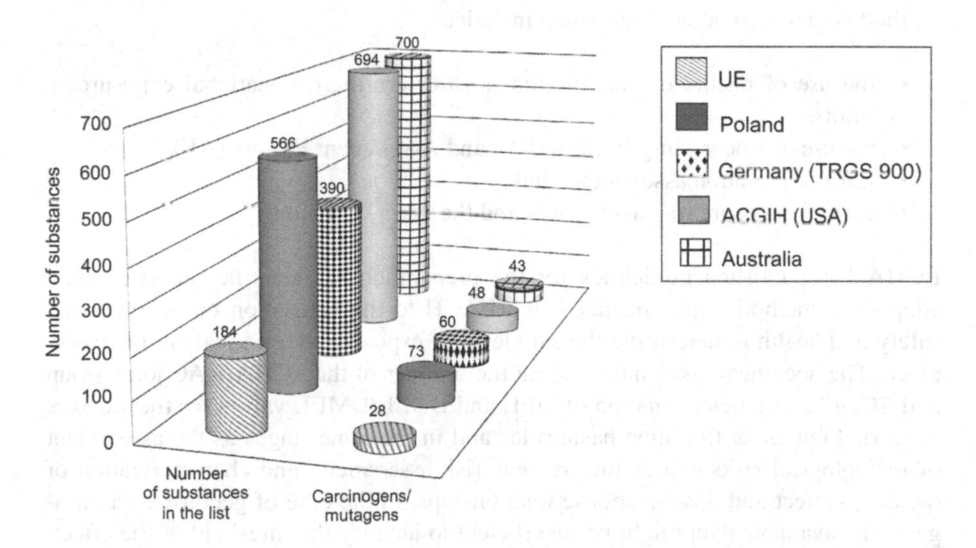

FIGURE 4.1 A summary of the number of carcinogenic/mutagenic substances for which limit values have been established in comparison with the total number of substances included in the appropriate lists in the European Union, Germany, Poland, ACGIH, and Australia.

According to the REACH Regulation [EC 2006], all hazardous substances at the workplace should be registered. Each of these substances should have appropriate technical documentation, supplemented by a chemical safety report (for substances placed on the EU market > 10 t/year), which includes relevant exposure scenarios describing the safe use of the substance throughout its life cycle. Annex I to REACH introduced new health-based reference values above which people should not be exposed. This level of exposure is a Derived No Effect Level (DNEL). Because there are different types of exposure and target groups (worker, consumer), different types of DNELs have been defined. In the workplace, eight types of DNELs are important (local and systemic effects in acute and long-term inhalation exposure and through the skin). Generally, two DNEL values are considered most important in risk management: inhalation and dermal. DNEL values may be derived only when there are quantitative dose–response dependencies in reference to particular adverse effects. In most cases of carcinogenic and mutagenic effects, a Derived Minimal Effect Level (DMEL) has to be determined. This parameter is not defined in the legal context, but only in guidelines.

In 2015, the Commission turned to the Risk Assessment Committee (RAC) of the European Chemicals Agency and the SCOEL to perform a comparative assessment of methodologies used by those committees to determine DNEL/DMEL values for workers or occupational exposure limits OEL/BOELV. The first report of the comparison of methodology used in the determination of derived no-effects levels (DNEL) with methodology used for occupational exposure limits (OEL) and DNEL values in the case of dermal exposure, with the "skin" notation, was published in February 2017 [ECHA 2017a]. The second report, on scientific assessment of non-threshold substances, mainly genotoxic carcinogens, was published in December 2017 [ECHA 2017b]. On the basis of both reports, a number of differences in the methodologies were identified, which included:

- the use of results of human studies to determine occupational exposure limits,
- the use of Uncertainty Factors (UF) and Assessment Factors (AF),
- sensory irritation assessment, and
- dermal exposure risk assessment and the "skin" notation.

ECHA has published guidelines for the preparation of scientific reports, which adapts the methodologies included in REACH to the legislation on occupational safety and health to determine the safe level of exposure to chemicals in the workplace. The document takes into account the findings of the ECHA/RAC joint group and SCOEL. The determination of OEL and DNEL/DMEL values for the workers is carried out using the same basic rules and in the same stages as the assessment of toxicological risks – literature review, risk assessment, and characterization of the dose–effect and dose–response relationships. In the case of genotoxic carcinogens, the available data might be insufficient to identify the threshold of the effect. Both committees used similar methodologies for these substances, assuming a linear relationship between exposure and effect and using the T25 methodology (daily substance dose in mg/kg body weight, which causes the risk of cancer to increase by 25% in lifelong exposure) and Benchmark Dose (BMD). In general, the SCOEL methodology and its fundamental rules for the determination of thresholds, based on the mode of action (MoA) of carcinogens, turned out to be appropriate to use within the REACH framework with certain adaptations. These include: (a) a detailed explanation of the uncertainty factors used, indicating that the proposed level of exposure may include residual risk, (b) omission of the grouping of carcinogens determined in SCOEL, (c) determination of the point of departure (PoD) to establish exposure limits with their Assessment Factors (AF), and (d) the use of allometric scaling and different correction factors, as described in the ECHA guidelines [ECHA 2019b].

Since 2019, the Committee for Risk Assessment (RAC) of the European Chemicals Agency (ECHA) has been carrying out a scientific assessment of the relationships between health effects of hazardous chemical substances and occupational exposure levels. The RAC Committee may also make recommendations regarding the notation of "skin" indicating that it must be protected. Other notations, such as "allergy", are also possible. Moreover, the RAC Committee can also set biological limit values (BLV) or biological guidance values (BGV). The OEL/BOELV values are determined in a multi-stage process (Figure 4.2). The EU Commission consults the suggested BOELVs for carcinogenic/mutagenic agents with social partners, as per the provisions of the Treaty on the Functioning of the European Union concerning social policy. ECHA begins work on the proposition of a binding value after receiving a request from the Commission to prepare a scientific report to be considered by the RAC for substances selected from the priority substance lists DG EMPL (stages 1 and 2).

In 2019, the Commission requested ECHA to assess lead and its compounds, and diisocyanates. Lead is the main substance toxic to reproduction. Diisocyanates are substances to which a large number of workers are exposed and which cause a large number of incidences of asthma.

Pronk [2014] presented the results of the review and comparison of methodology used by SCOEL, ECHA, and four member states to establish OEL values for non-threshold carcinogens. It has been established that there are many similarities, but also some differences. One of the similarities was that methodologies for determining OEL values are based on similar rules. All the organizations/institutions use similar general criteria of the quality and adequacy of data selected for establishing the limits. Most of the organizations preferred to use data from e.g. epidemiological studies on humans than animal studies, but in most cases the results of epidemiological studies were not available or sufficient to establish limit values. Differences observed in the OEL values for non-threshold carcinogens were caused by use of different levelsof cancer risk. Another source of the differences was the choice of concentration/doses in animal exposure, which caused adverse effects, and uncertainty factors used in the extrapolation of the results from animal studies to humans. If, on a later stage, other factors are considered, such as socio-economic or technical feasibility, they can also lead to differences in the resulting occupational exposure limits.

In the European Union, if the European Commission establishes a binding occupational exposure limit value (BOELV), the member states must establish their own binding occupational exposure limit value at the BOELV level or lower. There are plenty of examples of non-binding or recommended occupational exposure levels,

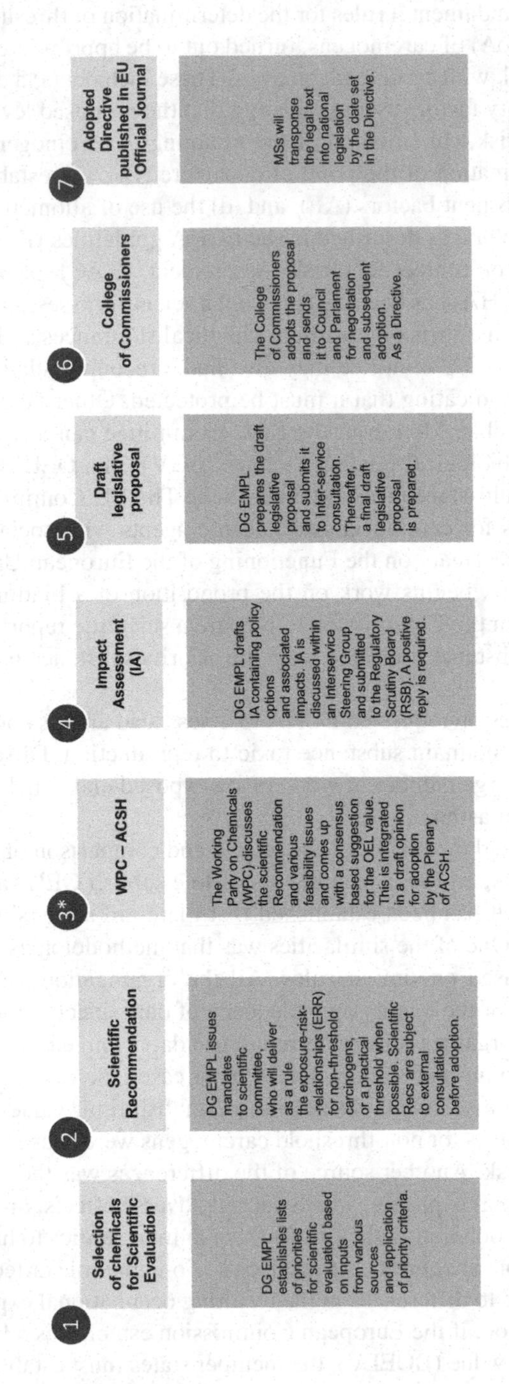

FIGURE 4.2 The procedure for the establishment of occupational limit values for chemical substances by ECHA (access: https://echa.europa.eu/pl/oel).

such as recommended exposure limits set by the NIOSH (REL), ACGIH (TLV-TWA®), Occupational Alliance for Risk Science (OARS), Workplace Environmental Exposure Levels (WEEL, earlier developed under AIHA), and SCOEL's indicative occupational exposure limit values (IOELV) [Deveau et al. 2015].

This distinction between non-binding and binding limits can, however, disappear, as some regulatory authorities adapt OEL values within the framework of their operations. For example, TLV-TWA® ACGIH are *de facto* legally binding in a number of Canadian provinces, European countries, and many other countries around the world. Moreover, there is a distinction between "health-based" and "regulatory adjusted" values, with the latter including technical and economic feasibility. Some OEL values – such as REL established by NIOSH – can be a hybrid of health and technical considerations. However, different organizations cannot clearly define when the OEL values are the hybrid of health-based and regulatory adjusted values; the lack of clarity of these rules can cause difficulties in the implementation of decisions in risk management.

Examples of these alternative OEL values include temporary OEL values, obtained for internal business purposes or by trade associations, or suppliers, which are used to bridge an information gap when there are no available OEL values from a recognized body [Deveau et al. 2015].

Hammerschmidt and Marx [2014] emphasize that even though DNEL and DMEL values are not linked to any legal consequences in relation to the workplace, they serve as reference values in the context of risk assessment. In the ECHA database, DNEL and DMEL values are presented in the information on structure. DMEL values have been established only for a small number of substances.

Table 4.1 presents a list of binding values for 27 chemical substances included in EU directives with the values set in Germany, Poland, and the United States (ACGIH), with DMEL/DNEL values included in the GESTIS database and in the ECHA database.

The determination of DMEL values is essentially based on the assumption of linear dose–response relationship between cancer and exposure to a carcinogenic substance. The selected DMEL values are used for cases of exposure included in exposure scenarios. In the case of proper long-term systemic effects, five DN(M)EL values can be established, depending on the routes of exposure and exposed populations. In most cases, the occupational DNEL/DMEL values should be established for dermal and inhalation exposure. DMEL values are reference values linked to risk, which should be used to better focus risk management measures. DMEL-based approach is useful in the development of chemical safety assessment to estimate remaining/residual risk [ECHA 2019b].

As a different methodology is used to establish DMEL (inhalation exposure) values and the OEL values, the DMEL values are smaller than the corresponding OEL values estimated for the same health effects. This may indicate that the OEL value does not indicate an adequate level of protection in the event of occupational exposure to a carcinogen, as required by REACH. Calculations lead to new, smaller values that require different risk management measures and operational conditions than before [ECHA 2019b].

TABLE 4.1
Substances with EU Binding Occupational Exposure Limit Values (as of June 2019): Maximum Admissible Concentration in Poland [Dz. U., poz. 1286, 2018], MAC Levels in Germany [TRGS 505, 517, 519, 553, 900, 910], TLV-TWA® in ACGIH (US) [2019], DNEL/DMEL Levels in Germany [IFA, GESTISDatabase 2019], and ECHA [ECHA 2019a]

Substance	EU Directive 2004/37/EC	EU Limit Value	Germany	Source	Poland MAC Value	ECHA DMEL/DNEL for Workers	GESTIS DNEL	ACGIH [2019]
Acrylamide [79-06-1]	2017/2398/EC	0.1 mg/m³	AC: 0.07 mg/m³ TC: 0.15 mg/m³	TRGS 910	0.07 mg/m³	0.07 mg/m³ Carc. 1B Muta. 1B	–	0.03 mg/m³ A3
Arsenic acid and its salts, as well as inorganic arsenic compounds	2019/983/EC	0.01 mg/m³ (IF)	AC: 0.00083 mg/m³ (IF) TC: 0.0083 mg/m³ (IF)	TRGS 910	0.01 mg/m³	0.006 mg/m³ DNEL Carc. 1A	Arsenic: 0.004 mg/m³ Arsenic acid: 0.006 mg/m³	0.01 mg/m³ A1
Asbestos	2009/148/EC	100,000 F/m³	AC: 10,000 F/m³ TC: 100,000 F/m³	TRGS 517, 519, 910	0.1 F resp/cm³			0.1 F/cm³ A1
Benzene [71-43-2]	2004/37/EC	3.25 mg/m³	AC: 0.2 mg/m³ TC: 1.9 mg/m³	TRGS 910	1.6 mg/m³	not established Carc. 1A		0.5 ppm A1
Beryllium [7440-41-7] and inorganic beryllium compounds	2019/983/EC	0.0002 mg/m³ (IF)	AGW: 0.00006 mg/m³ (RF) 0.00014 mg/m³ (IF)	TRGS 900	0.0002 mg/m³	not established Carc. 1B		0.00005 mg/m³ IF; A1
Bromoethylene (vinyl bromide) [593-60-2]	2017/2398/EC	4.4 mg/m³	AGW: 4.4 mg/m³	TRGS 900	0.4 mg/m³	not established Carc. 1B		0.5 ppm A2
1,3-Butadiene [106-99-0]	2017/2398/EC	2.2 mg/m³	AC: 0.5 mg/m³ TC: 5 mg/m³	TRGS 910	4.4 mg/m³ 2.2 mg/m³*	2.21 mg/m³ Carc. 1A Muta. 1B		2 ppm A2

Cadmium [7440-43-9] and its inorganic compounds	2019/983/EC	0.001 mg/m³ (IF)	AC: 0.00016 mg/m³ (RF) TC: 0.001 mg/m³ (IF)	TRGS 910	0.001 mg/m³*	not established Carc. 1B	0.004 mg/m³* local	0.01 mg/m³ 0.002 mg/m³ RF; A2
Chromium (VI) compounds [–]	2017/2398/EC	0.005 mg/m³ (as chromium)	0.001 mg/m³ (IF)	TRGS 910	0.001 mg/m³ 0.005 mg/m³*	–		0.0002 mg/m³ A1
Diesel engine exhaust emissions [–]	2019/130/EC	0,05mg/m³ (EC)	AGW: 0,05 mg/m³ (EC)	TRGS 900	0.5 mg/m³ RF 0.05 mg/m³ (EC)*	–	Diesel fuel: 68.3 mg/m³ systematic	Diesel fuel: 100 mg/m³ A3
Epichlorohydrine [203-439-8]	2019/130/EC	1.9 mg/m³	AC: 2.3 mg/m³ TC: 8 mg/m³	TRGS 910	1 mg/m³	DNEL: 1.52 mg/m³ Carc. 1B	1.52 mg/m³ local, systematic	0.5 ppm A3
1,2-Epoxy-propane (propylene oxide) [75-56-9]	2017/2398/EC	2.4 mg/m³	AGW: 2.4 mg/m³	TRGS 900	9 mg/m³ 2.4 mg/m³ *	not established Carc. 1B Muta. 1B	2.4 mg/m³ local	4.8 mg/m³ (2 ppm) A3
Ethylene dibromide [203-444-5]	2019/130/EC	0.8 mg/m³			0.01 mg/m³	0.0005 mg/m³ Carc. 1B		– A3
Ethylene dichloride [203-458-1]	2019/130/EC	8.2 mg/m³	AC: 0,8 mg/m³ TC: 4 mg/m³	TRGS 910	8 mg/m³	6.6 mg/m³ Carc. 1B		41 mg/m³ (10 ppm) A4
Ethylene oxide [75-21-8]	2017/2398	1.8 mg/m³	AC: 0.2 mg/m³ TC: 2 mg/m³	TRGS 910	1 mg/m³	1.8 mg/m³ Carc. 1B Muta. 1B		1.8 mg/m³ (1 ppm) A2
Formaldehyde [50-00-0]	2019/983/EC	0.37 mg/m³	AGW: 0.37 mg/m³	TRGS 900	0.37 mg/m³	DNEL: 9 mg/m³ Carc. 1B	0.375 mg/m³ local 9 mg/m³ systematic	0.1 ppm A1
Hardwood dusts [–]	2017/2398/EC	2 mg/m³ (IF)	2 mg/m³ (IF)/ 5 mg/m³ (IF)	TRGS 553	3 mg/m³ (all woods) 2 mg/m³ (all woods)*	–		0.5 mg/m³ IF(western and red cedar) A4 all other species 1 mg/m³ IF

(Continued)

TABLE 4.1 (CONTINUED)
Substances with EU Binding Occupational Exposure Limit Values (as of June 2019): Maximum Admissible Concentration in Poland [Dz. U., poz. 1286, 2018], MAC Levels in Germany [TRGS 505, 517, 519, 553, 900, 910], TLV-TWA® in ACGIH (US) [2019], DNEL/DMEL Levels in Germany [IFA, GESTISDatabase 2019], and ECHA [ECHA 2019a]

Substance	EU Directive 2004/37/EC	EU Limit Value	Germany	Source	Poland MAC Value	ECHA DMEL/DNEL for Workers	GESTIS DNEL	ACGIH [2019]
Hydrazine [206-114-9]	2017/2398/EC	0.013 mg/m³	AC: 2.2 µg/m³ TC: 22 µg/m³	TRGS 910	0.013 mg/m³	not established Carc. 1B		0.013 mg/m³ (0.01 ppm) A3
Lead [7439-92-1] and inorganic lead compounds	98/24/EC	0.15 mg/m³	0.1 mg/m³	TRGS 505	0.05 mg/m³	not established Lead Carc. 1A		0.05 mg/m³ A3
4,4′-Methylene-bis(2-chloro-aniline) [101-14-4]	2019/983/EC	0.01 mg/m³			0.02 mg/m³ 0.01 mg/m³*	0.000776 mg/m³ Carc. 1B		0.01 ppm IFV A2
4,4′-Methylene-dianiline [101-77-9]	2019/130/EC	0.08 mg/m³	AC: 0,07 mg/m³ TC: 0,7 mg/m³	TRGS 910	0.08 mg/m³	0.0148 mg/m³ Carc. 1B		0.1 ppm A3
2-Nitropropane [79-46-9]	2017/2398/EC	18 mg/m³	AC: 0.18 mg/m³ TC: 1.8 mg/m³	TRGS 910	18 mg/m³	not established Carc. 1B		36 mg/m³ (10 ppm) A3
Refractory ceramic fibres [–]	2017/2398/EC	0.3 F/ml	AC: 0.01 F/ml TC: 0.1 F/ml Aluminium silicate fibres	TRGS 910	0.3 F/cm³			0.2 F/cm³ A2
Respirable Crystalline silica dust [–]	2017/2398/EC	0.1 mg/m³	0.05 mg/m³	Issued by the BMAS, GMBl. (2016) No 31, p. 623	0.1 mg/m³			0.025 mg/m³ A2
***o*-Toluidine [95-53-4]**	2017/2398/EC	0.5 mg/m³	Manufacture/ use only in enclosed systems AGW: 0.5 mg/m³	GefStoffV, Annex II (6) TRGS 900	3 mg/m³ 0.5 mg/m³*	not established Carc. 1B		– A3

Trichloroethylene [79-01-6]	2019/130/EC	54.7 mg/m^3	AC: 33 mg/m^3	TRGS 910	50 mg/m^3	DNEL: 54.7 mg/m^3 Carc. 1B	54.7 mg/m^3 systematic	54.7 mg/m^3 (10 ppm) A2
Vinyl chloride [75-01-4]	2017/2398/EC	2.6 mg/m^3	AGW: 2.6 mg/m^3	TRGS 900		7.7 mg/m^3 Carc. 1A		2.6 mg/m^3 (1 ppm) A1

Explanations:
*Poland – the concentration value has been proposed for inclusion in the list (regulation of January 9, 2020).
MAC (Poland) – Maximum Admissible Concentration.
A1 (ACGIH) – Confirmed Human Carcinogen.
A2 (ACGIH) – Suspected Human Carcinogen.
A3 (ACGIH) – Confirmed Animal Carcinogen with Unknown Relevance to Humans.
Carc. 1A (CLP) – Carcinogen Category 1A.
Carc. 1B (CLP) – Carcinogen Category 1B.
Muta. 1B (CLP) – Germ cell mutagens Category 1B.
AC – acceptable concentration.
TC – tolerable concentration.
EC – elemental carbon.
RF – respirable fraction.
IF – inhalable fraction.
IFV – inhalable and vapour.
F – fibres.
AGW – occupational exposure limit (GefStoffV Section 2 (8).
DNEL – derived no-effectlevel, poziom niepowodujący zmian.
DMEL – Derived minimal effect level is defined as a level of exposure below which the risk levels of cancer become tolerable.
TRGS 505 – Die TechnischenRegelnfürGefahrstoffe505 "Blei".
TRGS 517 – Die TechnischenRegelnfürGefahrstoffe 517. Asbestos.
TRGS 519 –TechnischeRegelnfürGefahrstoffe. Asbest - Abbruch-, Sanierungs- oderInstandhaltungsarbeiten.
TRGS 553 –TechnischeRegelnfürGefahrstoffe. Holzstaub.
TRGS 900 – Die TechnischenRegelnfürGefahrstoffe 900 "Arbeitsplatzgrenzwerte".
TRGS 910 – Die TechnischenRegelnfürGefahrstoffe 910

4.7 CARCINOGEN/MUTAGEN MIXTURES

The determination of exposure limit values for mixtures of carcinogens/mutagens, arising, for example, in technological processes, depends on several aspects:

- Have the carcinogenic components of the mixture been identified? In this case, it seems appropriate to establish an OEL value for each of them.
- Has the mixture as a whole been classified as carcinogenic? In this case, the derivation of the OEL value seems appropriate only if measuring methods are available to determine the concentration of the mixture as a whole or its characteristic component in the air of the working environment. In the latter case, however, the composition of the mixture needs to be stable [Wriedt 2016; Table 4.2].

Some of the mixtures, such as diesel exhaust, respirable fraction of crystalline silica, or wood dusts, are treated in the same way as other chemicals with established OEL values. Other mixtures are chemical mixtures to which a large number of workers

TABLE 4.2
List of complex carcinogenic mixtures with variable composition of constituents [Wriedt 2016]

Complex mixture	Classification/ Inclusion in Annex I of CMD	Substance(s) listed in Annex, Table 1	Comments
Mineral oils, used		Benzo(a)pyrene	Process-generated
N-Nitros-amines		*N*-Nitrosodiethanolamine, *N*-Nitrosodiethylamine, *N*-Nitroso dimethylamine, *N*-Nitroso di-n-propylamine	Process-generated
Petroleum and coal stream substances and mixtures		Benzene, Benzo(a)pyrene, 1,3-Butadiene	Placed on the market
Polychlorinated biphenyls	IARC: 1	Polychlorinated biphenyls	Legacy substances
Polychlorinated dibenzopara-dioxins	IARC: 3	2,3,7,8-Tetrachlorodibenzo-para-dioxin	Process-generated
Polychlorinated dibenzofurans	IARC: 3	2,3,4,7,8-Pentachloro-dibenzofuran	Process-generated
Polycyclic aromatic hydrocarbons (PAHs)	CMD, Annex I	Benzo(a)pyrene	Process-generated
Rubber dusts and fumes	IARC: 1	1,3-Butadiene, *N*-Nitrosodiethanolamine, *N*-Nitrosodiethylamine, *N*-Nitroso dimethylamine, *N*-Nitroso di-n-propylamine	Process-generated

may be exposed. This group includes polycyclic aromatic hydrocarbons (PAHs) and their nitrogenous compounds, used mineral oils, polychlorinated dibenzodioxins, and dibenzofurans, as well as nitrosamines.

The term "polycyclic aromatic hydrocarbons" (PAHs) refers to a large class of chemicals which contain carbon and hydrogen and are composed of two or more condensed aromatic rings. PAHs are very common environmental pollutants, as they are created during incomplete combustion of materials such as coal, crude oil, natural gas, wood, or waste, or during the pyrolysis of other organic materials, such as tobacco or grilled meat. The NTP report on carcinogens [2016] lists 15 single PAHs. In the past, benzo[*a*]pyrene was often used as a marker to PAH exposure. It is, however, currently possible to measure concentrations of a number of PAHs individually. There are no identified epidemiological studies on exposure to particular PAHs. It is difficult to assess the carcinogenicity of PAHs components of these mixtures due to potential chemical interactions and the presence of other carcinogens in the mixtures. In 2005, IARC has re-assessed the PAHs' carcinogenicity. NTP listed 15 major components of PAHs as presumed human carcinogens based on sufficient evidence of carcinogenicity from experimental animal studies [NTP 2016]. In some countries, no PAHs occupational exposure limits were established, and the monitoring of worker exposure is carried out through biological monitoring. ACGIH recommends the determination of 1-hydroxypyrene in urine, for which the Biological Exposure Index was established to be 2.5 μg/L of urine (sample taken at the end of the last work shift of the work week) [ACGIH 2019]. Similarly, in Great Britain, only a biological monitoring guidance value (BMGV) was established, on the level of 4 μmol of 1-hydroxypyrene/mol creatinine in urine (sample taken at the end of the work shift). In Poland, a Maximum Admissible Concentration (MAC, OEL) was set for PAHs, on the level 0.002 mg/m³ with regard to 9 PAH compounds with carcinogenic properties, taking into account their potency. For this purpose, the concept of exposure indicator (W_{WWA}) was introduced and calculated from the formula:

$$W_{wwa} = (c_1 \cdot k_1) + (c_2 \cdot k_2) + (c_3 \cdot k_3) + (c_4 \cdot k_3) + (c_5 \cdot k_3) + (c_6 \cdot k_3) + (c_7 \cdot k_4) + (c_8 \cdot k_4) + (c_9 \cdot k_4).$$

The values of carcinogenic potency coefficients (k) in relation to the 9 PAHs range from 1 for benzo[*a*]pyrene to 5 for dibenz[*a,h*]anthracene. Whereas C1 – C9 express the concentrations of individual PAHs obtained from measurements. The adopted MAC value was equivalent to that of benzo[*a*]pyrene in Poland. The value of W_{WWA} ≤ 0.002 mg/m³ should protect workers against the additional risk of cancer [Sapota 2002].

Occupational exposure to diesel exhaust is present everywhere, where diesel-driven vehicles and equipment are used, e.g. in mining, construction, agriculture, forestry, waste management, and transportation. The level of emissions and composition of exhaust emissions from diesel engines depend, among others, on the type, age, operating, and maintenance conditions of the engine, composition, and physical properties of the fuel as well as the exhaust gas purification techniques used.

They are multi-component mixtures of several hundreds of chemicals, resulting from incomplete combustion of fuel and engine oil, as well as the present modifiers and contaminants. Exhaust toxicity is connected to the presence of compounds proven to be toxic and carcinogenic, i.e. PAHs and their nitro-derivatives, benzene and nitrobenzene, toluene, xylenes, formaldehyde, acetaldehyde, carbon, nitrogen, and sulfur oxides. Due to the "rich" composition of exhaust gases and the complexity of biological systems with which they interact, it is difficult to predict their toxic activity.

In occupational exposure, exhaust fumes are absorbed in the respiratory system. Over 90% of particles emitted from engines have a diameter smaller than 1 μm and therefore it is respirable fraction. The chronic poisoning is usually observed in people professionally exposed to diesel engine exhaust gas for at least several years. Results of epidemiological studies indicate the presence of a relationship between occupational exposure to diesel exhaust and an increased incidence of certain groups of cancers, particularly lung cancer and bladder cancer. IARC has classified diesel engine exhaust under Group 1, as carcinogenic to humans [IARC 2014].

Both NIOSH and NTP consider diesel engine exhaust to be potential occupational carcinogens (substances labeled "Ca" and "R") [NTP 2016; NIOSH 2017].

Commission for the Investigation of Health Hazards of Chemical Compounds in the Work Area of the German Research Foundation (DFG) classified diesel engine exhaust emissions under Category 2 – substances considered carcinogenic to humans, for which there is sufficient data from animal experiments [DFG 2018].

Researchers assessing the genotoxicity of exhaust emissions from diesel engines have observed that the emissions contain, in addition to mutagens, substances that damage chromosomes. Results of animal studies suggest that exposure to diesel engine exhaust emissions may affect male fertility.

Elemental carbon is a characteristic component of diesel engine exhaust emissions; therefore it was adopted as an indicator for the whole diesel exhaust exposure. Accepting the risk assessment estimated by Vermeulen et al [2014] and DECOS [2019] for OEL – 0.5 mg/m^3, the risk of lung cancer incidence is 1 of 1,000 exposed people, while for concentration 0.05 mg/m^3 (BOELV value) the risk is reduced to 1 case of 10,000 occupational exposed people.

Polychlorinated dibenzo-*p*-dioxins (PCDDs) and polychlorinated dibenzofurans (PCDFs) are formed as by-products during various industrial processes, combustion, accident, etc. Occupational exposure to PCDDs/Fs is present in those branches of industry in which processes generating them are present. Potential routes of human exposure to these compounds are oral, inhalation, and dermal.

Epidemiological studies are the basis of the assessment of carcinogenic activity of dioxins (including 2,3,7,8-TCDD) and furans in humans. Cohorts include subjects occupationally exposed to chlorophenols, phenoxyacetic herbicides, and mixtures of polychlorinated dibenzodioxins and furans. Unfortunately, presently available data do not allow for a relationship to be established between cancer incidence (and cancer-related death) and the level of exposure. Their carcinogenic activity might be caused by their relatively long half-life, particularly in humans, which results in permanent activation of aryl hydrocarbon receptor. PCDD/F have been assessed for carcinogenicity by IARC. According to IARC, there is sufficient evidence of carcinogenicity in humans only for 2,3,7,8-TCDD (CAS: 1746-01-6) and 2,3,4,7,8-pentaCDF (CAS: 57117-31-4); therefore these compounds were classified under group 1 of carcinogens. Other PCDDs/Fs are included in group 3 – substances that cannot be classified as carcinogenic to humans. As a basis for the establishment of the MAC (OEL) value for PCDDs and PCDFs mixture in Poland, the Interdepartmental Commission for Maximum Admissible Concentrations and Intensities for Agents Harmful to Health in the Working Environment has adopted the results of risk assessment of an additional incidence of liver cancer in workers occupationally exposed to 2,3,7,8-TCDD conducted in 2017. The risk was estimated to be 1×10^{-4} for 40 years of exposure to the compound in the concentration of 18 pg/m^3. In the case of combined exposure, the content of polychlorinated dibenzo-*p*-dioxins and furans in the tested samples (air, food, soil, etc.), and their limit values are expressed as a so-called toxicity equivalent (TEQ). For mixtures of PCDDs and PCDFs, a value of 18 pg WHO_{2006}-TEQ/m^3 was adopted. The result expressed in pg WHO-TEQ/m^3 is not a *de facto* concentration, but an expression of the toxicity of congeners of dioxins and furans in the sample in reference to a model dioxin, i.e. TCDD. A common feature of these compounds is the same mechanism of action mediated by the intracellular Ah receptor and the ability to elicit a similar spectrum of biological responses. Toxicity equivalent (TEQ) is a normalized value, calculated as the sum of the products of the concentrations of individual congeners and their corresponding toxic equivalency factors (TEFs), which are numerical values (dimensionless values) with a value from 0 to 1. TEFs express relative toxicity of particular congeners in comparison with the most toxic compound, i.e. 2,3,7,8-tetrachlorodibenzo-*p*-dioxin (2,3,7,8-TCDD, commonly known as TCDD), which has the value of 1. [Skowroń and Czerczak 2013; Szymańska et al. 2020].

4.8 KNOWN SUBSTANCES – A NEW RISK

The IARC Advisory Group, at a meeting in March 2019, considered over 170 chemical substances/mixtures nominated for re-evaluation of carcinogenicity and classification. In short summaries on each substance/mixture, information was included on human exposure, results of epidemiological studies, and results of research on carcinogenicity in experimental animals, to discuss the mechanisms of carcinogenicity in accordance with the assessment approach presented in the preamble to the IARC monograph [IARC 2019a].

Substances/mixtures were given priority for carcinogenicity/mutagenicity assessment in 2020–2024 based on (i) human exposure data and (ii) the range of available evidence of a carcinogenicity assessment (i.e. availability of relevant evidence for carcinogenicity in humans, in experimental animals, and carcinogenicity mechanisms).

In the list of high-priority substances, which had not been evaluated by IARC, the following substances present in the work environment are included:

- bromodichloroacetic acid (disinfecttion by-products in disinfected water used for showering, bathing, swimming, or drinking; rationale: human bladder cancer),

- Cupferron (*N*-nitroso-*N*-phenylhydroxylamine ammonium salt) (used for the separation and precipitation of metals, such as copper, iron, vanadium, and thorium; rationale: various cancers in mice and rats),
- glycidamide (used as intermediate in organic synthesis, e.g. in the production of dyes and plasticizers; rationale: cancers in test animals),
- malachite green (triphenylmethane dye used in the fishing industry as a fungicide; rationale: liver cancers in test animals),
- vinclozolin (fungicide, commonly used in the past for certain fruits, nuts, grapevine, vegetables, ornamental plants, and as a wood preservative; rationale: cancers in test animals).

Medium priority for the assessment of carcinogenicity/mutagenicity is given to the following substances:

- alachlor (used as herbicide on corn and soybeans; rationale: laryngeal cancer and leukemia in human),
- gentian violet (used in human medicine against fungi and intestinal parasites; rationale: thyroid cancers in test animals),
- C.I. Direct Blue 218 (a copper chelated dye used for cellulose, acetate, nylon, silk, wool, tissue, papers, and textile goods with a urea-formaldehyde finish; rationale: cancer in test animals),
- diphenylamine (industrial antioxidant, fungicide, insecticide, used in the production of dyes and pesticides, as a stabilizer for nitrocellulose explosives and celluloids, and as an antioxidant on apple leaves; rationale: cancers in test animals),
- 1,2-diphenylhydrazine (was used as an industrial intermediate, mainly in the manufacture of benzidine dyes; rationale: cancers in test animals),
- mancozeb (widely used fungicide, significant exposure for production workers, applicators, residents of the areas where it is used, and consumers; rationale: tumors of the central nervous system in human)
- *o*-benzyl-*p*-chlorophenol (broad-spectrum biocide in disinfectant solutions, used in hospitals and households for general cleaning and disinfecting, occupational exposure occurs by dermal absorption; rationale: kidney cancer in mice)
- biphenyl (used in a number of products and processes, including as a fungicide and an ingredient of chemicals used in agriculture; rationale: cancer in test animals),
- pendimethalin (dinitroaniline herbicide with unlimited applications for crops, lawns, and gardens by farmers, professional applicators, and homeowners; rationale: human lung cancer, cancer in test animals),
- terbufos (organothiophosphate herbicide used widely in agriculture; rationale: in humans prostate cancer, non-Hodgkin's lymphoma),
- tris(chloropropyl) phosphate (organophosphate flame retardant and plasticizer, and is used in a variety of industrial and household products; rationale: thyroid cancer in humans, structurally similar compounds are carcinogenic to rodents).

The following substances will be subsequently assessed by IARC:

- 2-hydroxy-4-methoxybenzophenone (UV filter used in sunscreens, and, to a degree in industrial products, such as plastics and coatings; rationale: a significant increase in the incidence of adenoma or thyroid cancer and uterine polyps or sarcoma in test animals),
- butyl methacrylate (used in the industry to produce polymers for car coatings, varnishes, enamels, glues, oil additives (greases), dental products and textile emulsions, and leather and paper finishing; rationale: cancers in test animals),
- cinidon ethyl (dicarboximide herbicide for use on cereals, e.g. wheat and rye; rationale: liver and parathyroid tumors in test animals),
- furmecyclox (furamide fungicide and wood preservative; rationale: cancers in test animals),
- isophorone (widely used solvent and chemical intermediate; rationale: cancer in test animals),
- methanol (a number of commercial applications; rationale: cancer in test animals),
- *S*-Ethyl -*N*-*N*- dipropylthiocarbamate (a thiocarbamate herbicide used to selectively control annual and perennial grass weeds and some leaves in citrus, beans, corn, potatoes, and pineapples; rationale: increased risk of non-Hodgkin's lymphoma, leukemia, cancer of the large intestine and pancreas in farmers) [IARC 2019c].

Clapp et al. [2008] discussed in detail new evidence on environmental and occupational causes of cancer in a 2008 study. Despite the weakness of some individual studies, they concluded that the publications have strengthened combined evidence for particular types of exposure linked to an increased risk of cancer, such as:

- leukemia after exposure to 1,3-butadiene;
- lung cancer caused by air contaminants, including PAHs;
- non-Hodgkin's lymphoma (NHS) as a result of exposure to pesticides and solvents;
- prostate cancer as a result of exposure to pesticides, PAHs, and metalworking fluids or mineral oils [EU-OSHA 2014].

4.9 SUMMARY

Occupational cancers currently pose a significant problem and will remain so, as people are and will be exposed to carcinogens at the workplace. EU actions aimed at establishing binding limit values for these substances, applicable for all EU countries, will have an impact on reducing the number of cancer diseases in Europe and will ensure an equal level of worker protection. It is estimated that the introduction of limit values for 50 carcinogenic substances will allow us to avoid deaths of 100,000 workers in the next 50 years.

Therefore, clear priorities are needed to prevent serious health hazards associated with carcinogens/mutagens in the workplace. Of course, occupational cancer is the largest individual threat, taking into account the number of deaths in the European Union and in other developed countries.

The objective of activities being carried out at the European level, and on a global scale, is to eliminate occupational cancers. This can be achieved by gradually decreasing occupational exposure to substances, particularly to carcinogens/mutagens, and to related processes and workplaces, as well as workplaces known to cause or contribute to the initiation of occupational cancers. This requires extensive European and international cooperation to identify and market new policies and practices for carcinogenic/mutagenic substances [Takala 2015].

According to a survey conducted by EU-OSHA, the first steps to be taken in relation to carcinogens/mutagens are legal provisions and setting binding limit values. Then, it pointed out the monitoring of exposure to carcinogenic/mutagenic substances at workplaces in various countries, information campaigns, ending with practical solutions and controls [Roadmap on Carcinogens Conference 2019].

There are still new threats to carcinogens/mutagens that include, for example, nanomaterials (carbon nanotubes), some of which have recently been classified by IARC as carcinogens or endocrine disruptors.

OEL values are developed by many organizations around the world and are an indispensable criterion for assessing health risk in the workplace, but have only been developed for a small proportion of chemicals. Because there are many organizations that establish OEL values around the world, each with their own approach, these values can vary significantly. It is necessary to extend international harmonization efforts to promote similarities and transparency in the approaches used by various organizations around the world [Deveau et al. 2015].

Despite significant progress, further researches on occupational cancer are still necessary. Epidemiological evidence is insufficient or completely missing for most of the more than 1,000 agents assessed by IARC. Many other factors present in the workplace have not been tested for carcinogenicity. It is also necessary to identify by geographical regions the number of workers exposed and to provide quantitative exposure data as a basis for hazard identification, exposure–response assessment, and risk assessment.

The consequences of occupational cancer and its impact on society go beyond deaths and morbidity. They also include reduced quality of life, loss of productivity at work, and increased healthcare costs. In addition to individual emotional suffering and pain associated with the disease, this leads to economic costs for the whole of society. It is estimated that, in the European Union alone, healthcare costs and production loss costs amount to 4–7 billion euro annually. After adding welfare loss, premature deaths, and cancer diagnostics, the total annual cost is estimated at an average of 334 (242–440) billion euro [Musu and Vogel 2018].

Duplication of limit values for carcinogenic/mutagenic substances should be avoided and the role of OEL/BOELV limit values should be clarified in the context of derived no-effect levels (DNELs) or derived minimum effect/effect levels (DMELs) in the REACH regulation.

Harmonization of the limit values for carcinogenic mutagenic substances has and can have many advantages. The OEL development process is complex and lengthy and requires a lot of economic investment. Intensive international cooperation efforts could reduce the need for many OEL regulators for carcinogens/mutagens or could encourage the division of labor between organizations, thereby preventing duplication of personnel and financial resources that arise when many organizations achieve limit values for these chemicals alone. The harmonization of rules of establishing limit values for these substances can also reduce the confusion and economic ineffectiveness, which can particularly affect international companies, which are required to observe a number of legally binging OEL/BOELV values. Inconsistent practices in the determination of limit values for carcinogenic/mutagenic substances can also cause discrepancies in worker protection between different countries.

Harmonization might be particularly beneficial to workers in smaller companies/countries. Properly conducted harmonization can also lead to higher clarity and use of the best practices to establish OEL values for carcinogenic/mutagenic substances [Deveau et al. 2015].

REFERENCES

ACGIH [American Conference of Governmental Industrial Hygienists]. 2019. *TVLs and BEIs based on the documentation of the threshold limit values for chemical substances and physical agents & biological exposure indices*. Cincinnati, OH: American Conference of Governmental Industrial Hygienists.

BAuA [Bundesanstalt für Arbeitsschutz und Arbeitsmedizin]. 2013. The risk-based concept for carcinogenic substances developed by the committee for hazardous substances. www.baua.de/dok/3581564 (accessed January 10, 2020).

Blair, A., L. Marrett, and L. B. Freeman. 2011. Occupational cancer in developed countries. *Environ Health* 10(Suppl 1):S9. doi: 10.1186/1476-069X-10-S1-S9.

Bolt, H. M., and A. Huici-Montagud. 2008. Strategy of the scientific committee on occupational exposure limits (SCOEL) in the derivation of occupational exposure limits for carcinogens and mutagens. *Arch Toxicol* 82(1):61–64. doi: 10.1007/s00204-007-0260-z.

Brandys, R. C., and G. M. Brandys. 2008. *Global occupational exposure limits for over 6,000 specific chemicals*. 2nd ed. Hingdale, IL: Occupational and Environmental Health Consulting Services Inc.

Clapp, R. W., M. M. Jacobs, and E. L. Loechler. 2008. Environmental and occupational causes of cancer: New evidence, 2005–2007. *Rev Environ Health* 23(1):1–37.

DECOS. 2019. Diesel engine exhaust. Health-based recommended occupational exposure limit. https://www.healthcouncil.nl/documents/advisory-reports/2019/03/13/diesel-engine-exhaust (accessed December 19, 2019).

Demers, P.A., C.E. Peters and A-M. Nicol. 2008. Priority Occupational Carcinogens for Surveillance in Canada. CAREX Canada, School of Environmental Health, University of British Colombia (accessed April 1, 2020). https://www.carexcanada.ca/CAREX_Canada_Occupational_Priorities_Report.pdf.

Deveau, M., C. P. Chen, G. Johanson et al. 2015. The global landscape of occupational exposure limits: Implementation of harmonization principles to guide limit selection. *J Occup Environ Hyg* 12:127–144. doi: 10.1080/15459624.2015.1060327.

DFG [Deutsche Forschungs gemeinschaft]. 2018. List of MAK and BAT values 2018: Maximum concentrations and biological tolerance values at the workplace. Report 54. Weinheim, Germany: Wiley-VCH. doi: 10.1002/9783527818402.

Dz. U. poz. 1286 (z późn. zm.). 2018. Rozporządzenie Ministerstwa Rodziny, Pracy i Polityki Społecznej z dnia 12 czerwca 2018 r. w sprawie najwyższych dopuszczalnych stężeń i natężeń czynników szkodliwych dla zdrowia w środowisku pracy. http://isip.sejm.gov.pl/isap.nsf/download.xsp/WDU20180001286/O/D20181286.pdf (accessed December 19, 2019).

EC [European Commission]. 1998. Council Directive 98/24/EC of April 7, 1998, on the protection of the health and safety of workers from the risks related to chemical agents at work (fourteenth individual Directive within the meaning of Article 16(1) of Directive 89/391/EEC). *Off J Eur Union*. Document 31998L0024. https://eur-lex.europa.eu/legal-content/EN/ALL/?uri=celex:31998L0024 (accessed January 10, 2020).

EC [European Commission]. 2004. Directive 2004/37/EC of the European Parliament and of the Council of April 29, 2004, on the protection of workers from the risks related to exposure to carcinogens or mutagens at work (Sixth individual Directive within the meaning of Article 16(1) of Council Directive 89/391/EEC) with amended. *Off J Eur Union*. Document 02004L0037-20140325. https://eur-lex.europa.eu/legal-content/en/TXT/?uri=CELEX%3A02004L0037-20140325 (accessed January 10, 2020).

EC [European Commission]. 2006. Regulation (EC) No 1907/2006 of the European Parliament and of the Council of December 18, 2006, concerning the Registration, Evaluation, Authorisation and Restriction of Chemicals (REACH), establishing a European Chemicals Agency, amending Directive 1999/45/EC and repealing Council Regulation (EEC) No 793/93 and Commission Regulation (EC) No 1488/94 as well as Council Directive 76/769/EEC and Commission Directives 91/155/EEC, 93/67/EEC, 93/105/EC and 2000/21/EC with amended. *Off J Eur Union*. Document 02006R1907-20140410. https://eur-lex.europa.eu/legal-content/EN/TXT/?uri=CELEX%3A02006R1907-20140410 (accessed January 10, 2020).

EC [European Commission]. 2008. Regulation (EC) No 1272/2008 of the European Parliament and of the Council of December 16, 2008, on classification, labelling and packaging of substances and mixtures, amending and repealing Directives 67/548/EEC and 1999/45/EC, and amending Regulation (EC) No 1907/2006. *Off J Eur Union*. Document 32008R1272. https://eur-lex.europa.eu/eli/reg/2008/1272/oj?eliuri=eli:reg:2008:1272:oj (accessed January 10, 2020).

EC [European Commission]. 2009. Directive 2009/148/EC of the European Parliament and of the Council of November 30, 2009, on the protection of workers from the risks related to exposure to asbestos at work. *Off J Eur Union*. Document 32009L0148. https://eur-lex.europa.eu/legal-content/EN/TXT/?uri=CELEX%3A32009L0148 (accessed January 10, 2020).

EC [European Commission]. 2017. Directive (EU) 2017/2398 of the European Parliament and of the Council of December 12, 2017, amending Directive 2004/37/EC on the protection of workers from the risks related to exposure to carcinogens or mutagens at work. Official Journal of the European Union. *Off J Eur Union*. Document 32017L2398. https://eur-lex.europa.eu/eli/dir/2017/2398/oj (accessed January 10, 2020).

EC [European Commission]. 2018. European Commission. Directorate-General for Employment, Social Affairs and Inclusion. Methodology for derivation of occupational exposure limits of chemical agents. The general Decision-Making Framework of the Scientific Committee on Occupational Exposure Limits (SCOEL) 2017. Luxembourg: Publications Office of the European Union. doi: 10.2767/435199.

EC [European Commission]. 2019a. Directive (EU) 2019/130 of the European Parliament and of the Council of January 16, 2019, amending Directive 2004/37/EC on the protection of workers from the risks related to exposure to carcinogens or mutagens at work. *Off*

J Eur Union. Document 32019L0130. https://eur-lex.europa.eu/legal-content/PL/TXT/?uri=CELEX:32019L0130 (accessed January 10, 2020).

EC [European Commission]. 2019b. Directive (EU) 2019/983 of the European Parliament and of the Council of June 5, 2019, amending Directive 2004/37/EC on the protection of workers from the risks related to exposure to carcinogens or mutagens at work. *Off J Eur Union*. Document 32019L0983. https://eur-lex.europa.eu/legal-content/PL/ALL/?uri=CELEX:32019L0983 (accessed December 19, 2019).

ECHA [European Chemical Agency]. 2017a. Joint task force ECHA Committee for Risk Assessment (RAC) and Scientific Committee on Occupational Exposure Limits (SCOEL) on Scientific aspects and methodologies related to the exposure of chemicals at the workplace. https://echa.europa.eu/documents/10162/13579/rac_joint_scoel_opinion_en.pdf/58265b74-7177-caf7-2937-c7c520768216 (accessed December 19, 2019).

ECHA [European Chemical Agency]. 2017b. Joint Task Force Committee for Risk Assessment (RAC) and Scientific Committee on Occupational Exposure Limits (SCOEL) on scientific aspects and methodologies related to the exposure of chemicals at the workplace. https://echa.europa.eu/documents/10162/13579/jtf_opinion_task_2_en.pdf/db8a9a3a-4aa7-601b-bb53-81a5eef93145 (accessed December 19, 2019).

ECHA [European Chemical Agency]. 2019a. C&L inventory database. https://echa.europa.eu/pl/information-on-chemicals/cl-inventory-database (accessed December 19, 2019).

ECHA [European Chemical Agency]. 2019b. Guidance on information requirements and chemical safety assessment, Chapter R.8: Characterisation of dose [concentration] – response for human health. http://echa.europa.eu (accessed December 19, 2019).

EPA [United States Environmental Protection Agency]. 2019. Integrated risk information system. https://www.epa.gov/fera/risk-assessment-carcinogenic-effects (accessed December 19, 2019).

EU-OSHA [Occupational Safety and Health Administration]. 2014. Exposure to carcinogens and work-related cancer: A review of assessment measures. https://osha.europa.eu/en/tools-and-publications/publications/reports/report-soar-work-related-cancer (accessed December 19, 2019).

Furuya, S., O. Chimed-Ochir, K. Takahashi et al. 2018. Global asbestos disaster. *Int J Environ Res Public Health* 15:1000. doi: 10.3390/ijerph15051000.

Hammerschmidt, T., and R. Marx. 2014. REACH and occupational health and safety. *Environ Sci Eur* 26:6. doi: 10.1186/2190-4715-26-6. http://www.enveurope.com/content/26/1/6 (accessed December 19, 2019).

IARC [International Agency for Research on Cancer]. 2014. *Diesel and gasoline engine exhausts and some nitroarenes*. Lyon, France: International Agency for Research on Cancer. https://monographs.iarc.fr/wp-content/uploads/2018/06/mono105.pdf (accessed December 19, 2019).

IARC [International Agency for Research on Cancer]. 2019a. Preamble. Lyon, France: International Agency for Research on Cancer. https://monographs.iarc.fr/wp-content/uploads/2019/07/Preamble-2019.pdf (accessed December 19, 2019).

IARC [International Agency for Research on Cancer]. 2019b. Monographs on the identification of carcinogenic hazards to humans. Agents classified by the IARC Monographs, Volumes 1–125. https://monographs.iarc.fr/agents-classified-by-the-iarc (accessed December 19, 2019).

IARC [International Agency for Research on Cancer]. 2019c. Report of the advisory group to recommend priorities for the IARC Monographs during 2020–2024. Lyon, France: International Agency for Research on Cancer. https://monographs.iarc.fr/wp-content/uploads/2019/10/IARCMonographs-AGReport-Priorities_2020-2024.pdf (accessed January 20, 2020).

IFA [Institut für Arbeitsschutz der Deutschen Gesetzlichen Unfallversicherung]. 2019. GESTIS, Database: International limit values for chemical agents. http://www.dguv.de/

ifa/GESTIS/GESTIS-Internationale-Grenzwerte-für-chemische-Substanzen-limit-values-for-chemical-agents/index-2.jsp (accessed December 19, 2019).

Jongeneel, W. P., P. E. D. Eysink, D. Theodori et al. 2016. Work-related cancer in the European Union. Size, impact and options for further prevention. RIVM Letter Report 2016-0010. https://www.rivm.nl/publicaties/work-related-cancer-in-european-union-size-impact-and-options-for-further-prevention (accessed December 19, 2019).

Kauppinen, T., J. Toikkanen, D. Pedersen et al. 2000. Occupational exposure to carcinogens in the European Union. *Occup Environ Med* 57:10–18.

Loomis, D., N. Guha, A. L. Hall, and K. Straif. 2018. Identifying occupational carcinogens: An update from the IARC Monographs. *Occup Environ Med* 75(8):1–11. doi: 10.1136/oemed-2017-104944.

Mengeot, M. A., T. Musu, and L. Vogel. 2014. *Preventing work cancers: A workplaces health priority*. Brussels: ETUI, European Trade Union Institute.

Miller, E. C., and J. A. Miller. 1981. Mechanisms of chemical carcinogenesis. *Cancer* 47:1055–1064.

Mol, A., and M. Stolarek. 2011. Indukowane mutagenami uszkodzenia DNA i mechanizmy ich powstania. *Postępy Biologii Komórki* 38(3):491–505. https://www.pbkom.eu/sites/default/files/artykulydo2012/38_3_491.pdf (accessed December 19, 2019).

Musu, T., and L. Vogel 2018. *Cancer and work. Understanding occupational cancers and taking action to eliminate them.* Brussels: ETUI, European Trade Union Institute. https://www.etui.org/Publications2/Books/Cancer-and-work-understanding-occupational-cancers-and-taking-action-to-eliminate-them (accessed April 1, 2020).

Niepsuj, A., S. Czerczak., and K. Konieczko. 2020. Substancje chemiczne i procesy technologiczne o działaniu rakotwórczym lub mutagennym w środowisku pracy w Polsce w latach 2013–2017. *Med Pr* . 71(2): 187–203. doi: 10.13075/mp.5893.00956.

NIOSH [National Institute for Occupational Safety and Health]. 2007. NIOSH pocket guide to chemical hazards. Cincinnati, OH: U.S. Dept. of Health and Human Services. https://www.cdc.gov/niosh/docs/2005-149/pdfs/2005-149.pdf (accessed December 19, 2019).

NIOSH [National Institute for Occupational Safety and Health]. 2017. Whittaker, C., F. Rice, L. McKernan et al. NIOSH chemical carcinogen policy. Cincinnati, OH: U.S. Department of Health and Human Services, Centers for Disease Control and Prevention, National Institute for Occupational Safety and Health, Publication No. 2017-100. https://www.cdc.gov/niosh/docs/2017-100/pdf/2017-100.pdf (accessed April 1, 2020).

NTP [National Toxicology Program]. 2016. 14th report on carcinogens. https://ntp.niehs.nih.gov/go/roc14 (accessed December 19, 2019).

Occupational and Safety Health Administration, United Sates Department of Labour (OSHA US) 2014. https://www.osha.gov/laws-regs/federalregister/2014-10-10 (accessed December 19, 2019). Table Z-1-A https://www.osha.gov/dsg/annotated-pels/tablez-1.html.

Pronk, M. E. J. 2014. Overview of methodologies for the derivation of occupational exposure limits for non-threshold carcinogens in the EU. RIVM Letter Report 2014-0153. https://www.rivm.nl/publicaties/overview-of-methodologies-for-derivation-of-occupational-exposure-limits-for-non (accessed December 19, 2019).

Roadmap on Carcinogens Conference "Working together to eliminate occupational cancer" Finland Ministry of Social Affairs and Health, Employment, Social Policy, Health and Consumer Affairs, EPSCO, November 27–28, 2019, Helsinki, Finland. https://roadmaponcarcinogens.eu (accessed December 19, 2019).

Safe Work Australia. 2018. Workplace exposure standards for airborne contaminants. https://www.safeworkaustralia.gov.au/system/files/documents/1804/workplace-exposure-standards-airborne-contaminants-2018_0.pdf (accessed December 19, 2019).

Sapota, A. 2002. Wielopierścieniowe węglowodory aromatyczne: Dokumentacja dopuszczalnych poziomów narażenia zawodowego. *Podstawy i Metody Oceny Środowiska Pracy* 2(32):179–208.

Schrenk, D. 2018. What is the meaning of "A compound is carcinogenic"? *Toxicol Rep* 5:504–511.

Skowroń, J., and S. Czerczak. 2013. Zasady ustalania dopuszczalnych poziomów narażenia dla czynników rakotwórczych w środowisku pracy w Polsce i w krajach Unii Europejskiej. *Med Pr* 64(4):541–563. doi: 10.13075/mp.5893.2013.0046.

Smith, M. T., K. Z. Guyton, C. F. Gibbons et al. 2016. Key characteristics of carcinogens as a basis for organizing data on mechanisms of carcinogenesis. *Environ Health Perspect* 124(6):713–721. doi: 10.1289/ehp.1509912.

Szymańska, J., B. Frydrych, and E. Bruchajzer. 2019. Spaliny emitowane z silników Diesla, mierzone jako węgiel elementarny: Dokumentacja proponowanych dopuszczalnych wielkości narażenia zawodowego. *Podstawy i Metody Oceny Środowiska Pracy* 4(102):43–103. doi: 10.5604/01.3001.0013.6378.

Szymańska, J., B. Frydrych, P. Struciński, W. Szymczak, A. Hernik, and E. Bruchajzer. 2020. Mieszanina polichlorowanych dibenzo-*p*-dioksyn i polichlorowanych dibenzofuranów: Dokumentacja proponowanych dopuszczalnych wielkości narażenia zawodowego. *Podstawy i Metody Oceny Środowiska Pracy* 1(103):71–142. doi: 10.5604/01.3001.0013.7815.

Takala, J. 2015. Eliminating occupational cancer in Europe and globally. Working Paper 2015.10. Brussels: ETUI, European Trade Union Institute. https://www.etui.org/content/download/21462/179550/file/WP+2015-10-Eliminating+occupational+cancer+Web+version.pdf (accessed January 20, 2020).

Takala, J. 2018. *Carcinogens at work: A look into the future*. Vienna: Austrian EU Presidency. https://roadmaponcarcinogens.eu/wp-content/uploads/2018/10/Takala.pdf (accessed April 1, 2020).

TRGS 505. 2007. Technische Regeln für Gefahrstoffe. Blei. https://www.baua.de/DE/Angebote/Rechtstexte-und-Technische-Regeln/Regelwerk/TRGS/pdf/TRGS-505.pdf?__blob=publicationFile&v=3 (accessed December 19, 2019).

TRGS 517. 2013. Technical rules for hazardous substances. Activities with potentially asbestoscontaining minerals and mixtures and products manufactured from same. https://www.baua.de/EN/Service/Legislative-texts-and-technical-rules/Rules/TRGS/pdf/TRGS-517.pdf?__blob=publicationFile&v=2 (accessed December 19, 2019).

TRGS 519. 2014. Technische Regeln für Gefahrstoffe. Asbest – Abbruch-, Sanierungs- oder Instandhaltungsarbeiten. https://www.baua.de/DE/Angebote/Rechtstexte-und-Technische-Regeln/Regelwerk/TRGS/pdf/TRGS-519.pdf?__blob=publicationFile&v=6 (accessed December 19, 2019).

TRGS 553. 2008. Technische Regeln für Gefahrstoffe. Holzstaub. https://www.baua.de/DE/Angebote/Rechtstexte-und-Technische-Regeln/Regelwerk/TRGS/pdf/TRGS-553.pdf?__blob=publicationFile&v=2 (accessed December 19, 2019).

TRGS 900. 2006. Technische Regeln für Gefahrstoffe. Arbeitsplatzgrenzwerte. https://www.baua.de/DE/Angebote/Rechtstexte-und-Technische-Regeln/Regelwerk/TRGS/pdf/TRGS-900.pdf?__blob=publicationFile&v=13 (accessed December 19, 2019).

TRGS 910. 2014. Technical rules for hazardous substances. Risk-related concept of measures for activities involving carcinogenic hazardous substances. https://www.baua.de/EN/Service/Legislative-texts-and-technical-rules/Rules/TRGS/pdf/TRGS-910.pdf?__blob=publicationFile&v=2 (accessed December 19, 2019).

Unifor Canada. 2019. *Occupational cancer fighting back makes a difference*. Toronto, ON: Unifor Health, Safety and Environment Department. https://www.unifor.org/sites/default/files/documents/document/unifor_occupational_cancer_eng_final-web_v2.pdf (accessed December 19, 2019).

Van Kesteren, P. C. E., N. G. M. Palmen, and S. Dekkers. 2012. Occupational exposure limits and classification of 25 carcinogens. RIVM Letter Report 320002001. http://www.rivm.nl/bibliotheek/rapporten/320002001.pdf (accessed December 19, 2019).

Vermeulen, R., D. T. Silverman, E. Garshick et al. 2014. Exposure-response estimates for diesel engine exhaust and lung cancer mortality based on data from three occupational cohorts. *Environ Health Perspect* 122:172–177.

Wriedt, H. 2016. *Carcinogens that should be subject to binding limits on workers' exposure*. Brussels: ETUI, European Trade Union Institute. https://www.etui.org/Publications2/Reports/Carcinogens-that-should-be-subject-to-binding-limits-on-workers-exposure (accessed December 19, 2019).

5 Assessment of Chemical Risk in the Work Environment

Małgorzata Pośniak

CONTENTS

5.1 INTRODUCTION

Chemical substances and their mixtures occur in the work environment in virtually all branches of the world economy. The technological processes in which they are produced, processed, or used are a source of pollution of the work environment, and sometimes also of the natural environment. Exposure to substances or their mixtures may cause various adverse changes in the state of the health of workers. The body's responses to these agents are primarily influenced by the dose absorbed into the body, as well as many other factors. The type and amount of adverse health effects depend, among others, on toxic and physicochemical properties, absorption routes, sex and age of the exposed person, the general state of health, as well as the intra-individual features of the exposed. It also depends on external factors—temperature, air humidity, and exposure time.

Assessment of occupational risk is the basis for the proper management of the hazards posed by different agents in the work environment, including chemicals and their mixtures that threaten workers' health. Quantitative assessment of occupational exposure and risk is a very complex process because of the possibility of the occurrence of multi-component mixtures of chemical substances in the work environment which cause various health effects. The scale of these effects depends on many factors, e.g. worker's load, health, and age and environmental factors such as temperature, pressure, humidity, etc., which complicates the risk assessment process. It requires the identification of all substances used in the workstation, the hazards associated with them, and the assessment of working conditions. The result of these actions is the basis for taking appropriate preventive measures to control and, if possible, eliminate such risks.

However, a complete elimination of risk in the case of exposure to harmful chemicals is problematic. It is due to the fact that it requires the cessation of the use of the hazardous chemical, which is a very difficult task, requiring costly research aimed at finding harmless or less harmful substances, the use of which will not affect the quality of manufactured products. It should be noted that the total elimination of chemical risks is, in principle, the elimination of chemical substances and their mixtures from production and processing, as well as the abandonment of the use of these substances even during basic activities necessary for human life such as personal hygiene, cleaning, or cooking. Contemporary societies, in both highly developed and economically backward countries, cannot live without the use of chemical products. Therefore, the only solution that can, to some extent, reduce the risks associated with the production, processing, and use of hazardous agents is to encourage employers to make every effort to consciously and correctly manage chemical risks in the workplace.

A particularly difficult issue in the chemical risk management process is the identification of the risks associated with the long-term use of new chemicals and substances, but which, thanks to the results of new toxicological and epidemiological studies and the use of modern measurement methods and techniques to assess their physicochemical, toxic, and exposure properties, become "new" factors, posing a threat to human health and often to human life and the environment. Understanding the threats arising from new processes and technologies, as well as changes in public opinion in relation to the already known threats, is a very important and difficult task. It is also necessary to pay special attention to chemical hazards leading to a rapid increase in the number of exposed people, and, above all, the increasing number of workers who have been diagnosed with diseases resulting from the effects of hazardous chemicals.

During research conducted by the EU-OSHA under the European Risk Observatory, the following definition was used to identify emerging chemical risks:

> Any occupational risks, both new and increasing risks, are so-called emerging occupational risks. By "new" it is meant that:
>
> - the risk was previously unknown and is caused by new processes, new technologies,
> - new types of workplace, or social or organizational change; or
> - a long-standing issue is newly considered as a new risk due to a change in social or public perceptions; or,
> - new scientific knowledge allows a long-standing issue to be identified as a risk.
>
> The risk is "increasing" if:
>
> - the number of hazards leading to the risk is growing; or
> - the likelihood of exposure to the hazard leading to the risk is increasing (exposure level and/or the number of people exposed); or
> - the effect of the hazard on workers' health is getting worse (seriousness of health effects and/or the number of people affected). [Brun et al. 2009]

Due to the complexity of the occupational risk assessment process posed by hazardous chemical agents, and, above all, by those newly identified or increasing threats, this chapter discusses issues related to methods and tools for use in the chemical risk assessment process. An attempt was made to assess the risks and key directions that should be taken to ensure the most reliable results of the chemical risk assessment in a work environment enabling proper risk management were indicated.

5.2 CHEMICAL RISK ASSESSMENT IN THE LIGHT OF STANDARDS AND REGULATIONS

It is the responsibility of every employer to carry out a risk assessment related to the presence of hazardous chemicals and their mixtures in the work environment. In the European Union, this is due to the requirements of the European Council Directives 98/24/EC and 2004/37/EC and the regional regulations implementing the requirements of these directives into the legislation of the Member States. The legal provision regulating in all the EU countries the issues of risk assessment created by those agents, including occupational risk, is the Commission Regulation (EC) No 1907/2006 of the European Parliament and of the Council of December 18, 2006, concerning the Registration, Evaluation, Authorization, and Restriction of Chemicals (REACH) [EC 1998; 2004; 2006].

The aim of the REACH regulation is to "improve the protection of human health and the environment from the risks that can be posed by chemicals, while enhancing the competitiveness of the EU chemicals industry". The basic activity enabling the achievement of this goal is carrying out risk assessment, including the risks posed by chemical substances in the workplace. Therefore, it is the responsibility of manufacturers and importers of chemical substances and their mixtures to carry out

> without prejudice to Article 4 of Directive 98/24/EC, chemical safety assessment and preparation of a chemical safety report for each substance on its own or as a component of a preparation or in a product, or for a group of substances, in accordance with para. 2–7 and REACH Annex I.

Annex I, "General provisions regarding the assessment of substances and the preparation of chemical safety reports", specifies methods for assessing and documenting the fact that the risks arising from the use of manufactured or imported substances are appropriately reduced during the production and use of these substances for their own use, and that users and employers, at a later stage of the supply chain, have the opportunity to properly reduce the risks. Since 2018, this entry also applies to nanoforms of chemical substances that have been registered and without prejudice to the requirements applicable to other forms of these substances [EC 2018].

Annex XII, "General provisions for downstream users for the assessment of substances and the preparation of chemical safety reports", sets out the way in which downstream users* (including employers), in the case of a use not included in the

---

* Art. 3 section 13. downstream user: means any natural or legal person established within the Community, other than the manufacturer or the importer, who uses a substance, either on its own or in a preparation, in the course of his industrial or professional activities. A distributor or a consumer is not a downstream user. Reimporter subject to exemption pursuant to the provisions of Art. 2 para. 7 lit. c) is considered as a downstream user.

provided safety data sheet, are to assess the risks arising from their use of the substance and its nanoforms, and prepare documentation showing that the risk is appropriately limited and the users down the supply chain can properly reduce this risk.

The US federal agencies that develop or implement health and safety regulations regarding workplace health and safety are Occupational Safety and Hygiene Administration (OSHA), Mining Safety and Hygiene Administration (MSHA), and the Environmental Protection Agency (EPA). Safety in working with chemical substances, including substances with nanoforms, is primarily covered by OSHA. The rules of conduct regarding occupational exposure to these factors are regulated, among others, by OSHA regulation: OHS services—29 USC 654, section 5(a)1, communication in case of danger—29 CFR 1910.1200, providing of personal protective equipment—29 CFR 1910.132, respiratory protection—29 CFR 1910.134, hazardous substances in laboratories—29 CFR 1910.1450, 79 FR 61383 OSHA Request for Information: Chemical Management and Permissible Exposure Limits (PELs) [OSHA 2012; 2014].

The EPA pursuant to the Toxic Substances Control Act (TSCA) of 1976 regulates the assessment of the risk posed to the environment and people (including workers) by chemical substances and their nanoforms.

According to the European and American regulations, occupational risk assessment related to the presence of chemical substances and their mixtures, including nanoforms, is a process of testing chemicals' dangerous properties and the conditions in which people work with them, in order to determine the hazards and possible harm that may appear, including the possible occurrence of individual sensitivity. This assessment should include hazards to the workers' health arising from the passage of the substance through the respiratory system, the skin, and the digestive tract, as well as the threats resulting from the substances' physicochemical properties. Its basic goal is to thoroughly understand the specific properties and potential hazards caused by chemical substances and their mixtures, and above all, to take appropriate actions aimed at reduction of the hazards [EC 1998; OSHA 2014].

Correct chemical risk management is possible thanks to the implementation of the legal provisions' requirements, as well as principles specified in the standards and standardization documents of the International Organization for Standardization (ISO), the European Committee for Standardization (CEN), or the regional standardization organizations, e.g. the German Committee for Standardization (KAN) or the Polish Committee for Standardization (PKN), regarding the assessment of exposure to chemical agents and dusts, including nano-objects, at the site and the associated risk. The key standards and technical documents in the chemical risk assessment are presented in Table 5.1.

In accordance with the legislation, guides, and standardization documents, the employer should assess and document the occupational risk associated with the work performed, apply the necessary preventive measures to control the risk, and inform workers about the occupational risk associated with their work [EC 1998; 2006; Lentz et al. 2019].

In cases where complete elimination of hazards is not possible, which is often the case in the presence of chemical substances and their mixtures at the workplace, appropriate technical and organizational solutions should be used, including

TABLE 5.1
Summary of EN and ISO Standards or Technical Documents Regarding the Risks Associated with the Presence of Chemical Substances in the Workplace

Range of Standards	ISO Standards and Technical Documents	CEN Standards
General strategy requirements for measuring inhalation exposure to chemicals to demonstrate compliance with occupational exposure limit values (OELs). These requirements apply to all stages of the measurement procedures, regardless of the physical form of the chemical agent (gas, vapor, aerosol), measurement procedures with separate sampling methods and analytical methods, and direct reading devices.	ISO 20581:2016 Workplace air—General requirements for the performance of procedures for the measurement of chemical agents	EN 689+AC:2019—Workplace exposure—Measurement of exposure by inhalation to chemical agents—Strategy for testing compliance with occupational exposure limit values
General rules for assessing dermal risk at workplaces. Standards provide guidance on dermal exposure assessment and commonly used dermal exposure measurement methods. The standards are designed to help occupational health professionals and researchers develop a strategy for assessing dermal exposure.	ISO-14294:2011 Workplace atmospheres—Measurement of dermal exposure—Principles and methods	CEN/TR 15278:2006—Workplace exposure—Strategy for the evaluation of dermal exposure CEN/TS 15279:2007—Workplace exposure—Measurement of dermal exposure—Principles and methods
Technical documents recommend using a control band approach to limit the risks associated with occupational exposure to nanoparticles and their aggregates and agglomerates (NOAAs), even if data on their toxicity and quantitative exposure assessment is limited or not available. Methods for managing occupational risk associated with inhalation exposure.	ISO/TS 12901-1:2012—Nanotechnologies—Occupational risk management applied to engineered nanomaterials—Part 1: Principles and approaches ISO/TS 12901-2:2014—Nanotechnologies—Occupational risk management applied to engineered nanomaterials—Part 2: Use of the control banding approach	No EN standards

The technical document provides guidelines for assessing the potential occupational risk associated with NOAA, enabling the identification of exposure route, exposed parts of the body and potential consequences of exposure in relation to dermal absorption, local effects, and inadvertent ingestion	ISO/TS 21623:2017—Workplace exposure—Assessment of dermal exposure to nanoobjects and their aggregates and agglomerates (NOAA)	No EN standards
The standard provides guidelines for measuring small particle aerosol fractions, in particular, in order to determine the numerical concentration of ultra-fine aerosols and nanoaerosols at workplaces using condensation particle counters (CPC).	No ISO standards	EN 16897:2017 (E) Workplace exposure—Characterization of ultrafine aerosols/nanoaerosols—Determination of the number concentration using condensation particle counters
The standard specifies the use of different indicators to measure inhalation exposure to NOAA during the basic and comprehensive assessment, as described within the EN 17058 standard.	No ISO standards	EN 16966:2019 Workplace exposure—Measurement of exposure by inhalation of nanoobjects and their aggregates and agglomerates—Metrics to be used such as number concentration, surface area concentration, and mass concentration
The standard provides guidelines for workplace exposure assessment by inhalation of nanoobjects as well as their aggregates and agglomerates (NOAA). Provides guidelines for the sampling and measurement strategies to be adopted and methods for assessing data.	No ISO standards	EN 17058:2019 Workplace exposure—Assessment of exposure by inhalation of nanoobjects and their aggregates and agglomerates
This technical document defines the principles of determination methods of OELVs and occupational exposure bands (OEBs) for nanomaterials and their aggregates and agglomerates (NOAAs).	ISO/TR 18637:2016 Nanotechnologies—Overview of the available frameworks for the development of occupational exposure limits and bands for nanoobjects and their aggregates and agglomerates (NOAAs)	No EN standards
The standard provides methods for determining the size and shape distribution of nanoparticles, by acquiring and evaluating scanning images from an electron microscope, as well as obtaining and reporting accurate results.	ISO/DIS 19749:2018 Nanotechnologies—Measurements of particle size and shape distributions by scanning electron microscopy	No EN standards

collective protection measures limiting the impact of these threats on the health and safety of workers. If the application of these solutions is not sufficient, the employer should provide workers with personal protective equipment suited to the type and level of hazards. REACH supports employers in activities related to chemical risk management by introducing a total ban on the use of the most dangerous substances or introducing quantitative restrictions based on Annex XVII: "Restrictions on the production, placing on the market, and use of certain hazardous substances, preparations, and articles".

The Directive 98/24/EC and regional regulations implementing its requirements in the European Union countries contain detailed data that should be taken into account when performing a risk assessment. In cases of exposure to several chemical substances, the joined risks of all agents should be assessed. The chemical risk assessment should also cover periods of work with an increased exposure, such as renovations and repairs of machinery and equipment used at the assessed workstations. It should also be carried out again if changes have been made to the composition of the used substances, technological processes, and when the medical knowledge regarding the impact of the agent on health has progressed. The risk associated with the occurrence of chemical substances and their mixtures appears both as a result of direct contact between a chemical agent and the human body, and as a result of the energy generated by a chemical reaction, such as fire or explosion. Due to the possibility of the two risk mechanisms arising from the presence of chemical agents, in accordance with the recommendations of the above directive, employers should use methods that ensure the assessment of occupational risks associated with inhalation and dermal exposure to chemical agents, as well as the possibility of an accident resulting from the presence of hazardous chemical agents [EC 1998].

5.3 MEASUREMENT METHODS FOR RISK ASSESSMENT

5.3.1 Inhalation Exposure

The main absorption route of chemical substances present in the work environment, in the form of gases, vapors, and aerosols, into a worker's body, is the respiratory system. Therefore, the basic element of occupational chemical risk assessment is inhalation exposure, and the most reliable and frequently used methodology is the quantitative measurement of concentrations of these factors in the workplace air, calculation of exposure indicators, and the determination of the relationship to the occupational exposure limits (OELs) at the workstation. Due to the possibility of exposing a worker to several or even a dozen or so chemical substances at a single workplace, and the variability of their determined concentrations over time, as well as the complexity of technological processes, this is the most expensive method, which is a serious limitation in its use.

5.3.1.1 Chemical Risk Assessment Criteria

The main obstacle in using the risk assessment method based on exposure measurement is the very limited number of chemicals for which the OELs are determined as assessment criteria. This is due to the systematic influx into the market of new chemicals, for which the established committees or government organizations responsible for the determination of the occupational exposure limit values for chemical substances in the workplace, in the European countries, as well as in the United States and Canada or Australia, are not able to determine legally obligatory OEL values. At the same time, they are not able to verify the existing values in accordance with the latest scientific achievements [OSHA 2014]. This situation is a result of the time-consuming and costly process of establishing OELs, requiring data analysis by large research teams. In many countries, e.g. in the United States, OELs, determined by the Occupational Safety and Health Administration (OSHA), the European indicative (IOELVs), or the binding (BOELVs) values, require careful analysis of toxicological and epidemiological results, quantitative health risk assessment along with other elements, such as availability of selective analytical methods for measuring exposure at a level adapted to the determined values, plus technical feasibility and economic feasibility [Lentz et al. 2019]. Therefore, the percentage of chemicals used with legally established OELs, despite over 70 years of practice in using these values for the occupational exposure assessment, is very low. The occupational exposure limit values based on the results of toxicological and epidemiological studies are established for only 1,341 substances [Deveau et al. 2015]. For example, in the United States, limit values have been established for 694 chemicals [ACGIH 2019]; in the European Union, for 184 substances [EC 2000; 2009; 2017a; 2017b; 2019a; 2019b]; and in Poland, for 566 substances in Ordinance of Ministry of Family, Labour and Social Affairs [Dz. U. poz. 1286, 2018]. In the absence of legally established OELs, the occupational exposure limits for a given substance may also be used as criteria for the occupational risk assessment, known as "derived no-effect level", so-called DNEL/DMEL, required pursuant to the regulation related to European registration, evaluation, authorization, and restriction of chemicals [EC 2006], set by chemical product manufacturers. DNELs are required for all compounds manufactured, used, or imported in the European Union in quantities exceeding 10 tonnes, regardless of data availability. They should be used with caution, as there is no peer review or public consultations in the process of their determination. In addition, although these values should be derived in accordance with detailed guidelines, there are no competence requirements for individuals responsible for determining them [Deveau et al. 2015]. A list of these values for 6,951 chemicals is available in the Gestis database [IFA 2019]. These numbers are still small compared to approximately 85,000 chemicals that are used worldwide [Krimsky 2017].

Lack of legally applicable OELs for thousands of chemicals makes it impossible to assess the occupational risk in relation to these values. This created a need to develop guidelines for chemicals that lack exposure limits. For this reason, researches from the National Institute for Occupational Safety and Health (NIOSH) have proposed the introduction of an Occupational Exposure Bands (OEB) process that will enable for the management of chemical risks posed by substances that do not have set OEL values [Lentz et al. 2019]. The process of creating OEBs is to use data on the toxicity of a chemical to determine ranges of concentrations in the air at the workplace that will provide the basis for the limitation of exposure to protect workers' health. NIOSH does not recommend using OEB for chemicals with determined OELs. Additionally, the process of determining occupational exposure bands, as well as the use of the resulting OEB values in the assessment and control chemical risks, is not legally required. Their use in practice is completely voluntary. The strategy of occupational exposure bands proposed by NIOSH is primarily aimed at filling the gap in access to criteria enabling risk assessment for many thousands of chemical substances and taking appropriate prevention measures to ensure safe working conditions, and thus to protect employees' health. The OEB process is defined as a systematic process of determining potential inhalation exposure ranges or exposure categories, using qualitative or quantitative health hazard information in relation to selected health effects' endpoints, and its purpose is to enable risk assessment and occupational risk management [Lentz et al. 2019].

The basis for the OEB determination process for individual substances proposed by NIOSH is the selection and assessment of information on critical health hazards, in order to classify the chemical substance, according to the posed hazard, to the appropriate OEB, from among five exposure categories A–E (band A—substances of low harmfulness, band E—the most harmful substances). Thus, OEBs determine the range of potential toxicity, expressed in the range of concentrations in the workplace air, for gases and vapors in ppm, and for particles in mg/m^3 (Table 5.2).

Thanks to the consistent and documented process of characterization of hazardous chemicals in accordance with the OEB process, the right risk management decisions can be made in the absence of OELs. This process can also be used to prioritize chemicals for which OEL values need to be determined. The process of creating bands makes it possible to identify threats to the worker's health that should affect decisions in the field of medical care. Classification into relevant occupational exposure bands does not take into account the impact of technical or economic feasibility on the exposure limitation, as well as the feasibility of selective exposure measurement at the level of the proposed concentrations.

Although the authors of the NIOSH report [Lentz et al. 2019] specify the air concentration ranges in the work environment for each exposure band that can be used in risk management, the concept of OEB differs from the concept of control bands (CB), where there is a direct connection of risk bands with appropriate prevention measures [HSE a; b] Paik et al. 2008; Zalk et al. 2011; Zalk and Heussen 2011]. For OEB, only data based on the information regarding the influence on human health or toxicological studies are used to determine the overall level of hazard potential and associated air concentration range for chemicals with similar hazard profiles. The process of creating OEB has been designed to use the results of scientific research and requires more complex input data, and thus ensures that more accurate results

TABLE 5.2
Range of Chemical Agents' Concentrations in the Air in Relation to Occupational Exposure Bands [Lentz et al. 2019]

OEBs	A	B	C	D	E
Gases and vapors, ppm	> 100	> 10 to 100	> 1 to 10	> 0.1 to 1	≤ 0.1
Particles/dusts, mg/m^3	>10	> 1 to 10	> 0.1 to 1	> 0.01 to 0.1	≤ 0.01

are obtained. The unique feature of the OEB process is primarily its three-level system that allows users with varying levels of expertise to use the process. The procedure for using each of the three levels is described in detail in the NIOSH report [Lentz et al. 2019] and is schematically presented in Figure 5.1.

Tier 1 is a qualitative approach to determine the scope of OEB, which can be carried out by occupational medicine doctors. Its rating is based on the Globally Harmonized System of Classification and Labeling of Chemicals (GHS) or the European CLP system, sometimes called EU-GHS. Chemicals that have the potential for serious or irreversible health effects at relatively low doses are in the D or E band, while substances that can cause reversible health effects at higher concentrations are in the C band.

Tier 2 is a semi-quantitative procedure to be carried out by professional hygienists with intermediate toxicological knowledge, based on the assignment of OEB based on the analysis of toxicological data derived primarily from reliable sources. In tier 2, users assign OEB based on the key resulting from literature sources. Chemicals are assigned to bands A–E. Level 2 users should be trained in the OEB process.

Tier 3 is an expert quantitative assessment that can be performed by toxicologists or experienced professional hygienists with advanced knowledge in the field of toxicology. This level includes a quantitative, comprehensive assessment of scientific information and requires the analysis and evaluation of all available data to determine the allocation of exposure bandwidth.

Information and detailed instructions, for individuals determining OEB at different levels, regarding the procedure for determination of the OEB concentration range for chemicals with different toxicity potential, can be found in the NIOSH report and in the "e-Tool" tab, available on the NIOSH website:wwwn.cdc.gov/NIOSH-OEB/. The e-Tool provides users with a fully automated way to classify chemicals at Tier 1 and a simplified method for classifying chemicals at Tier 2. The Tier 1 band is

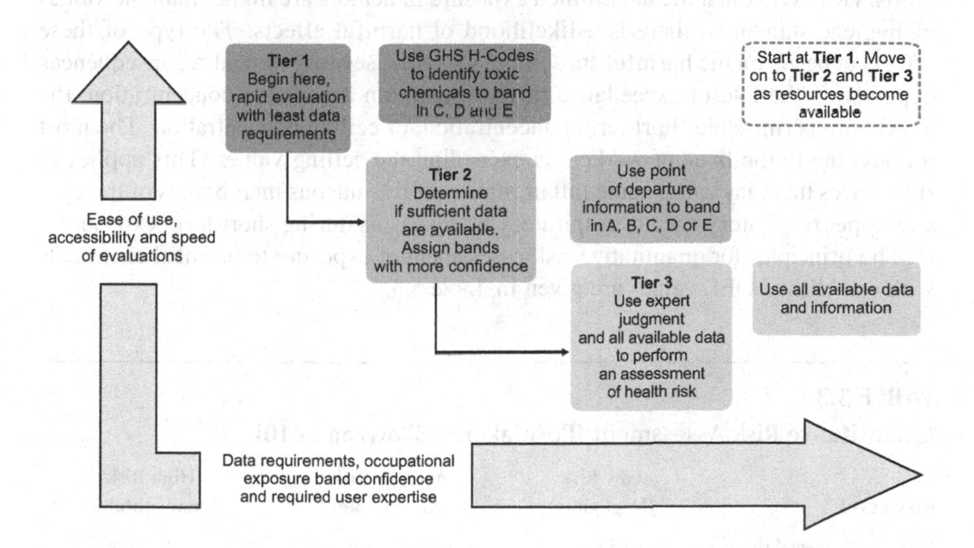

FIGURE 5.1 Three tiers of the occupational exposure banding process [Lentz et al. 2019].

generated automatically if the evaluated chemical is in the existing e-Tool database. The Tier 2 process requires obtaining relevant data from recommended sources. It can be time-consuming to correctly classify chemicals into the appropriate OEB band.

The procedure for determining the occupational exposure band developed at NIOSH is a tool particularly important in case of the chemicals creating the so-called emerging risks. It allows establishing the range of acceptable occupational exposure limits for these substances, based on information on all nine standard toxicological health endpoints, i.e. carcinogenicity, reprotoxicity, target organ toxicity, genotoxicity, respiratory sensitization, skin sensitization, acute toxicity, corrosivity and irritation to the skin, as well as eye damage and irritation. In addition, all exposure routes—inhalation, dermal, and oral—are included in the OEB determination process. The designated hazard categories are associated with quantitative exposure ranges.

5.3.1.2 The Principle of Quantitative Chemical Risk Assessment

General principles of occupational risk assessment, including the risk related to exposure to chemical substances, are provided by guides and standards, e.g. ISO 20581:2016, EN 689+AC:2019. These documents recommend, in cases where it is possible, the estimation of occupational risk based on the values characterizing the exposure, i.e. the OELs and the results of measurements of chemical concentrations in the work environment, and, in fact, in the workers' breathing zone.

The probability of occurrence of a harmful health effect resulting from the impact of these substances on a worker practically does not occur when the values of exposure indicators are less than or equal to the values of hygiene standards. Only in the case of carcinogenic compounds, cancer may occur with a probability of 10^{-6}–10^{-3}, at concentrations of this type of substances in the air at the level of hygienic standards. However, when the determined exposure indicators are higher than the values of hygiene standards, there is a likelihood of harmful effects. The type of these effects depends on the harmful substance, while the severity of health consequences depends on the fold of exceedance of the maximum acceptable concentration, the maximum permissible short-term concentration, or ceiling concentration. The most threatening to the lives of workers is exceeding the ceiling values. This applies to substances that may cause acute inflammation of the mucous membranes of the eyes and upper respiratory tract or respiratory tract spasms during short-term exposure.

The principles for quantitative risk assessment of exposure to harmful chemicals with established OEL values are given in Table 5.3.

TABLE 5.3
Quantitative Risk Assessment [Pośniak and Skowroń 2010]

Risk Level	Low Risk (Negligible)	Average Risk (Acceptable)	High Risk (Unacceptable)
Risk Assessment Criteria	<0.5 OELs	≥ 0.5 to 1 OELs	> 1 OELs

The adopted principle of occupational risk assessment does not apply to substances that are carcinogenic, mutagenic, and toxic to reproduction. When such substances are present in the work environment, the risk to all workers is always high if the exposure indicators are equal to or greater than 0.1 OELs. However, when the concentrations in the air are less than 0.1 OELs, the risk is estimated as average.

Due to the particular sensitivity of adolescents and pregnant and lactating women, which is reflected in the relevant provisions regulating their work [EC 1992; 1994; 2014a], the risk assessment for these groups of workers also deviates in some cases from the adopted principle. Occupational risk for these groups of workers exposed, among others, to carcinogenic, mutagenic, and reprotoxic substances, should be estimated as big, independently from their concentrations in the workplace air. Also, the risk is always great for young workers when working with sensitizing, corrosive, explosive, and flammable substances [EC 1994; 2014a].

5.3.1.3 Inhalation Exposure—Nanoparticles

Quantitative assessment of workers' exposure to nanoparticles of chemical substances emitted from nanomaterials and the associated risk is a very difficult issue. This is mainly due to the lack or incompleteness of data on the toxic effects of the engineered nanomaterials (ENMs) to explain the mechanism of their harmful effects on living organisms. Toxicological studies to date do not provide conclusive data on the harmful effects of engineered nanomaterials on human health and the environment. It is known that the toxicity of a substance depends mainly on its chemical structure. However, the particle size of the same chemicals can significantly change their toxic properties. The results of the most well-established studies indicate that the toxicity of insoluble particles of similar structure increases in inverse proportion to the diameter of the molecule and in direct proportion to their surface. In the work environment reality, in most of the cases, the numerical concentration of nanoparticles is high in comparison to the mass concentration. Additionally, many reports indicate that not the mass of particles present in the air, but their number and surface area are parameters that can be much more relevant in the case of health effects. Research results to date suggest that the toxicity of the nanoparticles depends primarily on their surface, and not on the number or mass in a given volume of air. Animal studies have shown that the specific surface area and low solubility of ENMs can cause inflammation in the lungs [Braakhuis et al. 2014]. Therefore, the parameter measured to assess the exposure to engineered nanoparticles should be surface concentration, not mass concentration. The numerical concentration is also considered as a more appropriate parameter alternative to the mass concentration, as indicated by the established recommended exposure limits (REL) based on the numerical concentration of the particles [van Broekhuizen et al. 2012].

No final decision has yet been made regarding the main parameter of nanoparticles, which should be the basis for occupational exposure assessment. However, studies of exposure to nanoparticles in the workplace air confirm the need for multidirectional measurements [Eastiake et al. 2016; Asbach et al. 2017; Debia et al. 2018]. Measurement strategies include measuring number, mass and surface concentrations of particles from nanotechnology processes and the background, using direct-reading instruments (DRI), morphological assessment and determination of the chemical composition of nanoparticles using scanning electron microscopy and energy-dispersive X-ray spectroscopy (SEM + EDS), as well as measurements of the concentration of nanomaterials in air samples taken in the workers' breathing zone using appropriate analytical techniques.

Due to the limited toxicological and epidemiological data, legally obligatory maximum allowable concentrations in the workplace air for nanomaterials have not been established yet. They are based on the lowest doses of these substances that do not cause harmful effects in humans, the so-called NOAEL values. Despite that, however, occupational recommended exposure levels (RELs) are determined for nanoparticles, their aggregates and agglomerates (discussed in Chapter 2), which can be used as criteria for assessing the occupational risk associated with exposure to engineering nanomaterials (ENMs). Another valuable source that can be used when choosing or establishing REL values is the ISO/TR 18637:2016 standardization technical document. It contains a review of the established REL values in several countries around the world, as well as the available methods and procedures for the determination of the occupational exposure limits and occupational exposure bands (OEB) for the produced nanoobjects, and their aggregates and agglomerates (NOAA), which can be used in the process of occupational risk assessment and decision-making in the field of occupational risk management.

Eastiake et al. [2016] proposed a measurement procedure to assess the occupational exposure to NOAA. This method called NEAT 2.0 is recommended to be used to characterize the inhalation exposure of workers in nanotechnology production plants. Its main purpose is qualitative assessment of the exposure to NOAA during the performance of professional activities in relation to the RELs established for the work shift (average time-weighted concentrations), performed in accordance with the principles of EN 689:2019 and ISO 2058:2016. It is recommended that samples on properly selected filters in the worker's breathing zone (PBZ) be taken throughout the work shift. In cases where there is an interest in the determination of exposure information related to a specific task, additional air samples can be taken in PBZ only during that particular task. To supplement the data from the PBZ samples, direct-reading instruments (DRI particle counters) are used to assess the work shift duration weighted average concentration of particles. Data from the DRI measurements provide information on peak emissions that could correspond to the exposure to NOAA. This data, in conjunction with the additional characteristics of nanoparticles, is used to introduce possible organizational changes in the work and strategies for using technical risk control measures. The NEAT 2.0 method recommends collecting integrated background data in real time throughout the entire sampling period, which allows for accurate determination of the concentration level of nanoobjects not derived from processes related to the production or use of ENMs. The basis of this method for quantitative exposure assessment is the collection of the PBZ samples for chemical analysis on filters, as well as the characterization (e.g. shape, size) and confirmation of the occurrence of ENMs of interest in the tested air by the SEM+EDS method.

NEAT 2.0 recommends parallel measurements using three portable, real-time DRI systems. The DRI results characterize process emissions by determining the number or mass concentration and the dimensional distribution of airborne particles.

DRIs are non-specific monitors and count all aerosol particles suspended in the air, not just those derived from the evaluated processes. Therefore, the NEAT 2.0 method recommends taking PBZ samples for analysis using more selective, time-based, laboratory-based methods to confirm and quantify NOAA exposure. In the case of exposure to carbon nanotubes (CNTs), carbon nanofibers (CNFs), or TiO_2, criteria and guidelines for occupational exposure assessment have been established. The NIOSH 7300 method can be used for the quantitative determination of TiO_2 in aerosols collected on filters. Concentrations of CNTs or CNFs can be determined using the NIOSH 5040 method [NIOSH 2019].

A similar strategy was used by Debia et al. [2018] during a study of occupational exposure to, among others, single-wall carbon nanotubes (SWCNTs), copper nanometals, CNFs, lithium iron phosphate, titanates, zinc NPs, multi-wall carbon nanotubes (MWCNTs), and nanoclays. Direct-reading instrument (DRI) kits were used to study the background and dimensional distribution of particles and to measure the number and surface of ENM particles at workplaces. Air samples for the NOAA laboratory analysis were taken in the PBZ breathing zone of the workers. Similar to the NEAT 2.0 method, the structure and chemical composition of ENMs was determined by transmission electron microscopy (TEM) or scanning electron microscopy (SEM). Quantitative determination of elemental carbon was performed using a thermal-optical analyzer, according to the NIOSH 5040 procedure. Nanometals were determined using inductively coupled plasma mass spectrometry analysis (ICP-MS) [NIOSH 2019].

A review article by a team of European researchers also confirms that the occupational exposure assessment of airborne nanomaterials at workplaces should include concentration measurement (numerical, surface) using direct-reading instruments and simultaneous sampling of individual PBZs for morphology and chemical composition tests using SEM + EDS as well as quantitative determination of NOAA in individual samples by instrumental methods, among others, ICP-MS [Asbach et al. 2017].

A comprehensive assessment of occupational exposure to NOAA allows for the collection of information that can be used to assess the level of exposure of employees to NOAA, identify sources of nanomaterial emissions and periods of increased emissions, and assess the effectiveness of exposure reduction and product handling practices. Integrated filter-based sampling is used to identify and quantify workers' exposure to ENMs. On the other hand, particle measurements using DRIs are used to examine background nanoparticle concentrations and NOAA concentration distribution during a working shift or measurement period, and assess the technical preventive measures used, e.g. ventilation.

5.3.2 Dermal Exposure

Dermal exposure is defined as the amount of a chemical substance that is present and absorbed by the outer layers of the skin and that can show various effects, i.e. systemic effects and/or local effects, on the skin surface. The assessment of occupational dermal exposure is one of the elements of risk assessment associated with the use of chemical substances and their mixtures at work. It is very important due to the fact that many substances irritate or sensitize the skin and mucous membranes, and some of them are absorbed through the intact skin, increasing the dose of the substance absorbed by inhalation into the body during professional activities. Quantifying dermal exposure is a very difficult task. There are no established permissible concentrations of chemicals deposited on the skin during work, so it is recommended to use qualitative methods to determine the risk associated with the effects of chemicals on the skin of the worker [Pośniak and Galwas 2007; Dobrzyńska and Pośniak 2018].

Explanation of the issue of the dynamic interaction of chemical environmental pollutants with human skin paved way for establishing conduct principles when assessing dermal exposure. Numerous publications on this topic [Brouwer et al. 2016; Jankowska et al. 2017a; McNally et al. 2019], and international and European standards, as well as the pre-normative documents listed in Table 5.2 (ISO-14294:2011 CEN/TR 15278:2006 CEN/TS 15279:2007), indicate ways and tools for assessment and limitation of occupational risk related to dermal exposure to chemicals, including those in the form of nanoparticles and their aggregates and agglomerates (ISO/TS 21623:2017). Conducting tests to assess the amount of substance deposited on the workers' skin and the amount absorbed by the intact skin under occupational exposure conditions is necessary for many substances. The results of these tests are, primarily, guidelines for the correct selection of personal protective equipment, and in the case of substances absorbed through the skin—help to assess the dose absorbed via this route, which sometimes significantly increases the dose of the substance absorbed by inhalation into the workers' bodies during professional activities.

Measurement methods of chemical substances used for dermal exposure assessment are being studied more and more. This issue was also addressed by the CEN Technical Committee TC 137, by developing a standardization document CEN/TS 15279:2007, in which the principles and methods of measurement for the quantitative assessment of occupational dermal exposure were proposed. A list of these methods is presented in Table 5.4.

The choice of method for measuring a chemical substance on the surface of the skin or work/personal protective measures depends on its physicochemical properties and deposited quantities. The dermal exposure assessment based on mass measurement currently faces various difficulties, while attempts are being made to use the results of these measurements to calculate and assess the amount of the substance absorbed by the body, due to the lack of appropriate limit values. The measurement results should be interpreted in relation to the strategy described in CEN/TR 15278:2007. Other methods for assessing dermal exposure include the artificial skin method and the biomonitoring of chemical substances or their metabolites in exhaled air, urine, and blood of employees, or other biological material.

The problem of assessing occupational skin exposure to NOAA during the production, processing, and use of nanomaterials is to some extent solved by the normative technical document ISO/TS 21623:2017. This document is primarily targeted at work hygienists and occupational safety specialists, and its task is to assist in recognizing the effects of occupational skin exposure and introducing appropriate prevention measures.

For the assessment of skin exposure to NOAA, an approach has been presented that assumes a gradual assessment of the workplace situation in a systematic way,

TABLE 5.4
Measurement Methods for Dermal Exposure Assessment of Harmful Chemical Substances

Measurement Principles/Sampling Technique	Sampling Method
Interception technique, i.e. interception of agent mass transport by use of a collecting media (e.g. α-cellulose paper, cotton, silk, flannel, paper impregnated with lanolin, aluminum foil, polypropylene, or polyurethane foam) placed at the skin surface or replacing work clothing during sampling time	Patche whole body
Removal technique, i.e. removal of the agent mass from the skin surface(i.e. the skin contaminant layer) at any given time	manual wipe tape-stripping hand wash hand rinse
Direct measurement, i.e. in situ detection of the agent or a tracer at the skin surface e.g. by image acquisition and processing systems, at a given time.	detecting of UV fluorescence of agent or added tracer as a surrogate for agent by video imaging. Attenuated Total Reflection FTIR, or using a light probe

Source: CEN/TS 15279:2007—Workplace exposure—Measurement of dermal exposure—Principles and methods

taking primarily into account the potential of exposure in relation to the likelihood of nanoparticle release and skin function disorders [Brouwer et al. 2016].

Stage 1 is the assessment of the workplace, taking into account toxicological hazards with respect to NOAA components and screening for the potential risk associated with skin exposure to insoluble NOAA and the performed work.

During Stage 2, based on observations of employees' behavior, it is indicated where there is a potential for dermal exposure (or accidental ingestion). These observations provide data on the frequency of contact with materials and surfaces of exposed body parts. At this stage, the use of DeRmal Exposure Assessment Method (DREAM) is recommended. Stage 3 of the assessment includes additional observations of the employees' behavior, which enable the evaluation of the correct use of working clothes and/or protective clothing and gloves, as well as contacts around the body (and thus accidental ingestion). Stage 4 involves a quantitative estimation of dermal exposure to NOAA. One method of characterizing and quantifying the skin exposure to NOAA, in the context of potential skin absorption, requires complex sampling (e.g. tape lifting methods) and detection methods (e.g. using an electron microscope to analyze the samples collected from the skin). In the event of high NOAA risk, which may dissolve in sweat, it is also recommended to assess the level of surface contamination (benches, tools, etc.) in the workplace, using the surface abrasion method. In addition, it is recommended to assess the exposure to these substances using biological monitoring (if available, e.g. As, Cr, Co, Ni in urine) for exposed workers as part of risk assessment [Marie-Desvergne et al. 2016; Larese et al. 2015; Li et al. 2016]. Stage 5—Assessment and review—involves setting risk priorities, formulation of an action plan for risk mitigation, documentation of the findings and the action plan (where applicable), and informing the involved employees about results and activities. The introduction of new equipment, substances, and procedures may lead to new threats. Therefore, risk assessment is reviewed at appropriate intervals and updated as necessary, including the assessment of skin exposure to NOAA (feedback loop).

Carrying out risk assessment according to the staged procedure contained in the standard is not easy and requires the collection and careful analysis of data on the used raw materials, process conditions, collective and personal protective measures used, and, above all, the knowledge of dermal exposure assessment tools and the ability to use them in practice. Much of this data can be found in the information annexes to the technical document ISO/TS-15279:2007. They contain data on industries associated with the use of nanomaterials or nano-enabled products, necessary to determine if the dermal exposure to NOAA occurs at the workplace, as well as methods to assess the skin condition of workers.

5.4 MEASUREMENTLESS METHODS FOR RISK ASSESSMENT

The use of quantitative methods for occupational exposure and risk assessment is limited due to the small number of occupational exposure limit values. Chemical risk assessment based on measurements of chemical substance concentration in the air, unfortunately, is a time-consuming and costly task, and generates health risks for the experts performing these measurements. Many small and medium enterprises do not have financial resources to carry out measurements. One way to deal with such problems is to conduct exposure assessments using models that are currently free and fairly easily available. Risk assessment results obtained using such models have greater uncertainty than results of exposure measurements, but may be useful when measurements are not possible for various reasons or to complement them. It is often necessary to integrate the actual (measured) and estimated, using non-measuring models, exposure values, or to rely solely on the model values to assess occupational exposure and risk. The models cover exposure variability better than measurements because the number of performed measurements is usually too low, due to the involved costs. The development of exposure assessment models began in the 1990s, and the first chemical risk assessment model developed in 1996 is still the basis for the development of new models. This model was based on a theory that occupational exposure may depend on three factors: internal emission of a substance, methods of dealing with a chemical substance and the impact of preventive measures, such as local mechanical ventilation or personal protective equipment [Landberg 2018].

In Europe, the use of measurementless models for occupational exposure assessment became necessary after December 2006, because of the adoption by the European Parliament and the EU Council of the Regulation (EC) No 1907/2006, which aims at systematic supervision of chemicals and provides necessary information for chemical risk assessment for most substances on the market in the community [EC 2006]. The main principle of the models' operation is a qualitative or measurement-free way of assessing the risks posed by dangerous chemical substances, especially those newly marketed, as well as those for which toxicological and epidemiological data is limited.

5.4.1 Control Banding Methods

The method of chemical risk management, the so-called Control Binding (CB), was developed in the 1970s of the last century, as a simplified tool for conducting qualitative risk assessment and taking measures aimed at protecting people in their workplace and living environment. CB has developed from a series of qualitative risk assessment strategies that began to appear in the 1970s. In the early 1990s, this method was implemented in the pharmaceutical industry to manage chemical risks at the workplace. In order to enable employers to comply with their obligation to assess the risks arising from the Directive 98/24 EC [EC 1998], the Health and Safety Executive (HSE) developed the CB method, known as Control of Substances Hazardous to Health Essentials [HSE a; b]. This tool was primarily developed for employers in the micro-, small-, and medium-sized enterprise sectors to support their risk assessment activities for all chemicals used in the workplace and to implement appropriate prevention measures to protect workers' health and ensure occupational safety. CB is a strategy for qualitative occupational risk assessment, which aims to ensure the safety of workers using new chemicals, where there is a lack of reliable toxicological and exposure data, and above all, the lack of data to determine acceptable levels of exposure in the workplace.

To determine the appropriate control risk strategy, exposure levels can be estimated based on the amount of used substance, the propensity to enter the air, volatility of liquids or dustiness of solids, and the frequency and duration of exposure. However, the potential threat is determined on the basis of the so called hazard statements (H phrases), which are used to classify the hazard associated with a given chemical substance in accordance with the recommendations of the CLP European Parliament Regulation [EC 2008], as well as the recommendations of the Globally Harmonized System of Classification and Labeling of Chemicals (GHS). CB methods are not used to predict the level of concentrations in the air, but only to know the risks posed by chemicals in the workplace and to plan and implement appropriate control measures.

The CB method is currently widely used all over the world as a practical tool for counteracting chemical hazards. It has been recognized by many national and international institutions, such as the International Labor Organization. Due to the rapid development of the CB model, opinions are increasingly emerging that the traditional risk-based measurement process will be less and less applicable in chemical risk assessment. This is due to the systematically growing number of potentially hazardous factors in the workplace with not fully recognized adverse effects on human health, for which it is not possible to establish acceptable exposure levels. The use of the CB method for chemicals creating new and increasing risks, to some extent currently solves the problem of assessing and limiting occupational risks for many employers, primarily small- and medium-sized enterprises. The principle of this method has been known for over 20 years and forms the basis for developing tools for chemical risk assessment, including nanomaterials.

5.4.2 Models for Qualitative and Quantitative Chemical Risk Assessment

Tools supporting the assessment of the risks of harmful chemical substances available on-line should meet the following criteria: they should be fast, easily accessible, cheap, simple to use, based on modern IT techniques, and, of course, meaningful and reliable, enabling credible estimation of the inhalation exposure without having to carry out measurement.

In the European Union, there is a wide range of models for exposure estimation and related occupational risks that can be used to predict the level of worker exposure without measuring the concentrations of chemicals at workstations. These models differ in objectives and their complexity. They are divided into two types: preliminary assessment models, the so-called first-level assessment models, and the more advanced and extensive second-level assessment models [ECHA 2016].

Tier 1 models are EASE, ECETOC TRA, EMKG-Expo-Tool, and MEASE.

The EASE model (Estimation and Assessment of Substance Exposure) is the first model to qualitatively assess the risk posed to humans and the environment by the existing and newly introduced substances, developed by the European Commission's Joint Research Center. Initially, it was the preferred model for the estimation of the worker inhalation exposure. It is a module implemented as part of the EUSES 2.1 model (The European Union System for the Evaluation of Substances), which is available free of charge and is friendly to all users. There is no information in the literature as to whether this model can be used with carcinogenic, mutagenic, or reprotoxic substances (CMR). The conducted validation tests show that the EASE program cannot be used uncritically instead of measurements, as it greatly overstates the results of concentration estimation. The EASE model is based on the results of real measurements collected in the countrywide National Exposure Database from 1986 to 1993, led by the Health and Safety Executive (HSE). It is believed that the reason for the overstatement of the anticipated concentration levels is that the concentrations of chemical substances in the air at work stations collected in this database were much higher at the end of the twentieth century than at present [Cherrie and Hugson 2005; Johnston et al. 2005; Creely et al. 2005; ECETOC 2012; 2017; Pośniak et al. 2015].

The EASE model was improved and became the basis for the development of other exposure prediction programs. The European Center for Ecotoxicology and Chemical Toxicology (ECETOC) developed the ECETOC TRA (Targeted Risk Assessment) model, based on the EASE model, and became a tool currently preferred by the European Chemicals Agency (ECHA) for estimating inhalation exposure of workers [ECETOC 2012; 2017]. The main difference between the EASE and the ECETOC TRA models is that in the case of EASE, the system is based on elements of a process category, while the ECETOC TRA uses application descriptors as a useful tool (a part of the information package), enabling the identification of relevant data for exposure estimation. ECETOC TRA is a general model for both respiratory and skin exposure, which does not require any specific training for the user. It is a model based on process category descriptors (PROCs) defined under REACH recommendations. Aerosol and mist exposure is not covered by ECETOC TRA and the results for these scenarios should be interpreted with caution. The results obtained by ECETOC TRAv3 represent the 75th percentile of the exposure distribution, and the results ECETOC TRA v2 represent the 90th percentile [Spinazzè et al. 2019].

To start the ECETOC TRA, regarding exposure assessment for workers, the following input data is required: specific substance data such as CAS number and name, molecular weight, physicochemical form, vapor pressure (volatility) of the liquid, and dustability of the solid. For each hazard category, a "generic reference exposure value" corresponding to the indicator reference value is determined, separately for inhalation and dermal route. The current version of ECETOC TRA allows for the use of both OELs and DNELs/DMELs (*Derived No Effect Level*) as a reference value. For the inhalation exposure, the estimated value is expressed in ppm or mg/m^3. The basic advantage of the ECETOC TRA model is its simple structure. However, it should be noted that the number of technological processes and operations covered by the model is very limited and insufficient for assessing risk at Tier 1, and the assessment does not take into account the amount of substances used in individual processes. Other comments and reservations relate to the choice of risk management measures. The ECETOC TRA model, as the only one among the analyzed models, allows for the estimation of the short-term concentration. This model is validated with regard to the estimated risk. The ECETOC TRA model can be used to assess exposure to carcinogenic, mutagenic, and reprotoxic substances (CMR) with extreme caution. This is due to the fact that specific CMR exposure may be outside the scope of the ECETOC TRA model (e.g. exposure to fibers). However, for simple substances such as benzene, the use of the ECETOC TRA model raises no objections [Pośniak et al. 2015].

The EMKG-Expo-Tool model is a part of a simple hazardous substance control system that was developed by the German Federal Institute for Occupational Safety and Health [BAuA 2014]. It is based on band risk management represented by the COSHH Essentials model for qualitative risk assessment in the workplace, assessing and confirming whether the risk is properly reduced. In contrast, the EMKG-Expo-Tool can be used as an instrument for the quantitative estimation of inhalation exposure levels and risk characterization (comparison of exposure levels with limit values such as OELs or DNELs). This model, as the only one among the tested models, takes into account the tonnage of the tested substance. It cannot be used to estimate exposure to mixtures of substances, because it treats the whole mixture as a pure substance. [Spinazzè et al 2019]. The EMKG-Expo-Tool model is not recommended for estimating exposure to CMR substances [Pośniak et al 2015; ECHA 2016].

The MEASE model combines the assumptions of the EASE and ECETOC TRA models with health risk assessment guidelines for metals. It is recommended for estimating occupational exposure to metals and inorganic substances by inhalation and dermal route. Like ECETOC TRA, MEASE utilizes use descriptors as a handy tool, a part of the information package, enabling the identification of an appropriate exposure estimation entry. Appropriate scenarios can be chosen from the selection lists by specifying the process category and selecting the REACH process descriptor (i.e. PROC). For PROCs 21–27a, the model estimates exposure based on measurement data from the metallurgy industry. It should be emphasized that the concentrations estimated by the MEASE model correspond to the concentration at the level of the 90th percentile, which means that 90% of the estimated concentrations are statistically lower or equal to the calculated value. The MEASE model is not recommended for use when estimating exposure to organic substances. However, this model, as one of the few, makes it possible to estimate exposure throughout the welding process, during which exposure to various carcinogens, such as chromium(VI) compounds, occurs [Jankowska et al. 2014; Lamb et al 2015].

Tier 2 models are STOFFENMANAGER® and ART. They are more powerful tools supporting enterprises in occupational risk assessment.

Stoffenmanager® (www.stoffenmanager.nl) is available as a free web application. In addition to the free version, a Premium Package is also available. The Stoffenmanager model was developed in the Netherlands by TNO Quality of Life to provide industrial plants with an effective risk assessment and mitigation tool. It is a model that enables quantitative estimation of inhalation exposure to vapors and dusts. It has been pre-validated by the authors for approximately 250 concentration measurement results. It turned out that Stoffenmanager maps real concentrations well; i.e. it is conservative enough. The Stoffenmanager® model, as the only one in the final estimation stage, creates an exposure distribution graph, which allows for reading of the concentration value for the 90th, 50th, or 25th percentile. Stoffenmanager has two specific groups of categories for estimating exposure to dusts generated as a result of processing a piece of material (a so-called massive object). However, they only cover raw materials of wood and stone. This model does not include categories related to the processing of metal elements. The Stoffenmanager model can be used to estimate exposure to CMR substances. Until recently, ECHA was in favor of the Stoffenmanager model developed in the Netherlands by TNO Quality of Life, as the leading model in the field of estimating first-tier exposure, but recently, the Stoffenmanager model is considered in the category of higher-level exposure assessment tools, i.e. intermediate tools between the first and second levels. This model requires more data and is more complicated [Marquart et al. 2008; Pośniak et al. 2015; ECHA 2016; Savic et al. 2017; 2018; Heussen and Hollander 2017].

Stoffenmanager® is an online tool designed to support small- and medium-sized enterprises (SMEs), but also large organizations in setting priorities and controlling the risks associated with the handling of hazardous substances in the workplace. Version 7.0. Stoffenmanager® is available in several languages, including Dutch, English, German, Finnish, Swedish, Polish, Spanish, and French. It is worth emphasizing that Stoffenmanager® is mentioned in the REACH guidelines [ECHA 2016] as a recommended risk assessment tool to ensure compliance with the EU legislation. This model enables the management of CMR substances as well as prioritizing the risks associated with inhalation and dermal exposure. In addition, it enables the assessment of inhalation risk (qualitative and quantitative methods) and comparison of results with limit values, as well as the calculation of the effects of the introduced preventive measures and their inclusion in the action plan, or automatic generation of workplace instructions. In Stoffenmanager®, the band risk assessment module is systematically improved. The previous method for selecting the hazard class based on product hazard statements based on the UK COSHH methodology has been verified and is currently based on a classification in accordance with the UN Globally Harmonized System (GHS). The program can also make a quantitative risk assessment using a quantitative and validated module for the assessment of aerosol and vapor exposure expressed in mg/m^3. The concentration is estimated for the so-called justified worst case, referred to as the 90th percentile. The exposure during the task can be compared with the OELs' value.

The next Tier 2 model is the ART, which is the most advanced tool for assessing exposure levels. ART is based on a mechanistic model combined with an empirical element associated with exposure databases. ART version 1.5 includes a Bayer element that can be applied to a mechanistic model to adjust the estimated exposure. The model provides estimates for the 75th or 90th percentile. The extension of the existing ART inhalation model is the dART model, a higher-order mechanical model for assessing dermal risk, which should be integrated with the existing ART programming platform. The model focuses on the exposure of hands to liquids with low volatility, i.e. liquid emitted into the air in the form of an aerosol (fog). It is a model that calculates the relative assessment of dermal exposure for a scenario tailored to measure dermal exposure in order to translate the results into quantitative exposure estimates [McNally et al. 2019].

The latest exposure and chemical risks estimation tool is the TREXMO model that integrates matrix exposure models—ART v.1.5, STOFFENMANAGER® v.4.0, ECETOC TRA v.3, MEASE v.1.02.01, EMKG-Expo-Tool, and EASE v.2.0. This tool is able to provide users with the most appropriate parameters to be used in a given exposure situation [Savic et al. 2016; 2019]. The purpose of this tool is to create one, user-friendly interface that helps users select appropriate parameters and use different exposure models to evaluate one scenario [Spinazzè et al. 2019].

5.4.3 Algorithm for Quantitative Risk Assessment of Carcinogenic, Mutagenic, Reprotoxic and Endocrine Substances

Based on the obtained research results of a project concerning a measurement-less assessment of occupational risk caused by carcinogenic and mutagenic substances implemented by the Central Institute for Labor Protection and the Institute of Occupational Medicine [Pośniak et al. 2015], as well as previous work in this area [Kupczewska-Dobecka et al. 2011; 2012; Jankowska et al. 2014; 2015], an algorithm of defined actions for the assessment of inhalation exposure to carcinogenic or mutagenic substances in the work environment was developed, the IT version of which was placed in the ChemPył knowledge base at the Central Institue for Labour Protection—National Research Institute (CIOP—PIB) website.

The algorithm (Figure 5.2) should facilitate and, in many cases, enable the identification of hazards related to the presence of carcinogenic/mutagenic substances in the work environment, especially in micro-, small-, and medium-sized enterprises. The use of this algorithm will provide valid data to keep records of work carried out in contact with carcinogenic/mutagenic substances and their mixtures, which should contribute to obtaining more reliable data on occupational exposure to carcinogens and mutagens, as well as statistics on occupational diseases and mortality caused by these dangerous factors.

The results of estimating the level of exposure to carcinogenic/mutagenic chemicals and their mixtures obtained with this algorithm, will, in practice, give employers

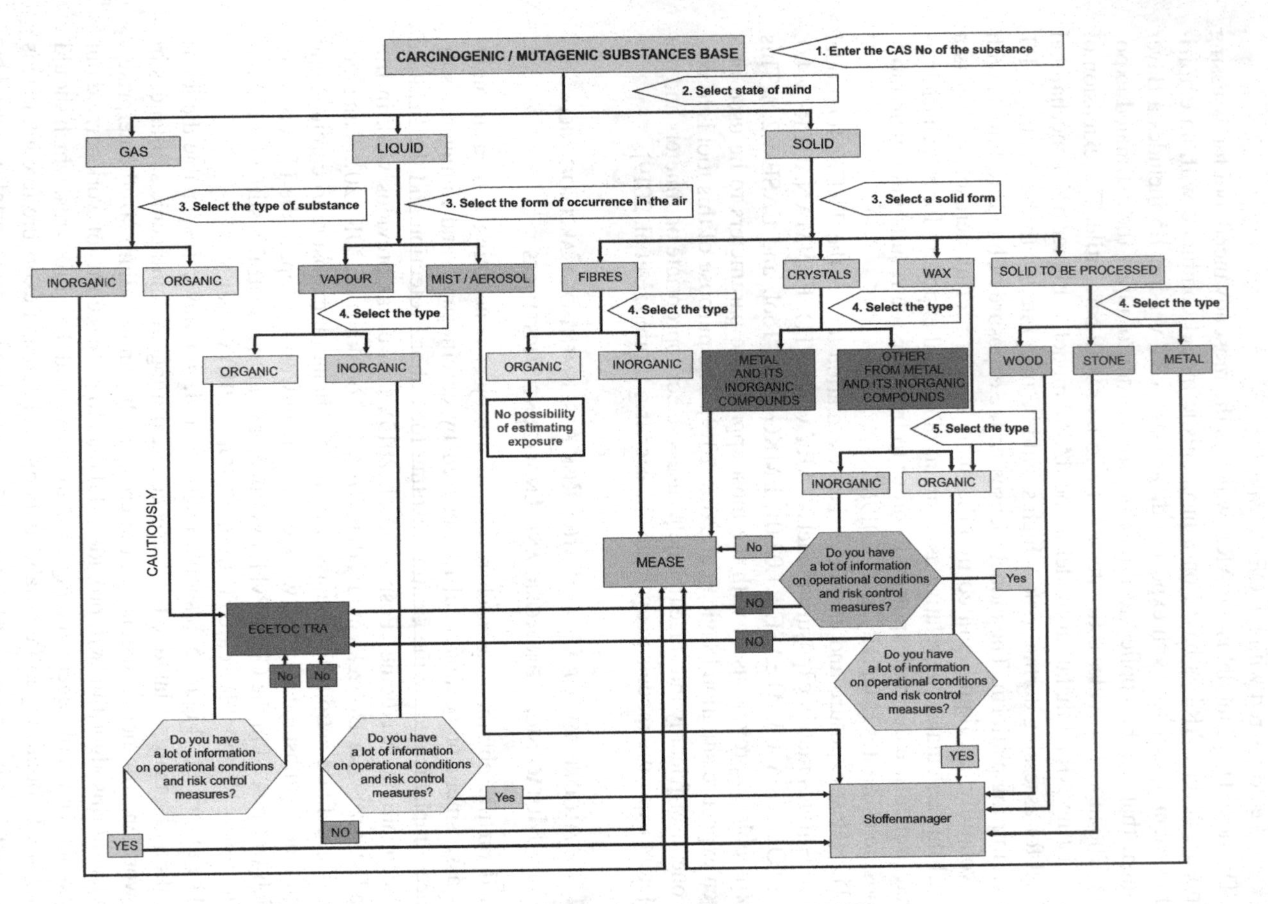

FIGURE 5.2 A diagram of steps during selection of a risk estimation model [Pośniak et al. 2015].

the basis to take appropriate actions (e.g. selection of appropriate risk mitigation measures, personal protective equipment), in order to minimize occupational risks, which, in the end, will contribute to the reduction of the number of occupational cancer diseases and mortality caused by them, reduction of the treatment costs, as well as the extension of professional activity of workers who have contact with carcinogens/mutagens at work.

5.4.4 Evaluation of Models for Occupational Exposure and Risk Assessment

The literature of recent years regarding the measurementless exposure and chemical risk assessment models focuses mainly on the publication of test results carried out to assess them [Riedmann et al. 2015; Savic et al. 2016; 2018; 2019; Lee et al. 2019a; 2019b; Spinazzè et al. 2019]. These studies often point to the poor quality of occupational exposure assessments' results for chemicals, obtained using REACH-recommended models. Therefore, validation of those models is the goal of many research projects.

Comparative studies of the ART and Stoffenmenager models carried out by Riedmann et al. [2015] suggested that ART may lead to more precise and effective actions, as a result of well-documented exposure scenarios. From the point of view of accuracy and stability, however, Stoffenmanager is a safer model. The results of the research conducted by the authors suggest that when the input data is uncertain or difficult to apply, the assessors should consider using the Stoffenmanager model, because it is the model through which the most reliable results are obtained compared to other models. In addition, Stoffenmanager also appears to be the most sustainable model for physical phenomena, such as emission from sources and dilution. However, the quality of the available data should ultimately determine the choice of the appropriate model.

Savic et al. [2017] have shown that different models can estimate exposure levels significantly different for the same exposure situations. This can lead to potentially dangerous conclusions about chemical risks. Assessment of the correlation and consistency of the ART, Stoffenmanager, and ECETOC TRA v3 models, based on 319,000 different exposure situations generated *in silico*, showed that the best correlated pair was Stoffenmanager–ART (R, 0.52–0.90), while the ART–TRA and Stoffenmanager–TRA were either lower (R, 0.36–0.69) or no correlation was found. Studies have shown that these models were more consistent when assessing vapor concentrations than aerosols and solids, in places close to the source of the emission than in distant places, as well as indoors than when exposed outdoors. Multiple linear regression analyses have shown a relationship between model parameters and relative differences between model predictions. Results of Savic et al.'s [2018] study highlight the need to use a multi-model strategy to assess critical exposure scenarios under REACH. In addition, in conjunction with occupational exposure measurements, they can also be used in future studies to improve the accuracy of exposure forecasting.

The latest research of Savic et al. [2019], comparing the results of exposure to chemicals performed by different assessors for the same SE exposure situations, using the Stoffenmanager, ART, and TREXMO models, showed large discrepancies between the results obtained by different people assessing the same exposure situations. The results of exposure level estimation using the TREXMO model were the most consistent.

Spinazzè et al. [2019] concluded, based on the review of literature, that assessment and validation studies of quantitative occupational exposure assessment models are still rare and only take into account a limited number of exposure scenarios. This review concerned Level 1 models: ECETOC TRA, MEASE, and EMKG-Expo-Tool, and Level 2 models: STOFFENMANAGER® and ART. Various studies describe ECETOC TRA as an insufficiently conservative model for Level 1, while the MEASE and EMKG-Expo-Tool seem to be conservative enough. STOFFENMANAGER® is the most balanced model. The ART model, with a medium conservativeness level, was recognized as the most accurate and precise. The results of the literature review showed that various methods of analysis were used to evaluate models, taking into account various reliability indicators (e.g. conservativeness, reliability, precision, accuracy, uncertainty), but most of the results refer to small-scale research. Accordingly, Spinazzè et al. [2019] recommend a meta-analysis of existing evidence regarding model reliability. In order to correctly assess the actual state of operation of the models, the planned tests should be conducted in a harmonized procedure, which will allow for consistent and comparable assessments. In addition to the reliability of the models, the compatibility of the results obtained by different assessors was tested. The results showed that inconsistencies between assessors can lead to very different risk assessment results. Therefore, it was proposed that the new TREXMO model should be a tool to support other models used for the measurementless assessment of occupational risk associated with exposure to harmful chemicals.

The assessment of Tier 1 models—ECETOC TRAv2 and TRAv3, MEASEv1.02.01, and EMKG-EXPO-TOOL—conducted by Lee et al. [2019a; 2019b] on 53 exposure situations (ES) to chemicals or during tasks, has shown that descriptors of process categories PROC 10 and PROC 15 in ECETOC TRA should be modernized, due to the low level of conservatism. In addition, it was recommended to re-evaluate the algorithms for liquids with medium and high vapor pressure, professional and industrial fields, and the variant without the use of local ventilation. In the case of EMKG-EXPO-TOOL, it was found that this tool was generally highly conservative when assessing exposure to volatile liquids with a vapor pressure of VP > 10 Pa, while it was less conservative when used for liquids with a high VP. It was recommended to re-assess the model in the case of exposure to very volatile liquids. In general, EMKG-EXPO-TOOL generated more conservative results than TRAv2 and TRAv3 for liquids with high vapor pressure. This finding stems, at least partly, from the fact that EMKG-EXPO-TOOL only considers pure substances and not mixtures of chemical substances. The Stoffenmanager®v4.5 and the Advanced REACH Tool (ART) v1.5, two higher-order exposure assessment tools, were evaluated by determining accuracy and reliability based on 282 exposure measurements from 51 exposure situations (ES). In this study, only the results of tests for liquids with a vapor pressure VP > 10 Pa with sufficient number of exposure measurements (n = 251 at 42 ES) were used. For VP > 10 Pa liquids, Stoffenmanager® v4.5 proved to be quite accurate and stable. In contrast, for the ART model, areas for improvement in some of the activities and input parameters were identified. Therefore, it is suggested that ART developers review the assumptions regarding exposure variability within the tool, to improve the quality of exposure estimation results. Additionally, further validation tests are required for both tools.

Heussen et al. [2019] commented on the research results on the evaluation of the TREXMO model and the integrated Stoffenmanager® 4.0, ART, MEASE, EMKG-EXPO-TOOL, EASE, and ECETOC TRAv3 models submitted by Savic et al. [2019]. They believe that all critical remarks about Stoffenmanager® result from the assessment of the outdated version of the model algorithms and the use of an unapproved Excel file, instead of the original Stoffenmanager® model. Therefore, the results of the research by Savic et al. [2019] do not provide information on differences in exposure estimation results obtained between Stoffenmanager users. The assessment should be repeated on the original Stoffenmanager® interface using the latest algorithms or on the version available on the new Stoffenmanager®Research website, operating in parallel to the website, www.stoffenmanager.com. Heussen et al. [2019] believe that the size of the surveyed group of people conducting the assessment should be increased. Only then can conclusions be drawn regarding the variability between the results obtained by different individuals performing exposure assessment using the Stoffenmanager® model.

5.4.5 Models for Occupational Risk Assessment of Nanomaterials

The use of nanomaterials in various areas of the national economy, on the one hand, contributes to the production of goods facilitating human life and improving their comfort; on the other hand, it can cause a threat to the health of employees engaged in the production processes, in the event of improper management of risk related to the exposure to hazardous chemical substances and their mixtures. Therefore, it is necessary to identify, assess, and limit the risks associated with workers' exposure to these hazardous materials. Estimated occupational risk assessment methods, designed specifically for nanomaterials, are helpful to employers and specialists in chemical risk management.

Currently, basic methods to solve the problem of occupational risks assessment posed by nanomaterials are, similar to the traditional chemical substances: bulk, control banding methods (CB methods), taking into account the physicochemical properties and the specificity of the interaction of nanoparticles with living organisms. These methods are increasingly used in the European Union, the United States, and Canada. In the last decade, several nano-specific CB methodologies have been developed for various types of work environments (e.g. small and large industrial sectors and laboratories), and each of these instruments has its strengths, weaknesses, and limitations [Paik et al. 2008; Ostiguy et al. 2010; Duuren-Stuurman et al. 2011; 2012; Schneider et al. 2011; EC 2014a; Dimeou and Emond 2017]. The Technical Committee (TC 229—Nanotechnology) of the International Organization for Standardization (ISO) has developed a standardized technical document ISO/TS 12901-2:2014 with a CB method for ENMs. The International Labor Organization, International Health Organization, as well as the European Agency for Safety and Health at Work recommend CB methods as, in fact, the only tool used to evaluate these new and incompletely known hazardous factors in the work environment. The most important of those tools are CB Nanotool, Precausiory Matrix, ANSES CB, NanoSafer, the Guidance, and Stoffenmanager Nano.

5.4.5.1 CB Nanotool-2

This method was developed by Paik S. and Zalk D. and has been widely used for 11 years to assess and control the occupational risk posed by ENMs [Paik et al. 2008; Pośniak et al. 2012]. CB Nanotool-2 is intended for use by both OSH experts and people without experience in chemical risk assessment. CB Nanotool-2 is a relatively simple tool based on rules established by COSHH Essentials. The basis of this method is the collection of data on the toxicological and physicochemical properties of the nanomaterials (NM) and parent substances (PM) used to obtain them, as well as information characterizing the potential exposure to nanoparticles. This data is used to determine the numerical values of the overall severity index, i.e. the potential harmful effects in the state of health of workers exposed to nanoparticles, and the total indicator of the likelihood of these effects. Based on 15 severity factors specifying the possibility of NM absorption by the respiratory system and deposition in its various regions, as well as local and systemic effects, the total severity index is determined. Seventy percent of the overall severity index relates to the assessment of nanomaterials, and the remaining 30% to the assessment of parent materials, because literature data indicates greater toxicity of nanomaterials. When determining the probability index, first of all, data on the number of exposed workers, and the time and frequency of exposure is taken into account, including the assessment of dustability for particles and the haze level for nanoparticles used in suspension. After determining the value of 15 indicators of the consequences severity, which may be the result of the impact of nanoparticles and parent material on the employee, the overall severity index of the consequences is determined as the sum of individual indicators. Similarly, the total probability index is determined as the sum of the factors. The maximum value of the combined severity index and probability index is 100. These indicators in the range of 0–25 indicate a low possibility of severe consequences and a very unlikely occurrence of occupational risk posed by nanomaterials, 26–50 indicate medium consequences and unlikely occupational risks, 51–75 point at high possibility of consequences and probable occupational risk, and 76–100 suggest very high possibility of consequences and highly likely occupational risk.

The level of occupational risks associated with exposure to ENMs is estimated on a four-point scale on the basis of the cumulative indicators of the severity of consequences and probability for a given operation.

5.4.5.2 Precautionary Matrix

Method published by the Swiss Federal Office of Public Health and the Federal Office for the Environment in the guidelines for preventive measures during the production and use of synthetic nanomaterials is addressed not only toward the employers and managers of OSH or to the services responsible for environmental protection, but also to the users of products containing nanomaterials [Höck et al. 2018]. Its use makes it possible to assess the needs of non-specific preventive measures, and also helps to identify risks in the use, production, and utilization of ENMs. This method calculates the indicator characterizing the preventive measures (V), taking into account the factors characterizing the nanoparticles (N), specific working conditions (S), and determining the potential hazard (W) and the level of exposure (E). The respective N, S, W, and E coefficients are assigned appropriate numerical weights using tables developed by the authors of the method, and after calculating the indicator characterizing the needs in terms of preventive measures, the rated workplace is classified into two categories: A ($V \leq 20$) and B ($V \geq 21$).

5.4.5.3 ANSES—CB

This method was dedicated by the French Agency for Food, Environmental and Occupational Health & Safety ANSES primarily to the work environment and people involved in the reduction of the chemical risk associated with the occurrence of nanomaterials in the work environment [Ostiguy et al. 2010]. As in the previous methods of "control banding", the level of risk is a derivative of the potential size of nanoparticle emissions and potential threats, but the risk is categorized under five levels, which correspond to the appropriate technical solutions of collective protection measures. ANSES uses five hazard bands that range from HB1 (very low; no significant health risk) to HB5 (very high; serious risk that requires an expert to perform a comprehensive risk assessment). Exposure ranges are assigned on the basis of emissions and can be grouped according to the following four levels: EP1 (solid), EP2 (liquid), EP3 (powder), and EP4 (aerosol). These five hazard ranges and four exposure ranges are directly linked to five levels of risk reduction (CL) that are ranked from the lowest CL1 (natural or general mechanical ventilation) to the highest CL5 (full isolation and maintenance by a specialist required).

5.4.5.4 Stoffenmenager Nano-tool

It is a method for the estimation of occupational risk associated with exposure to nanomaterials, developed at the Netherlands Organization for Applied Scientific Research—TNO. The basis for risk assessment in this method are the algorithms characterizing employee exposure to nanoparticles emitted from nanomaterials, and the threat resulting from their toxic effects.

Determining the level of exposure requires collecting data on the course of the evaluated process and working conditions at the assessed workplace. This information will allow the assigning of appropriate numerical values to the factors determining: time and frequency of nanomaterial use, collective and individual protection measures, concentration of nanoparticles conditioned by small or large distance from the source of nanoparticle emission, and background concentration, which are the basis for calculation of the indicator characterizing exposure to nanomaterials. Finally, based on this indicator, the end exposure indicator is assigned [Duuren-Stuurman et al. 2012]. In the Stoffenmanager Nano-tool method, hazard categories (from A to E, as in the CCOSH Essentiale) are determined on the basis of current literature data characterizing the toxicity of nanomaterials and their physicochemical properties. In the absence of any information on the harmful effects of nanomaterials, the categorization of hazards should be carried out by experts, based on their professional judgment. The final risk level is determined on the basis of a combination of five hazard categories (A–E) and four exposure categories. The hazard categories proposed by TNO for the most commonly used nanomaterials are very helpful for occupational risk assessors. Metal nanoparticles and their chemical substances with an aerodynamic diameter ≤ 50 nm, such as silver, iron, gold, lanthanum, tin, titanium dioxide, cerium dioxide, zinc oxide, cobalt and iron, silicon dioxide, aluminum trioxide and antimony, aluminosilicates and polymers, and dendrimers were classified under hazard Category D, and with a diameter > 50 nm—into hazard Category C. Hazard Category E was assigned to lead, quartz, and cristobalite nanoparticles, and all nanoparticles derived from parent material classified as carcinogenic, mutagenic, toxic for reproduction or sensitizing. This highest hazard category has also been assigned to other nanomaterials derived from materials with unexplored toxic properties.

The method developed by the TNO was the basis for the development of an IT tool under the name *Stoffenmenager Nano module 1.0.* This tool, which is easy to use and requires no expert knowledge, is generally available in Dutch and English on the TNO website (www.tno.nl). It is recommended by the Dutch Labor Inspectorate for risk assessment, primarily for entrepreneurs of small plants where nanomaterials are used. A case study published on the portal of the European Agency for Safety and Health at Work on the application of the *Stoffenmanager Nano* program [EU-OSHA 2012] by a small plant confirms that the "control banding" method has allowed employers using products containing nanoparticles to comply with the recommendations of the Dutch labor inspection, arising from the recommendations of the Directive 98/24 EC [EC 1998], in the area of occupational risk management associated with these specific chemical substances.

In all the presented methods of occupational risk assessment related to the exposure to nanoparticles emitted from nanomaterials, functioning under the name "control banding methods", the determined risk level is an indication for taking appropriate preventive measures, i.e. to achieve the main objective of occupational risk assessment.

5.4.5.5 The EC-nanotool

A very important tool for managing the risk associated with the occurrence of produced nanomaterials, and a guide developed by the Ministry of Employment, Social Affairs & Inclusion [EC 2014 b], intended for employers and employees. It provides clear and simple advice for the proper organization of safe workplaces, where new, not fully recognized factors present a threat to health and the environment. As in other methods, the procedure starts (Stage 1) with identifying all ENMs produced or used at a given workplace. If it is uncertain whether the substance or mixture used is a nanomaterial or whether it contains nanomaterial, it is advised to ask the supplier for this information, and if it is not possible to obtain this data, it should be assumed that ENMs are present. Next (Stage 2), in order to fully and better understand the risks arising from the occupational exposure to MNMs, data characterizing physicochemical and toxic properties should be collected. Information needs to be collected on the size distribution and content of MNMs in nanoproducts, characterizing nanoparticle shapes, their carcinogenicity, mutagenicity, reproductive toxicity (MNMs or parent products alone), water solubility, environmental stability, density, and physical state (including determination of, in case of solids, whether they are free nanoparticles or agglomerates or aggregates). The next stage (Stage 3) is to determine health hazards. Four hazard categories 1, 2a, 2b, and 3, have been specified in this method. In Stage 4, for each produced, processed, or used ENM, all activities during which there is a potential possibility of releasing nanoparticles while working with them should be identified, providing the amount of ENM used, the form in which they are emitted into the air (dust/fog), the duration of the activity, the frequency of activities performed, and the number of workers exposed. In the next stage, (Stage 5), the probability of exposure to ENMs is determined on the basis of process conditions and preventive measures against hazards. In the procedure discussed above, exposure can be classified into three categories. In Stage 6, appropriate prevention measures are selected.

5.4.5.6 Evaluation of Measurementless Models for Assessing the Occupational Risk Posed by Nanomaterials

Currently, CB methods are considered a large group of methods for detecting and categorizing the risks associated with the use of engineering nanomaterials [Brouwer 2012; Dimou and Emond 2017; Zalk et al. 2011; Zalk and Heussen 2011], primarily in research facilities for nanomaterials and plants using or producing ENMs, but in small quantities. The *ANSES*, *Stoffenmanager Nano*, and *Nanotool-2* methods are based on the classification of chemical substances and mixtures, relatively easy to find in safety data sheets and product data sheets. Unfortunately, these commonly used sources of information on hazardous chemicals for engineered nanomaterials do not contain complete information about their physicochemical properties and harmful effects on human health and the environment. A detailed assessment for 513 nanomaterials showed that most ENMs' safety data sheets do not contain sufficient data to assess hazards [Lee et al. 2013]. This condition will be systematically improved, due to the inclusion of nanomaterials in the registration obligation resulting from the amendments to the REACH regulation [EC 2018]. According to Derk H. Brouwer [2012], from the point of view of the level of detail and applicability in a wide range of professional activities, *ANSES models and Stoffenmanager Nano* seem to be the most suitable tools for assessing the risk associated with nanomaterials. CB *Nanotool-2 and ANSES* are rather recommended for use by experts with experience in risk assessment, while *Stoffenmanager Nano* is directed to users who do not need to deal with these issues professionally.

It should be noted that, despite various disadvantages, estimation methods are increasingly used to assess the occupational risk associated with the production and use of nanomaterials in the world. It is currently not possible to evaluate the results of using these tools because they have not been validated. The assessment results obtained with the use of CB *Nanotool-2* were compared at the stage of implementation of the method, with the assessment carried out by industrial hygienists, and good compliance was obtained [Zalk et al. 2009]. The authors of this CB tool, developed specifically for the needs of the qualitative assessment of the occupational risk posed by ENMs, 11 years after its development and practical application, have published studies aimed at the quantitative validation of the CB Nanotool [Zalk et al. 2019]. This tool is widely used to limit threats posed by ENMs by various institutions around the world, including the Safe Work Australia and Workplace Health and

Safety Queensland, which believe that it is currently the best method to manage risks related to the exposure to ENMs, primarily in the workplace. The *CB Nanotool-2* (version 2.0) validation tests were carried out at the Lawrence Beceley National Laboratory using the strategy developed by NIOSH, a so-called Nanoparticle Emission Assessment Technique (NEAT), using various techniques to quantitatively measure nanoparticles in the air at workplaces. Condensation and optical particle counters were used to perform real-time measurements in parallel with the study of nanoparticles collected on filters, using appropriate analytical techniques for their quantitative determination and confirmation of their presence in SEM+EDS microscopic analysis. In total, the results of quantitative measurements for 28 processes or activities using ENMs were compared with the CB Nanotool-2 qualitative risk assessment method. All quantitative results confirmed that the proposed preventive measures based on the qualitative assessment of *CB Nanotool-2* were sufficient to prevent exposure of workers to ENMs at levels exceeding the established exposure limits or background levels. The results of validation tests confirm the reliability of the results obtained with the use of *CB Nanotool-2* and recommend the use of this tool to reduce the exposure of employees to ENMs, in the absence of quantitative results of air monitoring.

In the coming years, systematic improvement in the methods for quantitative (measurements) exposure assessment of engineered nanomaterials will allow for the validation of all CB methods used so far. Additionally, the dynamic research conducted in many European scientific institutions, and, above all, in the United States, constantly provides new toxicological data on ENMs, as well as on occupational exposure to nanoparticles emitted from nanomaterials or the effectiveness of preventive measures used, and will fuel the improvement in qualitative methods of occupational risk assessment related to ENMs.

5.5 CHEMICAL RISK ASSESSMENT IN THE WORKPLACES—EXAMPLES OF GOOD PRACTICES

5.5.1 Reportoxic Susbtances

5.5.1.1 Occupational Risk Assessment of Exposure to Sodium Warfarin during the Production of an Anticoagulant Drug

Quantitative assessment of occupational exposure to sodium warfarin (WAS, CAS No. 129-06-6) during the production of an anticoagulant drug in Polish pharmaceutical plants was impossible due to the fact, that for WAS there were no established limit values in Polish legislation and European directives [EC 2000; 2009; 2017 a, b; 2019 a, b], as well as these values have not been specified by American hygienists [ACGIH 2019]. These values were also not found in the available safety data sheets developed by the manufacturers of the substance.

Sodium warfarin is a substance that poses a threat to human health, according to the classification criteria of the CLP Regulation (Table 5.5, Figure 5.3).

Using NIOSH Occupational Exposure Banding e-Toolwwwn.cdc.gov/NIOSH-OEB and taking into account reprotoxic properties, REL values for warfarin sodium have been established (Table 5.6).

TABLE 5.5
WAS Classification and Labeling CLP [EC 2008]

Classification	Acute toxicity, oral cat. 2., Harmful for reproduction cat. 1.B
Labeling—hazard statements	H300: Fatal if swallowed [Danger Acute toxicity, oral] H310: Fatal in contact with skin [Danger Acute toxicity, dermal] H330: Fatal if inhaled [Danger Acute toxicity, inhalation] H360: May damage fertility or the unborn child [Danger Reproductive toxicity] H372: Causes damage to organs through prolonged or repeated exposure [Danger Specific target organ toxicity, repeated exposure]

FIGURE 5.3 Warfarin sodium (WAS) pictograms.

TABLE 5.6
REL for Warfarin Sodium Established using NIOSH OEB e-Tool

Chemical Name: Sodium warfarin

Synonyms: 3-(a-Acetonylbenzyl)-4-hydroxycoumarin sodium salt

CAS#: 129-06-6

Overall Recommended Band

E

Particle Range: ≤ 0.01 mg/m³

Endpoint		Hazard Code	Hazard Category	Endpoint Band
Acute Toxicity	Oral	300	1	E
Skin Corrosion/Irritation				
Serious Eye Damage/ Eye Irritation				
Respiratory and Skin Sensitization				
Germ Cell Mutagenicity				
Carcinogenicity				
Reproductive Toxicity		360	1a	E
Specific Target Organ Toxicity		372	1	E
Overall Recommended Band				E

Then, a method was developed for determining WAS concentrations in the air of the workplace for assessing sanitary and hygienic conditions. The method consists of retaining WAS on a glass fiber filter, washing it out with distilled water, and analyzing the obtained solution using high performance liquid chromatography with spectrophotometric detection. The determination limit of the method is 0.001 mg/m³.

5.5.1.2 Chromatographic Determination Conditions

Based on the results of the tests, it was determined that the use of a polar column with hydrophobic silica gel of 5 µm granulation and 10 nm pore size, and modified with cyanopropyl functional groups, as well as mixtures of acetic acid, water, and acetonitrile as the mobile phase, enables the determination of warfarin sodium salt at 0.020 mg/ml. The analysis of the UV spectra of the test substance indicated that the most suitable analytical wavelength is $\lambda = 260$ nm.

Chromatographic conditions—mobile phase: acetic acid/acetonitrile/distilled water 1:25:74 (v/v); mobile phase flow: 1.5 ml/min; column temperature: -30°C; analytical wavelength: 260 nm. The tests were carried out using a liquid chromatograph with an Elite LaChrom DAD detector from Merck-Hitachi, equipped with a 25 cm Lichrospher 100 CN µm column of an internal diameter of 4.6 mm.

5.5.1.3 Air Sampling

Air samples should be taken in accordance with the principles specified in the EN 689:2018 norm. At the sampling point, the sampler for the inhalation fraction, equipped with a glass fiber filter, passing 480 1 of test air with a constant flow rate of 2 l/min. Samples stored in the refrigerator remain stable for 14 days. The results of validation of the developed WAS determination method are presented in Table 5.7.

TABLE 5.7
Validation Data of the WAS Determination Method in the Air at Workplaces

Validated Parameters	Value
Measuring range	0.16–16.0 µg/ml (0.001–0.1 mg/m³)
Limit of Detection (LOD)	0.015 µg/ml
Limit of quantification (LOQ)	0.046 µg/ml
Correlation coefficient (R)	0.9998
Total test precision, Vc	6.753
Total uncertainty of the method U_T [%]	14.351
Expanded uncertainty U [%]	28.702

TABLE 5.8
Measurement Results and Exposure Assessment

Employee	Exposure Time per Week [h]	REL [mg/m³]	Determined Concentration [mg/m³]	Exposure Indicator C_{wt} [mg/m³]	REL Multiplicity
Weighing					
Employee I	2	0.01	1.083	0.054	5.4
Employee II	2	0.01	2.284	0.115	11.4
Granulation and drying					
Employee I	6	0.01	0.155	0.023	2.3
Employee II	6	0.01	0.105	0.016	1.6
Mixing, sieving					
Employee I	6	0.01	0.100	0.02	2.0
Employee II	6	0.01	0.064	0.013	1.3
Tabletting					
Employee I	6	0.01	Not Detected	-	-
Portioning					
Employee	8	0.01	0.03	0.006	0.6

5.5.1.4 Measurements at the Pharmaceutical Plant during the Production of the Anticoagulant Drug

Inhalation exposure assessment has been carried out in accordance with the generally applicable principles specified in the EN 689+AC1:2019 standard. Measurements of WAS concentrations were carried out in the breathing zone of workers engaged in the production process, by individual dosimetry. To determine the concentrations of these substances in the air, in order to assess the occupational exposure, an analytical procedure, especially developed for the drug production process, was used as described above. The results of warfarin sodium salt measurements and exposure assessment during the production are presented in Table 5.8.

5.5.1.5 Result of Occupational Risk Assessment

The occupational risks posed were assessed as high for a worker performing work related to weighing, granulating, and drying, as well as mixing and sieving. On the other hand, when it was portioned, the risk was at a medium level.

5.5.1.6 Control risks

A glovebox was installed for work related to weighing, mixing, and sieving. The worker was equipped with respiratory protection equipment, overalls, gloves, and protective boots (Figure 5.4).

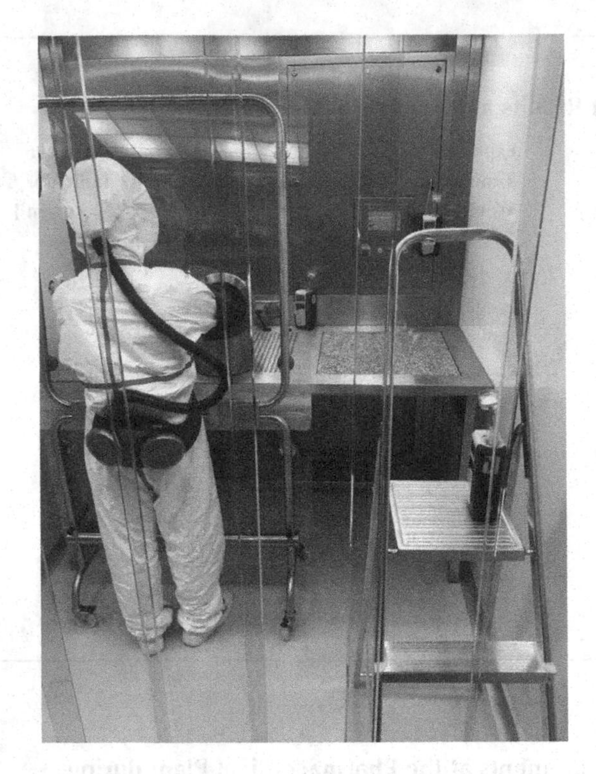

FIGURE 5.4 Work at the mixing and weighing workstation.

5.5.2 Engineering Nanomaterials

5.5.2.1 Occupational Risk Assessment of Exposure to Zirconium Oxide Nanopowder Calcia Stabilized, ZrO_2+8%CaO

NOAA exposure tests were conducted by Jankowska during the operation of the test workstation for the determination of the dustiness of the nanopowder—*Zirconium Oxide Nanopowder Calcia Stabilized, ZrO_2+8% CaO* [Jankowska et al 2017b]. The processes were carried out in a fume hood, and consisted of: transferring of the tested nanopowder into an open rotary drum in two processes, namely, 2g in the P-1 process and 6g in the P-2 process, carrying out a process with the nanomaterial in a hermetically sealed drum, pouring the nanopowder out from the open drum, and removing the residue with a brush. The volume of the air drawn from the fume hood was about 1,000 m^3/h.

5.5.2.2 Measuring Equipment

Measurement of NOAA concentrations for occupational exposure assessment was performed using a DiscMini Meter (Miniature Diffusion Size Classifier), Matter Aaerosol, enabling determination of the number concentration of 10–700 nm particles, in the range of 10^3–10^6 particles/cm^3, as well as the average particle size in the range of 10–300 nm with about 10–15% accuracy.

NOAA samples for microscopic and gravimetric tests were obtained on an MPS probe designed for acquiring air samples on TEM grids. The Ultra High Resolution FE-SEM

TABLE 5.9
Program of Tests [Jankowska et al. 2017b]

Symbol of Process	Information on Research	Test Duration
T-1	Background in the fume hood	10:20–10:58 (38min)
P-1	Pouring of the *ZrO_2+8% CaO* nanopowder into an open rotary drum (2g), carrying out the process with nanomaterial in a hermetically closed drum, pouring the nanopowder from an open drum and removing residues with a brush	10:59–11:06 (7min)
T-1,2	Tests between processes P-1 and P-2	11:07–11:29 (22min)
P-2	Pouring of the *ZrO_2+8%CaO* nanopowder into an open rotary drum (6g). Performing work to ensure the hermeticity of the drum. At this time, the drum with the powder inside was usually left open	11:30–12:54 (24min)
T-2	Tests after P-2 process	12:55–15:10 (135 min)

microscope was used to determine the shape of the particles, and the EDS-Bruker Quantacs 400 system was used to assess the chemical composition of the particles.

NOAA samples for microscopic studies were acquired with a volume flow of 1 dm^3/min, on the MPS sampler designed for obtaining air samples on TEM grids. Air samples using the sets (MPS + GilAir Plus) were obtained. To determine the shape and chemical composition of the particles, an Ultra High Resolution FE-SEM scanning electron microscope with the EDS-Bruker Quantacs 400 system was used. To determine the mass concentration of respirable fraction of the particles, the following measurement set was used: FSP10 probe (Cyclone Sampler), SG 10-2 pump (Personal Air Sampling Pump), GSA, with the air flow rate of 10 dm^3/h for 360 min.

5.5.2.3 Online Measurements and Sampling at the Work Station

NOAA exposure tests were carried out for real conditions of nanomaterials' application during the process of determining the dustiness of *ZrO_2+8%CaO* nanopowder by means of a small rotary drum. The test workstation was placed in a fume hood, which provided 1,000 dm^3/h air exchange and was equipped with HEPA filters.

The measuring system was located in a room, outside the fume hood, in the area of the breathing zone of the workers handling the conducted processes. The testing program is presented in Table 5.9.

5.5.2.4 Results of the Test and NOAA Exposure Evaluation for Occupational Risk Assessment

Number concentrations of NOAA (using the DiscMini device) were determined before, during, between, and after processes P-1 and P-2, which lasted 7 and 24 min, respectively. Measurements were carried out between processes for over 20 min, which helped the influence of the P-1 process on the concentration taken as the reference concentration before the P-2 process. The air temperature during the tests in the room was in the range of 24.3–26.8°C, the relative humidity of the air was in the

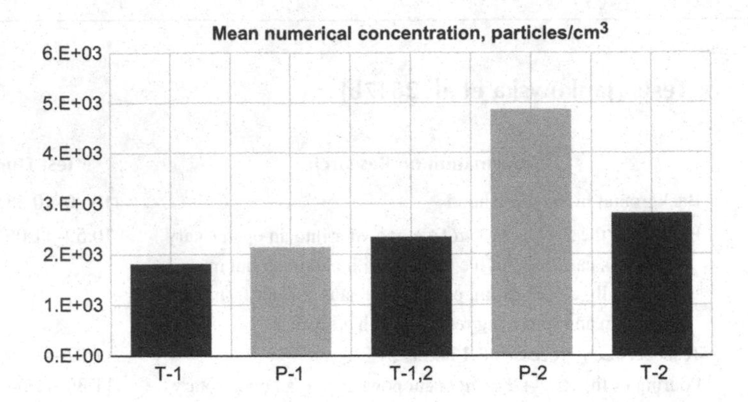

FIGURE 5.5 Average values of the numerical concentrations of particles obtained using DiscMini before, during, between, and after processes P-1 and P-2 [Jankowska et al. 2017b].

range of 22.9–24.7%, and the air speed was in the range of 0.00–0.02m/s. The results of average number concentrations before, during, between, and after processes P-1 and P-2 are depicted in Figure 5.5.

NOAA exposure indicators (Wp) for processes are presented and defined as ratios of the average number concentrations of particles during processes to the average number"background" concentrations or concentrations before the P-2 process in Table 5.10.

TABLE 5.10
Numerical Exposure Indicators for NOAA for Processes P-1 and P-2 [Jankowska et al. 2017b]

Information on Research	DiscMini1, particles/cm³	Wp
Background in the fume hood (T-1)	1 808	W_{P-1}=1.18
Pouring of *the* ZrO_2+8%*CaO* nanopowder into an open rotary drum (2g), carrying out the process with nanomaterial in a hermetically closed drum, pouring nanopowder from an open drum and removing residues with a brush (P-1)	2 136	
Tests between processes (T-1 and T-2)	2 330	W_{P-2}=2.09
Pouring of *the* ZrO_2+8%*CaO* nanopowder into an open rotary drum (6g). Performing work to ensure the hermeticity of the drum. At this time, the drum with the powder inside was usually left open **(P-2)**	864	

T—"background" or before the processes,
P—processes with ZrO_2+8%*CaO* nanopowder
Wp—NOAA exposure indicator
The numerical exposure factors for NOAA during processes P-1 and P-2 were 1.18 and 2.09, respectively.

5.5.2.5 Confirmation of the Presence of NOAA Particles in the Air from the Evaluated Process

Microscopic images and the chemical composition of particles taken from the air before ("background" T-1), during, and between processes P-1 and P-2, with *Zirconium Oxide Nanopowder Calcia Stabilized, ZrO_2+8%CaO*, are shown in Figures 5.6 and 5.7.

The results of the chemical composition analysis of the samples acquired during the processes with *ZrO_2+8%CaO* nanopowders confirmed that in the air, apart from the "background" particles, few particles of nanopowders used in the processes were suspended (Figure 5.7). No *ZrO_2+8%CaO* particles were found in samples obtained before nanopowder processes (Figure 5.6).

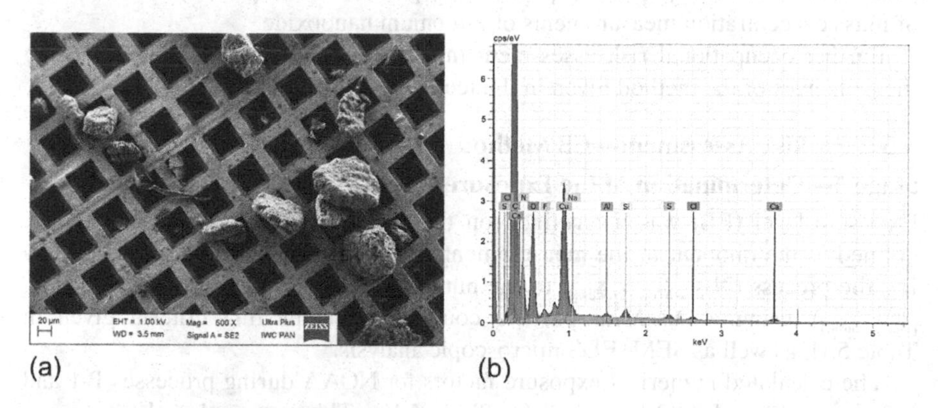

FIGURE 5.6 500x magnification microscopic images (a) and chemical composition—S, Cl, C, Ca, N, O, F, Cu, Na, Al, Si (b) of particles obtained from the air before ("background") the process with *Zirconium Oxide Nanopowder Calcia Stabilized, ZrO_2+8%CaO* [Jankowska et al. 2017b].

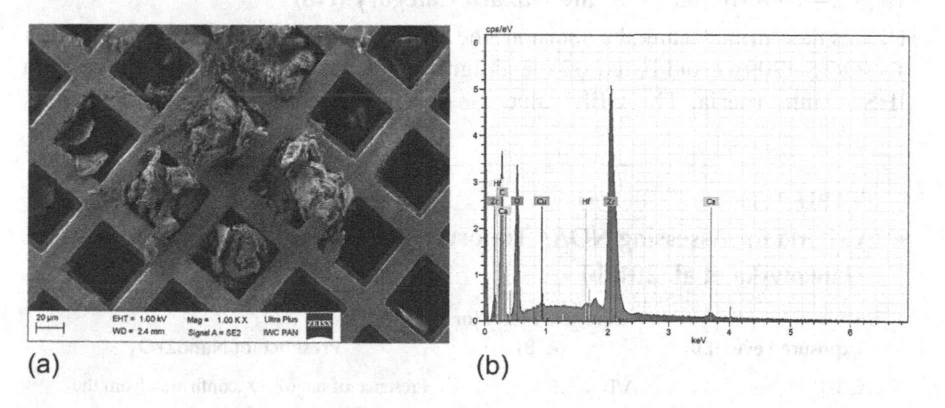

FIGURE 5.7 1.00KX magnification microscopic images (a) and chemical composition—Zr, Hf, C, Ca, O, Cu (b) of particles obtained from the air during and between processes P-1 and P-2 with Zirconium Oxide Nanopowder Calcia Stabilized nanopowder, *ZrO_2 +8%CaO* [Jankowska et al. 2017b].

Taking the criteria proposed by Brouwer et al. [2013] for the assessment, it can be concluded that during the P-2 process, the employee was highly likely exposed to NOAA of the tested nanomaterial, *Zirconium Oxide Nanopowder Calcia Stabilized, ZrO_2+8%CaO*, while during the P-1 process, exposure to these NOAA was only less likely.

5.5.2.6 Mass Exposure Indicator

The determined mass of particles on the measuring filter, in relation to the volume flow and time of sampling (360 min), constituted the basis for determination of the exposure index for the tested measuring cycle. During the measurement cycle with processes P-1 and P-2, the mass exposure index for the respirable fraction was 0.0175 mg/m^3. Due to the lack of a limit value for zirconium dioxide NOAA, the content of zirconium dioxide in the respirable fraction was not determined. It is currently not possible to assess occupational exposure during the reviewed processes, on the base of mass concentration measurements of zirconium nanooxide.

Further occupational risk assessment in this workstation was conducted using the principles of the method given in the technical document ISO/TS 12901-2:2014.

5.5.2.7 Risk Assessment—CB Method

Stage 1—Determination of the Exposure Level

Exposure level (EB) was determined on the basis of exposure indicators (A/B), defined as the quotient of the measurement results of the number of particles during the process ($N_{\text{DiscMini [Process]}}$) to the number of background particles ($N_{\text{DiscMini [Background]}}$), using a DiscMini particle counter and employing criteria given in Table 5.11, as well as SEM+EDS microscopic analysis.

The calculated numerical exposure factors for NOAA during processes P-1 and P-2 were 1.18 and 2.09, respectively (Table 5.10). This means that the exposure level for the process was EB-2 for P-1and EB-4 for P-2 (Table 5.8). The presence of nanoZrO_2 was confirmed by SEM+EDS (Figure 5.7).

Stage 2—Determination of the Hazard Category (HB)

HB was determined using the decision tree given in Chapter 7, Hazard Band Setting of ISO/TS 12091-1 and based on the assignment of risk groups in accordance with GHS health criteria. The OEL value for bulk Cyrconium and its compounds is

TABLE 5.11
Criteria for Assessing NOAA Exposure in the Work Environment [Jankowska et al. 2017b]

Exposure Level (EB)	Exposure Indicator (A/B)	Presence of NanoZrO_2
EB-1	A/B < 1.1	Presence of nanoZrO_2 confirmed from the SEM+EDS test process
EB-2	A/B ≥ 1.1 and < 1.5	
EB-3	A/B ≥ 1.5 and < 2	
EB-4	A/B ≥ 2	

TABLE 5.12
Risk Assessment Based on HB and EB According to ISO/TC 12901-2: 2014

Hazard Categories (HB)		Exposure Levels (EB)			
		EB-1	EB-2	EB-3	EB-4
	A	low	low	low	medium
	B	low	low	medium	high
	C	low	medium	medium	high
	D	medium	medium	high	high
	E	medium	high	high	high

TABLE 5.13
The Result of Risk Assessment in Relation to Processe s P-1 and P-2

Information on Research P-processes with ZrO_2+8%CaO Nanopowder	EB	HB	Risk Level
P-1	EB-2	C	medium
P-2	EB-4	C	high

5 mg/m^3. These compounds are not dangerous substances within the meaning of the Regulation (EC) No 1272/2008 and GHS. Based on these data and the decision tree, the hazard category C (HB–C) was determined for the *ZrO_2+8% Ca* nanopowder.

Stage 3—Risk Assessment

The level of occupational risk during the dustiness testing process of *Zirconium Oxide Nanopowder Calcia Stabilized, ZrO_2+8%CaO* was determined on the basis of the principles set out in chapter 7.6 of the ISO/TS 12901-2: 2014 technical document (Table 5.12).

It was established that during the work carried out to determine the dustiness of nanomaterials, the risk during the P-1 process was low, and that during the P-2 process was high (Table 5.13).

Stage 4—Applied Preventive Measures

The results of the assessment of occupational risk posed by the *ZrO_2+8%CaO* nanomaterial during work on the test workstation for the determination of the dustiness of nanomaterials indicate that the employee should be equipped with personal protective equipment: respiratory protection equipment with a P3 class filter, protective goggles, protective gloves, and a protective suit type 5—dust protection clothing.

5.5.2.8 Conclusion

Using the occupational risk assessment method for NOAA recommended by ISO/TC 12901-2:2014 and determining the level of exposure to NOAA, based on measurements of NOAA "background" concentrations and during the work carried out determining the dustiness of nanomaterials, it was established that the risk during the P-1 process was medium, and that during the P-2 process was high.

Due to the lack of a determined limit value for zirconium dioxide NOAA, occupational exposure cannot be assessed during the studied processes, based on mass concentration measurements.

5.6 SUMMARY

Correct assessment of the risk associated with exposure to chemicals and their mixtures posing threat to health and sometimes also human life in the workplace is a very difficult task. Despite several decades of intensive research focused on both measurement and measurementless methods for conducting exposure and chemical risk assessment, only methods based on the quantitative measurement of concentrations of individual chemical substances in the workplace can be considered credible, and the assessment criterion—allowable exposure level values. These methods are often considered the gold standard of exposure and chemical risk assessments [Landberg 2018]. First of all, due to the huge number of chemicals commonly used worldwide, but also for economic reasons, the use of such methods is currently very limited. Only for 1,351 chemical substances OEL values are determined on the basis of documented, reliable results of toxicological studies [Deveau et al. 2015]. The additional problem is the huge variability of exposure to chemical substances during various working shifts for different employees performing identical tasks, which often forces repeated measurements to ensure that the obtained measurement results are reliable.

The introduction of measurementless models in the 1990s for the qualitative and quantitative estimation of exposure to chemical substances was a way to solve the problems associated with measurements in the work environment. The intensive development of these methods took place in 2006, after the REACH Regulation was implemented in Europe. The chemical risk management process has been using a variety of models presented in this chapter for more than 15 years, especially for chemicals that do not have established OEL values.

In principle, all the latest literature data indicates that the biggest problem is to check the correctness of models' operations, their accuracy and stability, as well as the reliability of the obtained exposure estimation results. This can be achieved through their validation, consisting of the simultaneous determination of the estimated level of exposure using the model and comparison with measurement data obtained in the workplace for the tested chemical tested substances. The results of previous studies in this area are not conclusive. Most often, they recommend continuing validation studies of models, taking into account larger number of real measurement data, as well as larger groups of people performing chemical risk estimations with the same models. The main problem in the model validation process is the availability of reliable measurement databases, covering the concentrations of chemical substances in the work environment that can be used as reference for comparison with model concentrations. In the case of chemicals that create new and increasing risks, mostly concerning the nanomaterials, such databases, in principle, do not exist. Certain hopes are held for the NECID database created by PEROSH containing the results of NOAA measurements.

From the analysis of the available literature on the reliability of occupational exposure estimation models, some issues arose concerning the need to improve the models, their performance, or to extend their scope of application. The results of the occupational exposure assessment for chemical substances using the REACH recommended models are very often subject to a large error [Lee et al. 2019a; 2019b; Riedmann et al. 2015; Savic et al. 2016; 2017; 2018; 2019]. It is therefore necessary to constantly develop, regulate, and re-calibrate tools for exposure estimation and chemical risk assessment.

REFERENCES

ACGIH [American Conference of Governmental Industrial Hygienists]. 2019. *TVLs and BEIs based on the documentation of the threshold limit values for chemical substances and physical agents & biological exposure indicies.* Cincinnati, OH: American Conference of Governmental Industrial Hygienists.

Asbach, C., C. Alexander, S. Clavaguera et al. 2017. Review of measurement techniques and methods for assessing personal exposure to airborne nanomaterials in workplaces. *Sci Total Environ* 603:793–806. doi: 10.1016/j.scitotenv.2017.03.049.

BAuA [Bundesanstalt für Arbeitsschutz und Arbeitsmedizin]. 2014. EMKG-Expo-Tool 2.0: User Guide. https://www.baua.de/EN/Topics/Work-design/Hazardous-substances/REACH-assessment-unit/pdf/User-Guide-EMKG-Expo-Tool.pdf?__blob=publicationFile&v=2 (accessed January 15, 2020).

Braakhuis, H. M., M. V. D. Z. Park, I. Gosens et al. 2014. Physicochemical characteristics of nanomaterials that affect pulmonary inflammation. *Par Fibre Toxicol* 11. doi: 10.1186/1743-8977-11-18.

Brouwer, D. H. 2012. Control banding approaches for nanomaterials. *Ann Occup Hyg* 56(5):506–514. doi: 10.1093/anhyg/mes039.

Brouwer, D. H., B. van Duuren-Stuurman, M. Berges et al. 2013. Workplace air measurements and likelihood of exposure to manufactured nanoobjects, agglomerates, and aggregates. *J Nanopart Res* 15:2090. doi: 10.1007/s11051-013-2090-7.

Brouwer, D. H., S. S. Paan, M. Roff et al. 2016. Occupational dermal exposure to nanoparticles and nano-enabled products, Part 2: Exploration of exposure processes and methods of assessment. *Int J Hyg Environ Health* 219(6):503–512. doi: 10.1016/j.ijheh.2016.05.003.

Brun, E., I. Lenda, C. Blotiere et al. 2009. *Expert forecast on emerging chemical risks related to occupational safety and health.* Luxembourg: Office for Official Publications of the European Communities.

Cherrie, J. W., and G. W. Hughson. 2005. The validity of the EASE expert system for inhalation exposures. *Ann Occu Hyg* 49(2):125–134. doi: 10.1093/annhyg/meh096.

Creely, K. S., J. Tickner, A. J. Soutar et al. 2005. Evaluation and further development of EASE Model 2.0. *Ann Occup Hyg* 49(2):135–145. doi: 10.1093/annhyg/meh069.

Debia, M., G. L'Espérance, C. Catto et al. 2018. An assessment of methods of sampling and characterizing engineered nanomaterials in the air and on surfaces in the workplace. Study Research Project R-1009. https://www.irsst.qc.ca/media/documents/PubIRSST/R-1009.pdf?v=2020-01-03 (accessed December 31, 2019).

Deveau, M., C. P. Chen, G. Johanson et al. 2015. The global landscape of occupational exposure limits: Implementation of harmonization principles to guide limit selection. *J Occup Environ Hyg* 12:127–144. doi: 10.1080/15459624.2015.1060327.

Dimou, K., and C. Emond. 2017. Nanomaterials, and occupational health and safety: A literature review about control banding and a semi-quantitative method proposed for hazard assessment. *J Phys Conf Ser* 838:1–15. doi: 10.1088/1742-6596/838/1/012020.

Dobrzyńska, E., and M. Pośniak. 2018. Ocena ryzyka związanego z występowaniem substancji chemicznych na stanowiskach pracy: Metody bezpomiarowe w bazie CHEMPYŁ. *Przemysł Chemiczny* 97(4):633–638.

Duuren-Stuurman, B., S. R. Vink, D. H. Brouwer et al. 2011. Stoffenmanager nano: Description of the conceptual control banding model. TNO Report V9216. The Netherlands: Zeist. https://nano.stoffenmanager.com/public/factsheets/STMNano_%20Bevindingendocument.pdf (accessed January 14, 2020).

Duuren-Stuurman, B., S. R. Vink, K. J. M. Verbist et al. 2012. Stoffenmanager Nano version 1.0: A web-based tool for risk prioritization of airborne manufactured nano objects. *Ann Occup Hyg* 56(5):525–541. doi: 10.1093/annhyg/mer113.

Dz. U. poz. 1286 (z późn. zm.). 2018. Rozporządzenie Ministerstwa Rodziny, Pracy i Polityki Społecznej z dnia 12 czerwca 2018 r. w sprawie najwyższych dopuszczalnych stężeń i natężeń czynników szkodliwych dla zdrowia w środowisku pracy. http://isip.sejm.gov.pl/isap.nsf/download.xsp/WDU20180001286/O/D20181286.pdf (accessed December 19, 2019).

Eastlake, A. C., C. Beaucham, K. F. Martinez et al. 2016. Refinement of the nanoparticle emission assessment technique into the Nanomaterial Exposure Assessment Technique (NEAT 2.0). *J Occup Environ Hyg* 13(9):708–717. doi: 10.1080/15459624.2016.1167278.

EC [European Commission]. 1992. Council Directive 92/85/EEC of 19 October 1992 on the introduction of measures to encourage improvements in the safety and health at work of pregnant workers and workers who have recently given birth or are breastfeeding (tenth individual Directive within the meaning of Article 16 (1) of Directive 89/391/EEC). https://eur-lex.europa.eu/legal-content/EN/TXT/PDF/?uri=CELEX:31992L0085&from=EN (accessed January 15, 2020).

EC [European Commission]. 1994. Directive 94/33/EC of 24 June 1994 on the protection of young people at work. https://eur-lex.europa.eu/legal-content/EN/TXT/PDF/?uri=CELEX:31998L0024&from=EN (accessed January 15, 2020).

EC [European Commission]. 1998. Council Directive 98/24/EC of 7 April 1998 on the protection of the health and safety of workers from the risks related to chemical agents at work (fourteenth individual Directive within the meaning of Article 16(1) of Directive 89/391/EEC). https://eur-lex.europa.eu/legal-content/EN/TXT/PDF/?uri=CELEX:31998L0024&from=EN (accessed January 15, 2020).

EC [European Commission]. 2000. Commission Directive 2000/39/EC of 8 June 2000 establishing a first list of indicative occupational exposure limit values in implementation of Council Directive 98/24/EC on the protection of the health and safety of workers from the risks related to chemical agents at work. https://eur-lex.europa.eu/legal-content/EN/TXT/PDF/?uri=CELEX:32000L0039&from=en (accessed January 15, 2020).

EC [European Commission]. 2004. Directive 2004/37/EC of the European Parliament and of the Council of 29 April 2004 on the protection of workers from the risks related to exposure to carcinogens or mutagens at work (Sixth individual Directive within the meaning of Article 16(1) of Council Directive 89/391/EEC). https://eur-lex.europa.eu/legal-content/EN/TXT/PDF/?uri=CELEX:02004L0037-20140325&from=en (accessed January 15, 2020).

EC [European Commission]. 2006. Regulation (EC) No 1907/2006 of the European Parliament and of the Council of 18 December 2006 concerning the Registration, Evaluation, Authorisation and Restriction of Chemicals (REACH), establishing a European Chemicals Agency, amending Directive 1999/45/EC and repealing Council Regulation (EEC) No 793/93 and Commission Regulation (EC) No 1488/94 as well as

Council Directive 76/769/EEC and Commission Directives 91/155/EEC, 93/67/EEC, 93/105/EC and 2000/21/EC. https://eur-lex.europa.eu/legal-content/EN/TXT/PDF/?uri=CELEX:02006R1907-20140410&from=EN (accessed January 15, 2020).

EC [European Commission]. 2008. Regulation (EC) No 1272/2008 of the European Parliament and of the Council of 16 December 2008 on classification, labelling and packaging of substances and mixtures, amending and repealing Directives 67/548/EEC and 1999/45/EC, and amending Regulation (EC) No 1907/2006. https://eur-lex.europa.eu/eli/reg/2008/1272/oj?eliuri=eli:reg:2008:1272:oj (accessed January 15, 2020).

EC [European Commission]. 2009. Commission Directive 2009/161/EU of 17 December 2009 establishing a third list of indicative occupational exposure limit values in implementation of Council Directive 98/24/EC and amending Commission Directive 2000/39/EC. https://eur-lex.europa.eu/legal-content/EN/TXT/PDF/?uri=CELEX:32009L0161&from=EN (accessed January 15, 2020).

EC [European Commission]. 2014a. Directive 2014/27/EU of the European Parliament and of the Council of 26 February 2014 amending Council Directives 92/58/EEC, 92/85/EEC, 94/33/EC, 98/24/EC and Directive 2004/37/EC of the European Parliament and of the Council, in order to align them to Regulation (EC) No 1272/2008 on classification, labelling and packaging of substances and mixtures. https://eur-lex.europa.eu/legal-content/EN/TXT/PDF/?uri=CELEX:32014L0027&from=GA (accessed January 15, 2020).

EC [European Commission]. 2014b. Guidance on the protection of the health and safety of workers from the potential risks related to nanomaterials at work. 2014. Employment, Social Affairs & Inclusion. European Commission. https://osha.europa.eu/pl/legislation/guidelines/guidance-protection-health-and-safety-workers-potential-risks-related (accessed January 15, 2020).

EC [European Commission]. 2017a. Commission Directive (EU) 2017/164 of 31 January 2017 establishing a fourth list of indicative occupational exposure limit values pursuant to Council Directive 98/24/EC, and amending Commission Directives 91/322/EEC, 2000/39/EC and 2009/161/EU. https://eur-lex.europa.eu/legal-content/EN/TXT/PDF/?uri=CELEX:32017L0164&from=EN (accessed January 15, 2020).

EC [European Commission]. 2017b. Directive (EU) 2017/2398 of the European Parliament and of the Council of 12 December 2017 amending Directive 2004/37/EC on the protection of workers from the risks related to exposure to carcinogens or mutagens at work. *Off J Eur Union*. Document 32017L2398. https://eur-lex.europa.eu/eli/dir/2017/2398/oj (accessed January 10, 2020).

EC [European Commission]. 2018. Commission Regulation (EU) 2018/1881 of 3 December 2018 amending Regulation (EC) No 1907/2006 of the European Parliament and of the Council on the Registration, Evaluation, Authorisation and Restriction of Chemicals (REACH) as regards Annexes I, III, VI, VII, VIII, IX, X, XI, and XII to address nanoforms of substances. https://eur-lex.europa.eu/legal-content/PL/TXT/?qid=1545225466697&uri=CELEX%3A32018R1881 (accessed January 15, 2020).

EC [European Commission]. 2019a. Commission Directive (EU) 2019/1831 of 24 October 2019 establishing a fifth list of indicative occupational exposure limit values pursuant to Council Directive 98/24/EC, and amending Commission Directives 91/322/EEC, 2000/39/EC and 2009/161/EU. https://eur-lex.europa.eu/legal-content/EN/TXT/PDF/?uri=CELEX:32019L1831&rid=1 (accessed January 15, 2020).

EC [European Commission]. 2019b. Directive (EU) 2019/130 of the European Parliament and of the Council of 16 January 2019 amending Directive 2004/37/EC on the protection of workers from the risks related to exposure to carcinogens or mutagens at work. *Off J Eur Union*. Document 32019L0130. https://eur-lex.europa.eu/legal-content/PL/TXT/?uri=CELEX:32019L0130 (accessed January 10, 2020).

ECETOC [European Centre for Ecotoxicology and Toxicology of Chemicals]. 2012. ECETOC TRA version 3: Background and rationale for the improvements. Technical Report No. 114. Brussels: European Centre for Ecotoxicology and Toxicology of Chemicals. http://www.ecetoc.org/wp-content/uploads/2014/08/ECETOC-TR-114-ECETOC-TRA-v3-Background-rationale-for-the-improvements.pdf (accessed December 31, 2019).

ECETOC [European Centre for Ecotoxicology and Toxicology of Chemicals]. 2017. Targeted risk assessment: User guide of the integrated tool TRAM. Version 3.1. http://www.ecetoc.org/wp-content/uploads/2017/07/ECETOC_TRA_Integrated_Tool_ User_Guide_July2017.pdf (accessed December 31, 2019).

ECHA [European Chemicals Agency]. 2016. Guidance on information requirements and chemical safety assessment. Chapter R. 14: Occupational exposure assessment. Version 3.0. Helsinki, Finland: European Chemicals Agency. https://echa.europa.eu/documents/10162/13632/information_requirements_r14_en.pdf (accessed January 15, 2020).

EU-OSHA [European Agency for Safety and Health at Work]. 2012. Risk assessment by a small company using Stoffenmanager nano: Case study. https://test.osha.europa.eu/en/publications/risk-assessment-small-company-using-stoffenmanager-nano/view (accessed December 20, 2019).

Heussen, G. A. H., and A. L. Hollander. 2017. Stoffenmanager® exposure model algorithms within TREXMO and Stoffenmanager® lead to different outcomes. *Ann Work Expo Health* 61(5):604–606. doi: 10.1093/annweh/wxx018.

Heussen, G. A. H., M. Arnone, R. van der Haar et al. 2019. Response to Savic et al. on: Inter-assessor agreement for TREXMO and its models outside the translation framework. *Ann Work Expo Health*. doi: 10.1093/annweh/wxz094/5671603.

Höck, J., R. Behra, L. Bergamin et al. 2018. Precautionary matrix for synthetic nanomaterials. Version 3.1. Bern, Switzerland: Federal Office for Public Health and Federal Office for the Environment. https://www.bag.admin.ch/bag/en/home/gesund-leben/umwelt-und-gesundheit/chemikalien/nanotechnologie/sicherer-umgang-mit-nanomaterialien/vorsorgeraster-nanomaterialien-downloadversion.html (accessed January 15, 2020).

HSE [Health and Safety Executive]. a. COSHH essentials. Controlling exposure to chemicals: A simple control banding approach. https://www.hse.gov.uk/pubns/guidance/coshh-technical-basis.pdf (accessed January 15, 2020).

HSE [Health and Safety Executive]. b. The technical basis for COSHH essentials: Easy steps to control chemicals. http://www.coshh-essentials.org.uk/assets/live/CETB.pdf (accessed December 20, 2019).

IFA [Institut für Arbeitsschutz der Deutschen Gesetzlichen Unfallversicherung]. 2019. GESTIS, Database: International limit values for chemical agents. http://www.dguv.de/ifa/GESTIS/GESTIS-Internationale-Grenzwerte-für-chemische-Substanzen-limit-values-for-chemical-agents/index-2.jsp (accessed December 19, 2019).

Jankowska, A., S. Czerczak, M. Kucharska et al. 2015. Application of predictive models for estimation of health care workers exposure to sevoflurane. *Int J Occup Saf Ergon* 21(4):471–479. doi: 10.1080/10803548.2015.1086183.

Jankowska, A., S. Czerczak, and M. Kupczewska-Dobecka. 2017a. Bezpomiarowa ocena narażenia na działanie substancji chemicznych przez kontakt ze skórą w środowisku pracy. *Med Pr* 68(4):557–569. doi: 10.13075/mp.5893.00555.

Jankowska, E., P. Sobiech, B. Kaczorowska et al. 2017b. Ocena i kontrola ryzyka w odniesieniu do narażenia na nanoobiekty oraz ich aglomeraty i agregaty (NOAA) emitowane do powietrza podczas produkowania i stosowania nanomateriałów. Zalecenia. Warszawa: Centralny Instytut Ochrony Pracy – Państwowy Instytut Badawczy.

Jankowska, A., S. Czerczak, P. Maciaszek et al. 2014. Nowe narzędzie do oceny inhalacyjnego narażenia zawodowego na metale i substancje nieorganiczne. *Przemysł Chemiczny* 5(93):606–612.

Johnston, K. J., M. L. Phillips, N. A. Esmen et al. 2005. Evaluation of an artificial intelligence program for estimating occupational exposures. *Ann Occup Hyg* 49(2):147–153. doi: 10.1093/annhyg/meh072.

Krimsky, S. 2017. The unsteady state and inertia of chemical regulation under the US Toxic Substances Control Act. *PLoS Biol* 15(12):e2002404. doi: 10.1371/journal.pbio.2002404.

Kupczewska-Dobecka, M., S. Czerczak, and M. Jakubowski. 2011. Evaluation of TRA ECETOC model for inhalation workplace exposure to different organic solvents for selected process categories. *J Occup Med Environ Health* 24(2):208–217. doi: 10.2478/s13382-011-0021-3.

Kupczewska-Dobecka, M., S. Czerczak, and S. Brzeźnicki. 2012. Assessment of exposure to TDI and MDI during polyurethane foam production in Poland using integrated theoretical and experimental data. *Environ Toxicol Pharmacol* 34:512–518. doi: 10.1016/j.etap.2012.06.006.

Lamb, J., S. Hesse, B. G. Miller et al. 2015. Evaluation of Tier 1 exposure assessment models under REACH (eteam) project. Research Project F 2303. Dortmund: Federal Institute for Occupational Safety and Health. https://www.baua.de/EN/Service/-Publications/Report/F2303-D26-D28.pdf?__blob=publicationFile&v=4 (accessed January 15, 2020).

Landberg, H. 2018. The use of exposure models on assessing occupational exposure to chemicals. Doctoral dissertation. Lund: Faculty of Medicine, Lund University. https://portal.research.lu.se/portal/files/37105703/Kappa2017_12_21_a.pdf (accessed December 30, 2019).

Larese, F. F., M. Mauro, G. Adami et al. 2015. Nanoparticles skin absorption: New aspects for a safety profile evaluation. Inadvertent ingestion exposure: Hand- and object-to-mouth behaviour among workers. *Regul Toxicol Pharmacol* 72(2):310–322. doi: 10.1016/j.yrtph.2015.05.005.

Lee, E. G., J. Lamb, N. Savic et al. 2019a. Evaluation of exposure assessment tools under REACH, Part 1: Tier 1 tools. *Ann Work Expo Health* 63:218–229. doi: 10.1093/annweh/wxy091.

Lee, E. G., J. Lamb, N. Savic et al. 2019b. Evaluation of exposure assessment tools under REACH, Part 2: Higher tier tools. *Ann Work Expo Health* 63:230–241. doi: 10.1093/annweh/wxy098.

Lee, J. H., W. K. Kuk, M. Kwon et al. 2013. Evaluation of information in nanomaterial safety data sheets and development of international standard for guidance on preparation of nanomaterial safety data sheets. *Nanotoxicology* 7(3):338–345. doi: 10.3109/17435390.2012.658095.

Lentz, T. J., M. Seaton, and P. Rane. 2019. Technical report: The NIOSH occupational exposure banding process for chemical risk management. Cincinnati, OH: U.S. Department of Health and Human Services, Centers for Disease Control and Prevention, National Institute for Occupational Safety and Health, DHHS (NIOSH). Publication No. 2019-132. doi: 10.26616/NIOSHPUB2019132.

Li, Y., H. Yu, S. Zheng et al. 2016. Direct quantification of rare earth elements concentrations in urine of workers manufacturing cerium, lanthanum oxide ultrafine and nanoparticles by a developed and validated ICP-MS. *Int J Environ Res Public Health* 13(3):350. doi: 10.3390/ijerph13030350.

Marie-Desvergne, C., M. Dubosson, L. Touri et al. 2016. Assessment of nanoparticles and metal exposure of airport workers using exhaled breath condensate. *J Breath Res* 10:036006. doi: 10.1088/1752-7155/10/3/036006.

Marquart, H., H. Heussen, M. Le Feber et al. 2008. 'Stoffenmanager', a web-based Control Banding Tool using an exposure process model. *Ann Occup Hyg* 52(6):429–441. doi: 10.1093/annhyg/men032.

McNally, K., J. P. Gorce, N. Warren et al. 2019. Calibration of the dermal Advanced REACH Tool (dART) mechanistic model. *Ann Occup Hyg* 63(6):637–650. doi: 10.1093/annweh/wxz027.

NIOSH [National Institute for Occupational Safety and Health]. 2019. *NIOSH Manual of Analytical Methods (NMAM)*. 5th ed. Cincinnati, OH: National Institute for Occupational Safety and Health. https://www.cdc.gov/niosh/nmam/default.html (accessed December 30, 2019).

OSHA [Occupational Safety and Health Administration]. 2012. Hazard communication: Final rule. Washington, DC: U.S. Department of Labor. Occupational Safety and Health Administration. *Federal Register* 77(58):17574–17896. https://osha.gov/pls/oshaweb/owadisp.show:document?p_table=FEDERAL_REGISTER&p_id=22607 (accessed December 30, 2019).

OSHA [Occupational Safety and Health Administration]. 2014. 79 FR 61383 OSHA Request for information: Chemical management and permissible exposure limits (PELs). https://www.govinfo.gov/app/details/FR-2014-10-10/2014-24009 (accessed December 30, 2019).

Ostiguy, C., M. Riediker, P. Troisfontaines et al. 2010. Development of a specific Control Banding Tool for nanomaterials. ANSES. French agency for food, environmental and occupational health and safety. Request no. 2008-SA-0407. Expert Committee (CES) on Physical Agents. https://www.anses.fr/en/system/files/AP2008sa0407RaEN.pdf (accessed December 31, 2019).

Paik, S. Y., D. M. Zalk, and P. Swuste. 2008. Application of a pilot Control Banding Tool for risk level assessment and control of nanoparticle exposures. *Ann Occup Hyg* 52:419–428.

Pośniak, M., and J. Skowroń. 2010. Harmful chemical agents in the work environement. In *Handbook of occupational safety and health*, ed. D. Koradecka, 103–138. Boca Raton, FL: CRC Press/Taylor & Francis Group.

Pośniak, M., and M. Galwas. 2007. Ocena narażenia dermalnego. *Bezpieczeństwo Pracy. Nauka i Technika* 11:14–17.

Pośniak, M., E. Dobrzyńska, and M. Szewczyńska. 2012. Projektowane nanomateriały w środowisku pracy: Narzędzia do oceny ryzyka. *Przemysł Chemiczny* 91(4):588–593.

Pośniak, M., S. Czerczak, M. Kupczewska-Dobecka et al. 2015. Opracowanie narzędzi do bezpomiarowej oceny inhalacyjnego narażenia zawodowego na substancje rakotwórcze oraz zaleceń do profilaktyki chorób nowotworowych. Projekt realizowany w ramach umowy ZUS i CIOP-PIB. https://www.ciop.pl/CIOPPortalWAR/file/76865/Sprawozdanie_ZUS_Etap_2_2015n.pdf (accessed December 31, 2019).

Riedmann, R. A., B. Gasic, and D. Vernez. 2015. Sensitivity analysis, dominant factors, and robustness of the ECETOC TRA v3, Stoffenmanager 4.5, and ART 1.5 occupational exposure models. *Risk Anal* 35:211–225. doi: 10.1111/risa.12286.

Savic, N., D. Racordon, D. Buchs et al. 2016. TREXMO: A translation tool to support the use of regulatory occupational exposure models. *Ann Occup Hyg* 60(8):991–1008.

Savic, N., B. Gasic, and D. Vernez. 2017. Stoffenmanager® algorithm within version 6 differs from the published algorithm within old versions and TREXMO. *Ann Work Expo Health* 61:607–610.

Savic, N., B. Gasic, and D. Vernez. 2018. ART, Stoffenmanager, and TRA: A systematic comparison of exposure estimates using the TREXMO translation system. *Ann Work Expo Health* 62:72–87.

Savic, N., E. G. Lee, B. Gasic et al. 2019. Inter-assessor agreement for TREXMO and its models outside the translation framework *Ann Work Expo Health* 63:814–820. doi: 10.1093/annweh/wxz040.

Schneider, T., D. H. Brouwer, I. K. Koponen et al. 2011. Conceptual model for assessment of inhalation exposure to manufactured nanoparticles. *J Expos Sci Environ Epidemiol* 21:450–463. doi: 10.1038/jes.2011.4.

Spinazzè, A., F. Borghi, D. Campagnolo et al. 2019. How to obtain a reliable estimate of occupational exposure? Review and discussion of models' reliability. *Int J Environ R Public Health* 16(15):2764, 1–29. doi: 10.3390/ijerph16152764.

Van Broekhuizen, P., W. van Veelen, W. H. Streekstra et al. 2012. Exposure limits for nanoparticles: Report of an international workshop on nano reference values. *Ann Occup Hyg* 56(5):515–524. doi: 10.1093/annhyg/mes043.

Zalk, D. M., and G. H. Heussen. 2011. Banding the world together: The global growth of control banding and qualitative occupational risk management. *Saf Health Work* 2(4):375–379. doi: 10.5491/SHAW.2011.2.4.375.

Zalk, D. M., S. Y. Paik, and P. Swuste. 2009. Evaluating the Control Banding Nanotool: A qualitative risk assessment method for controlling nanoparticle exposures: *J Nanopar Res* 11:1685–1704.

Zalk, D. M., T. Spee, M. Gillen et al. 2011. Review of qualitative approaches for the construction industry: Designing a risk management toolbox. *Saf Health Work* 2(2):105–121. doi: 10.5491/SHAW.2011.2.2.105.

Zalk, D. M., S. Y. Paik, and W. D. Chase. 2019. A quantitative validation of the Control Banding Nanotool. *Ann Work Expo Health* 63(8):898–917. doi: 10.1093/annweh/wxz057.

6 Control of Occupational Risks

Tomasz Jankowski

CONTENTS

6.1 SUBSTITUTION

The protective measures introduction strategy should primarily take into account substitution activities that involve the use of less harmful physical forms of chemical substances. The harmful indicator replacement method reduces investment and operating costs associated with technical protection equipment. In the case of nanomaterials, nanopowders are replaced with emulsions, suspensions and pastes. Chemicals, including nanomaterials, are often used because of their unique properties, so their replacement can be difficult in many cases.

However, even if the use of a given substance cannot be eliminated, it may be possible to use it in a form that minimizes occupational risks associated primarily with inhalation exposure to nano-objects. To limit the exposure to chemical substances, in particular nanomaterials, by respiration, technological processes that require the opening of nanopowder tanks may be given up. Techniques that generate smaller amounts of aerosols can be used during the processing and use of nanomaterials. An example of such activities may be to replace nanomaterial sawing with a cutting process or applying a layer during roller painting instead of spraying with aerosol.

In the studies by Zatorski, Salasinska [2016] and Salasinska et al. [2017], they obtained lower heat release rates (HRR) among polymer composites containing nanomaterials in the form of carbon nanotubes and titanium dioxide. The flammability of composites with a maximum proportion of polyhedral carbon nanotubes with a diameter above 8 nm and a length of 10–30 μm of 3% by mass (HRR_{max} = 398.16 kW/m^2) was less than the flammability of the filler, 6% titanium dioxide by mass with a grain size of 20 nm (HRR_{max} = 530.95 kW/m^2).

A number of new pieces of information are being created in the world for designers, users looking for less dangerous alternatives to newly introduced chemical substances and their mixtures in the processes of production, processing and application. Selected issues can be found, among others, in the document titled Green Nanotechnology Challenges and Opportunities, published by the American Chemical Society of Green Chemistry Institute [ACS 2011].

The US Environmental Protection Agency (EPA) indicates alternative ways to find less harmful chemicals, including nanomaterials [EPA 2015]. The selection of alternative methods is carried out as part of activities related to occupational risk management, in accordance with the Toxic Substances Control Act (TSCA) Work Plan.

6.2 COLLECTIVE PROTECTION EQUIPMENT

In enterprises related to the processing of nanomaterials, in medical, chemical and physical laboratories, machinery and technological processes that pose a threat of chemical agents emissions should be equipped with appropriate collective protection equipment (CPE).

The spread of pollutants emitted into the air at workplaces can be reduced by using various types of collective protection equipment, the use of which, in accordance with EU directives and international documents, is a priority in relation to the use of personal protection equipment (PPE).

Technical collective protection equipment are primarily air ventilation, air filtration and air conditioning installations. If assuming the type of indicator that causes air exchange in the room as a criterion, the ventilation division diagram can be presented as shown in Figure 6.1.

Lack of proper air exchange, especially in facilities related to the processing of nanomaterials in medical, chemical and physical laboratories, is the cause of the increase in concentration and accumulation of air pollution, including harmful gases, vapors and dust.

The purpose of ventilation, that consists of constant or periodic exchange of air in this type of room, is therefore to:

- improve the condition and composition of air at workplaces in accordance with hygiene (protection of human health) and technological requirements (need to obtain products with specific properties),
- regulate such parameters of the air environment in rooms as concentration of pollution, temperature, humidity, velocity and direction of air movement.

An indicator that causes natural air exchange is pressure difference which is caused by the difference in temperature inside and outside a building, wind energy and the distance between vents that supply fresh air and extract used air.

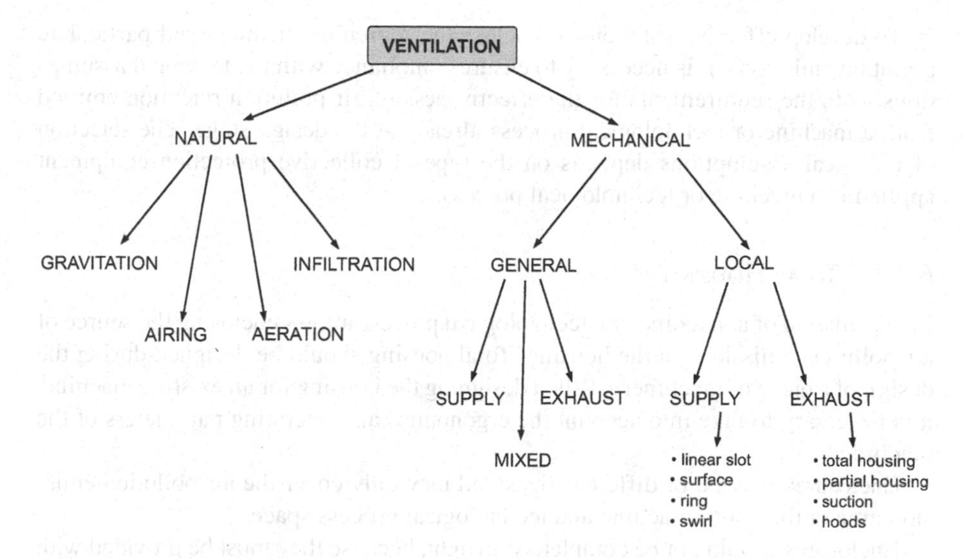

FIGURE 6.1 Types of air ventilation.

In general, natural ventilation is insufficient for safety reasons and mechanical ventilation is necessary.

Mechanical ventilation is an organized air exchange that is ensured by technical measures such as fans.

In production rooms, where chemical and particulate pollutants are emitted, negative pressure ventilation is usually used.

Mechanical ventilation of sufficient high efficiency, equipped with an efficient air distribution and filtration system, is also necessary in clean rooms. In these rooms, depending on requirements, the air should have an appropriate dust or microbiological cleanliness level. The supply of clean, pollution-free air and the extract of used, contaminated air reduces the risk of contamination of materials being tested. In rooms classified as the so-called clean, positive pressure ventilation is mainly used.

When it is necessary to create a microclimate in a room that constantly meets specific requirements, regardless of external weather conditions, air conditioning must be used, which not only provides the right amount of air, but also its required air quality (cleanliness, humidity, temperature, etc.).

The most favorable and recommended solution from the point of view of protection against the emission of air pollution to the machine environment and the technological process is total housing [Bell 2009]. The total housing of machinery and technological processes is not always possible. Then, apply:

- partial housings with working holes,
- local ventilation systems equipped with suction nozzles and hoods (stationary or adjustable).

Before removing the air into the atmosphere together with contaminants, housings and local ventilation systems should be connected with dust removal systems or air cleaner devices.

To develop effective solutions to enclose the source of chemical and particulate pollutant emissions, it is necessary to ensure compliance with the technical assumptions as to the requirements for the effectiveness of air pollution retention emitted from a machine or technological process already at the design stage. The selection of technical assumptions depends on the type of collective protection equipment applied to a machine or technological process.

6.2.1 Total Housing

Encapsulation of a machine or a technological process allows enclosing the source of air pollution emissions in the housing. Total housing should be designed during the design of a piece of machinery. When designing the housing for an existing machine, it is necessary to take into account the ergonomics and operating parameters of the machine.

Enclosures may be of different sizes and may only cover the air pollution emission area or the entire machine and technological process space.

Enclosures should not be completely airtight, because they must be provided with holes to put materials to be processed therein. Additionally, air flow must be provided in the air enclosure. Hence – there must be holes in the housing for air to enter and for contaminated air to be removed. As a result of the machine's operation or a technological process, different pressure zones are created inside the housing. Connecting the ventilation system to a housing allows creating a negative pressure in the entire enclosure, which prevents uncontrolled release of air pollutants into the environment.

Requirements for the effectiveness of reducing the risks of air pollution emitted during processing in the housing should take into account the appropriate selection of the following parameters:

- volume air flow of inflowing and air extracted from the housing,
- housing size and shape,
- method of connecting ventilation ducts.

The size and shape of the exhaust vents from the housing contribute to shaping the air velocity distribution in their interiors, which also affects the efficient functioning of the machine or technological process and the enclosure itself.

For material processing in the process of crushing, grinding, etc., in order not to take the material for the ventilation system of enclosures, the aim is to remove the least amount of dust. To reduce the amount of dust extracted into the ventilation system from housing, use:

- low air velocities in the plane of the ventilation duct openings,
- increased distance between the dust emission source and the air vent openings,
- special shape of surfaces of individual housing elements,
- additional internal partitions, covers, aprons, etc.

Automatic control of switching on and off of the machine, technological process and ventilation system is of crucial importance when reducing the risk of air pollution in the area protected by machine housing. Switching the installation on should be simultaneous with starting the material treatment in the machine protected by the housing. Installation should usually be shut down after a certain period after the machine has stopped, so that air pollution is removed from the enclosure before it is opened.

6.2.2 Partial Housing

Partial housing equipped with working holes provides workers with free access to machines being operated. Partial housing enables safe control of the technological process while removing polluted air from the ventilation system from the inside of the housing.

Partial housing limits the air pollution emission zone and prevents the effects of uncontrolled air flows caused by activities performed by workers, their movement in the environment and the work of other machines or technological processes.

The use of lightweight but durable materials for the construction of partial housing allows them to:

- be adapted to be placed directly in the area of the source of air pollution,
- be coupled with an air pollutant emission source, e.g. with portable cutting tools.

The basic technical parameters that characterize the effectiveness of partial housing in the process of risk reduction of air pollution are:

- volume air flow extracted from the housing (air flow rate adjustment to the velocity of air pollution flow from the source of emission),
- size and shape of the housing matched to the flow characteristics of air pollutants in the area of emission,
- method of connecting ventilation ducts,
- location of air pollutant openings in relation to their emission area.

Laboratory fume cupboards and laminar flow and biological safety cabinets (BSC) are a special type of partial housing.

Fume cupboards are used to partially close processes related to solid, liquid and gas substances in chemical and physical laboratories.

Laboratory fume cupboards should be designed so that:

- contaminated air with dangerous concentrations or air with a dangerous amount of pollutants did not get from the working chamber of the laboratory fume cupboard to a room,
- effective vapor removal that reduces the susceptibility to the formation of an explosive or dangerous atmosphere inside the fume cupboard workspace is ensured,

- the user is protected by a front sliding window against splashes of liquid and solid materials.

The laboratory fume cupboard is a partial housing that meets the requirements of EN 14175-2:2003 (Fume Cupboards – Part 2: Safety and performance requirements). A protective device is ventilated by forced air circulation through an adjustable opening vent, and in addition:

- it is equipped with an enclosure, the task of which is to limit the spread of contaminated air and prevent it from reaching the operators and other workers outside the protective device,
- it gives a certain degree of mechanical protection,
- it provides controlled extraction of polluted air.

Laboratory fume cupboards can be divided into:

- table tops, the working surface of which is at least 720 mm above floor level,
- low-level, the working surface of which is in the range from the floor level to 720 mm above the floor,
- transitive, the working surface of which is at the floor level or below,
- with variable air volume (VAV), ensuring variable air flow intensity depending on the opening of the sliding window,
- with forced filtration, capable of catching specific impurities and allowing air to be returned back to the laboratory.

An additional division criterion for laboratory fume cupboards is the position of the sliding window: vertical, horizontal or combined.

In accordance with the requirements of EN 14175-2:2003, the permissible dimensions of laboratory fume cupboards include:

- the total width should be a multiple of 100 mm, and the recommended widths are 1,200 mm and 1,500 mm,
- the total depth should be between 600 and 1,200 mm,
- the height of the workspace should not exceed 900 mm, and the recommended heights are 0 mm, 500 mm, 720 mm and 900 mm,
- other dimensions are allowed but must be agreed on between the user and the manufacturer.

The effectiveness of a laboratory fume cupboard can be presented in qualitative terms as the ability to retain and remove one or more pollutants released from a source inside the fume cupboard's workspace and the ability to minimize any interference caused by draughts, operators' movements or the passage of personnel in a room.

Hence – it is important to test the safety and effectiveness of the laboratory fume cupboard in accordance with the requirements given in the EN 14175 series and the ANSI/ASHRAE 110-1995 (method of testing performance of laboratory fume hoods).

The assessment of the effectiveness of laboratory fume cupboards at workstations is focused on conducting tests in various conditions to:

- assess safety and effectiveness of fume cupboards in laboratory conditions (type test),
- assess a specific laboratory fume cupboard in given environmental conditions and use the on-site test.

Of particular importance are tests of fume cupboards under operating conditions, as their operating parameters are strictly dependent on the effectiveness of general ventilation systems and local ventilation devices installed in a given laboratory room.

Research works are conducted in many research centers to increase the efficiency of laboratory fume cupboards. They were conducted by Bémer et al. [2002], to name a few, by the tracer gases method in laboratory conditions. They concerned:

- determining the effectiveness of laboratory fume cupboards. It was found that the efficiency of the extract operation can vary from 10% up to 84%, depending on the location of slits and extract holes in relation to the source of the tracer gas emission,
- research on the impact of construction and operating parameters of sources emitting gas tracers on the "mixing distance" of air with the tracer gas. It was found that when using multi-hole sources and high velocity of tracer emitting sources, a significant reduction in "mixing distance" could be achieved.

This research was conducted using three types of gas tracers, namely helium (He), nitrous oxide (N_2O) and sulfur hexafluoride (SF_6).

The number of air exchanges in the laboratory room was determined by the gas tracer method by Jung and Zeller [1994]. Two gas tracers were used in the research: sulfur hexafluoride and nitrous oxide. Based on the results obtained, it was found that the type of applied tracer gas does not affect the test results with good mixing of air with the tracer gas. Rules for generating the tracer gas when conducting tests in the laboratory room are also presented. It was found that the room background has a very significant effect on the results of measurements made using tracer gases.

In the research by Hampl and Niemelä [1986], Ojima [2002; 2007; 2009], Tseng et al. [2006; 2007] the gas tracer method was used to test the effectiveness of a laboratory fume cupboard in an industrial plant. The tracer gas (SF_6) was administered at a known, constant rate at the place of emission of real pollutants. The efficiency of the local exhaust ventilation was determined by measuring the concentration of the gas tracer at a selected point in the exhaust duct, and then the reference of the obtained concentration to the concentration of the tracer given at the source of pollutant emissions. The research determined the minimum "mixing distances" in relation to four types of exhaust ducts: straight duct, manifold, one-elbow duct and two-elbow duct.

Research on the performance assessment of laboratory fume cupboards, in accordance with the recommendations of the ANSI/ASHRAE 110-1995, was conducted, among others, by: Bell et al. [2002], Moore [2004], Ahn et al. [2008], Weale et al.

[2007], Tschudi et al. [2004a; 2004b], Varley [1993], Worrell et al. [2003], Saunders [1993] oraz Lieckfield Jr. and Poore [2001]. During the research, air velocity and gas tracer concentration measurements were made to assess the effect of the fume cupboard operating parameters on human protection in the work process. Human presence was simulated by a mannequin. Sulfur hexafluoride was used as a gas tracer in the tests.

The development of methods for testing the distribution of air flow in laboratory fume cupboards and air flow from the laboratory room to the fume cupboard working chamber has been observed in recent years.

Laboratory fume cupboards encapsulate the sources of emissions aerodynamically, and by removing air pollutants from the working chamber they greatly facilitate the general ventilation process [Tseng et al. 2010]. However, according to the research by Chen and Huang [2012; 2014], at high efficiency they do not eliminate this task completely, and therefore some part of air pollution gets from the emission source into the laboratory room. According to Huang, such uncontrolled emission of air pollutants is usually limited by increasing the volume of air extracted from the fume cupboard or by using general dilution ventilation in the laboratory [Huang et al. 2013]. In some research on air emissions from sources in fume cupboards, Tsai et al. [2010] used local air supplies that covered a specific zone of the laboratory room or supported local exhaust devices located there.

In the laboratories at the Central Institute for Labour Protection – National Research Institute (CIOP-PIB), new constructions of low-flow laboratory fume cupboard supported by air flow which allowed to increase the capture efficiency of pollutants diffusing towards the laboratory while increasing the degree of opening the fume cupboard sash [Jankowski 2011] were examined.

Innovative solutions in the scope of encapsulation of the working chamber by aerodynamizing the shape of the sliding window frame, sides and walls of the fume cupboard were applied.

This monitoring and air flow regulation system enabled directing the volume of air flow from the laboratory directly to the working plate of the chamber in which the source of air pollution is located. The structural changes of the fume cupboard allowed reducing the required amount of extracted air, which is needed to effectively remove its impurities, which will contribute to savings in the operating costs of laboratory rooms. A smaller air flow rate caused a reduction in its velocity in the fume cupboard inlet window to a range below 0.25 m/s. In the fume cupboards used so far, the minimum required air velocity in the inlet window was about 0.5 m/s. Lower volumes of extracted air in fume cupboards can contribute to savings of 15–20% in the operating costs of laboratory rooms. The research has also confirmed that the tested laboratory fume cupboards meet the performance requirements specified in standards EN 14175-2:2003 and EN 14175-3:2003 (Fume Cupboards – Part 3: Type test procedures) concerning limit value settings necessary for safe operation (Figure 6.2).

Parameters that characterize the assessment of efficiency of laboratory fume cupboards are determined based on three measurement methods:

- air flow visualization methods to observe the smoke flow in the opening plane of the sliding window and inside the laboratory fume cupboard,

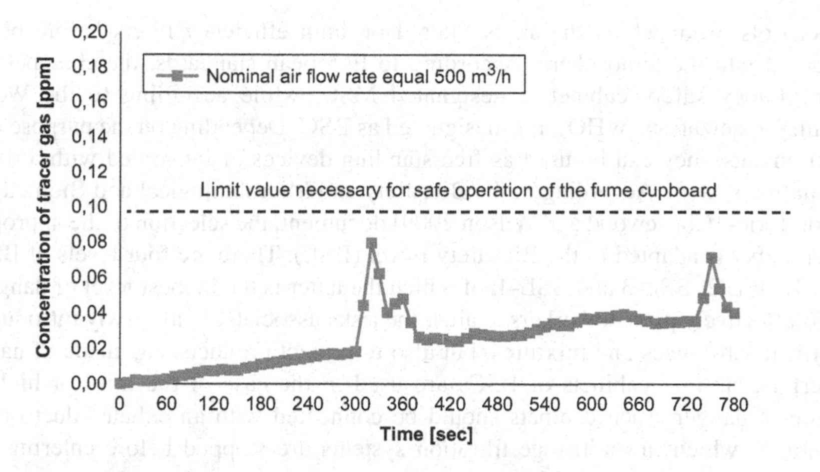

FIGURE 6.2 Changes in the containment indicator of a laboratory fume cupboard.

- anemometric methods to determine the air volume flow and pressure drop in the exhaust duct, air velocity in the front face and in front of the laboratory fume cupboard and air velocity in a room,
- gas tracer methods that allow measuring the containment indicator, robustness of the containment and the air exchange efficiency in a laboratory fume cupboard.

At CIOP-PIB, smoke flow observations are made using smoke generators: smoke pipes (Drägerwerk AG & Co. KGaA, Germany), while air flow velocity measurements are made using Testo 435-4 anemometers (Testo AG, Germany). Pressure drop in the exhaust duct is determined using an electronic differential micromanometer (IMG PAN, Cracow). In turn, the tracer gas method uses the MIRAN SapphIRe Model 100E single-beam infrared spectrophotometer (Thermo/Foxboro, USA).

Clean Laminar Cabinets and Biological Safety Cabinets are used for works that require a high degree of air purity, primarily in biological and medical laboratories. The cabinets allow the workplace to be insulated from the surroundings, thereby reducing the risk of air pollution spreading.

Clean laminar cabinets are used only to protect the material being processed inside. Due to the lack of workers' protection, they cannot be used for work with hazardous substances, both chemically and microbiologically. For this reason, they are not used as cabinets for the preparation of cytostatic drugs and as cabinets for working with biologically active materials. Clean laminar cabinets are used in technological processes that require a high degree of air purity, e.g. biotechnology, food production. Their principle of operation consists of drawing in air, then cleaning it from impurities and forcing it into the working part of the chamber towards the material. Air purification is made using high efficiency particulate air filters (HEPA) and ultra low penetration air filters (ULPA).

Biological Safety Cabinets (BSC) protect laboratory space together with workers and the material collected in the work space. BSC chambers are a ventilated enclosed place, designed to protect workers and the environment against hazardous bioaerosols, from which the air is filtered on high efficiency filters before being extracted into the atmosphere. According to European standards, the chamber for microbiology safety cabinet is designated MSC, while according to the World Health Organization (WHO), it is designated as BSC. Depending on the purpose and requirements, they can be used as free-standing devices or integrated with a room ventilation system. According to the Biosafety and Microbiological and Biomedical Laboratories [Chosewood and Wilson 2009] document, the selection of the appropriate chamber is adapted to the Biosafety Level (BSL). There are four levels of BSL: BSL-1, BSL-2, BSL-3 and BSL-4, of which the latter is the highest level of danger.

To effectively protect workers against the risks associated with newly introduced chemical substances and mixtures, but also to protect products, e.g. made of nanomaterials, laminar cabinets or BSCs are used in the case of the third or highest degree of danger. Such cabinets should be connected with an exhaust duct of air pollutants, which in multistage filtration systems are stopped before entering the atmospheric environment (Table 6.1).

TABLE 6.1
Biosafety Levels

Level of risk	Type of Rooms	Ventilation Requirements	Recommendations
BSL-1	Rooms in which microorganisms introduced with the material do not pose a risk of illness for workers	No special requirements for air ventilation.	Applying under pressure to the environment.
BSL-2	Rooms in which the introduced microorganisms can cause curable diseases	No special requirements for air ventilation.	Use of mechanical ventilation without recirculation.
BSL-3	Rooms in which the introduced microorganisms can be transported by air and absorbed in the respiratory tract of workers	The use of air curtains, unidirectional, single conduit mechanical under pressure ventilation and multistage air filtration, including HEPA filters, total or partial housing. The use of recirculation is prohibited.	Replacement of air filters at least once a year.
BSL-4	Rooms in which work is performed with materials that pose a high risk of losing workers' lives	The use of air curtains, disinfection locks, mechanical under pressure ventilation with HEPA air filters, enclosures connected directly to ventilation exhaust ducts. The use of recirculation is prohibited. Ventilation system designed only for a specific room.	Annual testing of the effectiveness of ventilation and air filtration systems.

Source: Chosewood 2009

6.2.3 Local Ventilation

When it is impossible to totally or partially encapsulate the source of air pollution, local exhaust ventilation (LEV) systems are used that are combined with a dust removal system or an air cleaning device [ACGIH 2019]. The task of local ventilation is to capture air pollution directly at the source of the emission and to prevent it from spreading in the room. The type of LEV used depends on both the location of the emission source and the direction and velocity of the air pollution spread. A disadvantage associated with the use of LEV is the need to place them directly in the area of dust emission sources. This is due to the small area of effective operation of the elements that capture air pollution (e.g. hoods, suction nozzles).

Placing the *hood* in the area where there is a threat of dust emission allows the removal of pollutants with a density lower than the density of air.

To eliminate or reduce the risk of chemical and particulate pollutant emissions, actions should be taken by:

- placing the hood directly above the emission source,
- choosing the shape and dimensions of the hood so that it can cover the entire pollution zone,
- avoiding the conduct of interference processes in the area of extracting pollutants from the emission source to the hood,
- the breathing zone of a machine operator should not be located on the pollutants' removal line from the source to the hood.

The above can be achieved by checking the following technical parameters:

- distance of the inlet plane of the hood from the surface of the source of pollution,
- geometrical dimensions of the hood,
- air flow velocity distribution in the area between the hood and the surface of the source of pollution,
- value of capture velocity,
- air velocity distribution in the inlet plane of the hood,
- volume of air flow extracted from the hood.

Suction nozzles can be placed close to the source of air pollution. This allows capturing most air pollutants in the workplace.

Suction nozzles, as well as housing, should be designed during the manufacture of machinery and planning of technological processes. It is necessary to provide verified data in the operating instructions concerning the value of the extracted air flow rate in relation to flow resistance.

When using suction nozzles, it is necessary to take into account the movability of the emission source. This is related to the effectiveness of the LEV, which mainly depends on the location relative to the source of air pollution. Therefore, two cases should be considered during the design of the suction nozzles:

- immovable source with constant extracted direction – immovable LEV is used,
- movable source – then the LEV is selected from three types:
  - immovable system with a high volume flow of exhaust air,
  - system moving along with the source without the ability to recognize the direction of particle ejection,
  - movable system positioning itself according to the direction of particle ejection.

A serious disadvantage associated with the use of suction nozzles is the occurrence of the phenomenon of rapid decrease in air velocity with an increase in distance from the plane of the inlet opening of the LEV.

Hence – local ventilation (suction nozzles) does not replace general ventilation of the room, but complements it. The key to high efficiency in capturing air pollution is a proper interaction between general and local ventilation.

The work to date is primarily related to increasing the effectiveness of local ventilation solutions and their adaptation to the threat of dust and chemical substances at workplaces. Currently, a number of studies focus on supporting the operation of local ventilation through the use of covers, tables with variable air flow direction (upper, lower and side extraction), general push–pull ventilation or positive displacement ventilation [Wang et al. 2012; Albuquerque et al. 2015].

As part of the research conducted by Jankowski [2019], it has been demonstrated that the side air supply support allows increasing the capture efficiency and reducing the time of air pollution flow from the working zone to the upper zone and the air filtration zone. A supportive flange diffuser was used at the suction nozzle. The air supply was directed directly into the area of effect of the LEV. Such a supply in the form of a flow supplied from the annular gap around the suction opening acts as a dynamic flange. Thus, it not only increases the range of the suction nozzle's effect as a capturing hood, but a directional effect also appears (Figure 6.3).

The use of the principles and control of technical parameters given below may eliminate or reduce the risk of air pollution emissions in the workplaces by:

- utilizing the flow of pollutants by placing the LEV by means of particle emissions,
- covering a maximum calcium emission zone of air pollution,
- matching the shape and dimensions of the suction nozzle to the area of the emission source,
- forcing sufficient particle capture velocity, adapted to the velocity of particles formed during material processing,
- determining the required value of the volume air flow extracted from the LEV,
- reducing the distance between the suction nozzle and the source of air pollution,
- lowering the required value of the extracted air volume by using covers and flanges,

FIGURE 6.3 Smoke flow visualization in the measuring plane of supply air device supporting the LEV.

- determining the air velocity distribution in the inlet plane of the suction nozzle,
- determining air velocity distribution in the area between the suction nozzle and the surface of the source of pollution.

6.2.4 Air Cleaning Devices

The air cleaning devices are designed for removing and purifying polluted air from the source of chemical emissions and harmful pollutants.

Generally, filtration and ventilation devices can be divided into wet and dry dust collectors. In addition to the general division, devices can be divided according to the phenomena used in them:

- inertia force (settling chambers),
- centrifugal force (cyclones),
- electrostatic force (electrostatic precipitators),
- materials used for filtration (fabric filters).

Settling chambers are one of the simplest dust collectors used at the beginning of the technological process of air purification. As a rule, settling chambers serve as preliminary dust collectors for cleaning air from particles with a falling velocity above 0.5 m/s. They are used in multistage air cleaning systems. Their operation is based on the use of the phenomenon of gravity. A condition for separating all grains in the chamber with a falling velocity in the purified air greater than the limiting value of the falling velocity in the air is to select the air velocity and the dimensions of the chamber so that the air flow is laminar. The process parameters of the settling chambers are:

- low flow resistance in the range of 20–50 Pa,
- low energy consumption,
- application for purification of hot gases without prior cooling.

Cyclones are the most common type of dust collectors. The principle of centrifugal force to separate grains from a swirled gas flow is used there. During a spiral movement, the dust particles are subjected to a centrifugal force causing them to move towards the walls. After touching the walls, dust grains lose their velocity and fall under the influence of gravity. The minimum particle size that can be separated in the cyclone depends on its design parameters and the physicochemical properties of the gas being purified, such as:

- volume air flow in the cyclone,
- number of revolutions of the gas flow in the cyclone,
- inner tube radius, cylindrical part, cylindrical layer height,
- dynamic gas viscosity,
- volumetric weight of dust particles.

Cyclones can work in single systems or can be combined into batteries. Cyclone and multi cyclone batteries are used to obtain the highest possible efficiency of grain cleaning, while cleaning large amounts of gases. Multi cyclones are a parallel connection of several dozen cyclones with small diameters placed in a common chamber. The multi cyclone design uses the phenomenon of increasing efficiency of purification while reducing the diameter of the cyclone. Cyclones are characterized by simple design, small size, low investment costs, significant flow resistances from 300 to 1,300 Pa and relatively fast wear of structural elements due to erosion.

Due to their high cleaning efficiency, *electrostatic filters* are commonly used for the separation of large amounts of gases. They work by the principle of electrostatic field interaction on solid or liquid particles suspended in gas. In a strongly uneven electric field created in a properly shaped system there are two types of electrodes that are electrically separated from each other into two opposite poles. Negative electrodes are shaped as thin rods, and at the positive pole has electrodes in the form of plates. The most commonly used electrode system is a system made up of a series of parallel plates at the same distance from each other and between them there are rows of thin rods. The application of high voltage to the emission electrodes causes the release of electrons, which move towards positive collecting electrodes, which results in the precipitation of further electrons from gas particles (corona discharge). Outside the corona discharge sphere, electrons flowing towards the collecting electrode negatively charge the gas particles, which in turn transfer the charge to dust grains, which are attracted to the collecting electrodes. They settle thereon and discharge. When the collecting electrodes are shaken, the dust grains fall to the bottom of the tank due to gravity. The cleaned gas is extracted through a pipe in the upper part of the electrostatic filter. The effectiveness of work of electrostatic filters depends on the strength of the electrostatic field and the physicochemical properties of dust and gas carrier. Electrostatic filters are characterized by a low pressure drop of 30–150 Pa, relatively low power consumption and high cleaning efficiency.

Fabric filters are among the most effective ways of cleaning air. The flow of polluted air through porous materials deposits particles in the filters. Fabric filters require large surfaces, but they are characterized by very high efficiencies of air purification depending on the dimensional distribution of particles.

Air cleaning devices use filtering fabrics manufactured using mainly three methods:

- needling,
- melt-blown,
- paper.

The spun-lace method is a technique used to produce fabrics by swirling the fibers in the fleece by means of geometrically defined water or air flows. This method belongs to the group of advanced techniques for producing fabrics. In the most general sense, the principle of fabric production is as follows: a fleece (fiber board with appropriate structure) is produced by a wet or dry method. This fleece is laid on a metal, supporting, metal sieve with an appropriate perforation. To interlink the fibers and obtain an appropriate fabric structure determining its functional characteristics, the fleece then passes through the three zones of the nozzle battery located above the surface of the fleece. In each zone, water jets of defined geometrical shape are pressed under high pressure towards the surface of the fleece. The water pressure in the first zone is about 294 Pa, and in the third from 539 to 882 Pa. In the middle zone it is an intermediate value between the pressures of the first and third zones. The distance between the top of the fleece and the surface of the nozzles is 5–7.5 cm. Fleece travel velocity is up to 30 m/min. After passing through all the spray passages, the fleece is inverted and the same operation takes place on the inverted fleece surface. At such high pressures, the extruded flows of water could cause uncontrolled movement of the fibers in the fleece, and as a result give a random fabric structure. To avoid this undesirable phenomenon, the top surface of the fleece is covered with a mesh of appropriate perforation and geometry of mesh arrangement.

The spatial orientation of fiber fragments as a consequence of the operation of their entanglement by water flows, increase of material density and friction forces between the fibers, structure of the fabric and the resulting physical and mechanical properties of the product depends on the apparatus and technological solutions of the production line, and the applied technological parameters. *The spun-lace method* has several equipment and technology variants that are used depending on which basic type of product is to be manufactured in the aspect of its application in a given machine park.

A characteristic feature of fabrics obtained by the *spun-lace* method is that they are made of staple fibers, without the use of additional binders, such as thermoplastic polymers, synthetic rubbers, acrylic polymers and polyurethanes. The presence of the binder phase always complicates the product structure, which is particularly undesirable for filtering fabrics. Spun-lace fabrics achieve cohesion as a result of friction forces between the fibers during the fiber entanglement operation by water flows. The structure of fabrics and their rheological properties depend on the type of fibers used (physicochemical and geometrical properties) and technological parameters of the production of fleece and fabrics.

Needled filtering fabrics are manufactured, among others, using polyester fibers (PES) with the following structural parameters:

- average fiber diameters in the range from 10 up to 400 μm,
- surface mass in the range from 150 up to 500 g/m^2,
- thicknesses of fabrics in the range from 5 up to 15 mm,
- mass of 0.003 g/cm^3.

The spun-lace techniques used in the country and in the world also allow producing layered composite fabrics, in which individual layers are made of fibers of different average diameter, using different intensities of needling. The above actions lead to the formation of fiber composites with various structures and porosities.

Pneumothermic fabrics obtained by the melt-blown method, also known as the technique of molten polymer blowing, allow producing filter materials from fibers with submicron diameters. If it is assumed that they have a circular cross section, then the average diameter of fibers, in the case of fabrics with high aerosol filtration efficiency, is in the range of 0.3–0.7 μm. Fabrics are also produced from fibers with a diameter distribution from 1 to several micrometers. The greater the differences in fiber sizes, the less uniform the structure of these fabrics. *The melt-blown method* allows producing fabrics with very high filtration properties.

In the *melt-blown* method, a polymer in the form of granules is moved from the hopper to the heated extruder cylinder. This way it is adjusted to the right viscosity before extrusion from the fiber-forming head. Compressed air passes from the regulator to the heater, in which it is dried and heated to the correct temperature. Then, it is directed to the fiber-forming head, and when it is lowered, the extruded polymer flows are blown into elementary fibers. Settling on the receiving device, they form a compact porous fleece.

The *melt-blown* method is a technologically flexible process. By selecting appropriate process conditions, fibers with a given diameter are obtained, and by choosing the operating conditions of the receiver, porosity values of the obtained fabric can be adjusted.

Fabrics produced by the *melt-blown* method are characterized by a compact structure, which provides them with a higher original packing density compared to fabrics made using the *spun-lace* method, without the need for additional process and technological operations.

Paper fabrics are primarily filter paper and different types of papers. These fabrics are made of thin glass, cellulose or chemical fibers and mixtures thereof. Paper fabrics have a compact spatial structure and are not mechanically strong. They are used for the production of special application filters, e.g. for the filtration of radioactive and biologically polluted air. Due to the high flow resistance of fabrics, a strong development of its surface is used already at the production stage. The method of obtaining paper fabrics determines their properties.

Purification of air in the aforementioned air cleaning devices allows moving it into the atmosphere or reintroducing into a room, while meeting the requirements of local regulations and standards concerning emissions and environmental protection. The advantages of air cleaning devices are the possibility of placing them in the immediate vicinity of the source of air pollution and operational savings that result from the use of recirculation air (especially in winter). Air recirculation should not be used in work rooms where it is possible to suddenly increase the concentration of harmful substances [Dz. U. 1997, No. 129, poz 844 z późn. zm.]. Technical parameters and actions that have an effect on reducing the risk of air pollution are:

- type of harmful substances (solid, liquid and gas),
- value of the volume of air flow extracted from the emission source area,
- size and shape of the inlet adapted to the characteristics of the emission area,
- method of connecting ventilation ducts,
- particle filtration efficiency.

6.2.5 General Ventilation

In the recommendations of the National Institute for Occupational Safety and Health (NIOSH), the American Society for Heating, Refrigerating and Air-Conditioning Engineers (ASHRAE), the American Conference of Governmental Industrial Hygienists (ACGIH) and the Environmental Protection Agency (EPA) in ventilated facilities, especially with technological processes, general mechanical ventilation should perform such a distribution of air that, in addition to maintaining the required conditions of thermal comfort, should interact with local exhaust ventilation encapsulating the source of air pollution [ACGIH 2019].

According to §148, item 4 of the Announcement of the Minister of Investment and Development of April 8, 2019, on the publication of a uniform text of the Regulation of the Minister of Infrastructure on technical conditions to be met by buildings and their location [Dz. U. 2019, poz. 1065],

> In a room where a technological process is a source of local emission of harmful substances with an unacceptable concentration or oppressive odour, hoods working with general ventilation should be used as this would enable meeting the quality requirements of the internal environment specified in the occupational health and safety regulations.

In rooms related to the treatment of nanomaterials, in medical, chemical and physical laboratories intended for treatment of nanomaterials, in medical, chemical and physical laboratories, local exhaust ventilation is usually supported by general supply or exhaust-supply ventilation. The organization of air distribution in the room has a significant effect on the quality of the air, including the concentration of its impurities. It is particularly important to properly shape the air flows in large rooms, e.g. large halls. Although the flow of air supplied to rooms required by law is often provided, wrong location of ventilation openings, e.g. a significant number near windows and walls, can lead to draughts in workplaces located close to these openings, while in the middle of a room proper air movement will not be provided and its pollution will accumulate.

Rooms where works with nanomaterials are performed should be equipped with mechanical ventilation ensuring that nano-objects and their aggregates and agglomerates (NOAA) are extracted from potential sources of emission and proper air distribution. It is possible thanks to the installation of properly co-operating general and local mechanical ventilation [Mierzwinski 2015]. Improper general mechanical ventilation may cause improper air distribution in the room and, as a consequence, formation of unventilated areas and accumulation of NOAA of significant concentrations in various room areas. The lack of local ventilation or the use of inefficient ventilation (e.g. fume cupboard) near the NOAA source may result in their spreading in the room air.

Basic parameters used to determine the type of ventilation in a room (negative pressure, balanced, positive pressure) and calculate the air exchange rate are the volume flows of the air entering and leaving the room. In a given room, different air distribution systems can be used, depending on whether the installed ventilation systems have been turned on or off, and the type of interaction of general and local mechanical ventilation.

Due to the air condition in a given room, after switching on/off the general mechanical ventilation system in the room the following can occur:

- negative pressure ventilation when the volume of air supply is lower than the volume flow of extracted air,
- balanced ventilation when the volume flows of supply air and exhaust air are similar,
- positive pressure ventilation when the volume of supply air is higher than the volume flow of extracted air.

The main difficulty in designing indoor ventilation systems is the need to maintain a constant value of negative pressure or positive pressure. The above equilibrium can be obtained by using an appropriate disproportion between supply and extract air volume flows. The automation and control system, whose executive component is the Variable Air Volume (VAV) controllers, is responsible for maintaining the balance of volume air flows and specific proportions between them [Szymanski et al. 2015].

6.2.6 Air Filtration

Air filters are an important element of collective protection equipment. Both in general ventilation systems and in air cleaning devices, single- or multistage filtration systems determined by hygienic or technological requirements determine the quality of the air extracted from or supplied to rooms.

Basic performance indicators of air filters are filtration efficiency and flow resistance. These parameters depend on:

- dust properties (particle size distribution, aerosol concentration, particle shape, electrostatic properties, chemical properties, dust wettability),
- air flow properties (temperature, humidity, velocity),
- structural parameters of air filter (filter design, properties of filter material used).

Filter efficiency is a parameter that determines its ability to clean the air of pollution particles with a given size distribution. The resistance of air flow through the filter has a significant effect on the selection of devices that introduce air into motion when flowing through the filter partition.

Depending on the required degree of purity of the air supplied or extracted from the rooms through the ventilation systems, different filtration systems are used that are designed based on data on operational parameters of air filters determined during the tests using standardized methods used for their classification.

Requirements for test methods and classification rules for air filters used in ventilation and air conditioning installations are specified in European standards.

According to the standards of EN 1822-1:2009 (High efficiency air filters (EPA, HEPA and ULPA) – Part 1: Classification, performance testing, marking) and EN 779:2012 (Particulate air filters for general ventilation – Determination of the filtration performance), established as part of the work of the TC 195 Technical Committee of the European Committee for Standardization (CEN), by class, air filters are divided into:

- coarse type G, medium type M, fine type F,
- Efficient Particulate Air filter (EPA), High Efficient Particulate Air filter (HEPA) and Ultra Low Penetration Air filter (ULPA).

The classification of G, M and F type air filters in accordance with EN 779:2012 is shown in Table 6.2. The G, M and F filters are classified based on average filtration efficiency, determined by the synthetic dust test and the aerosol test with a particle size of 0.4 μm.

G-type coarse filters are primarily used in ventilation and air conditioning systems in rooms with average air purity requirements, and as pre-filters before

TABLE 6.2
Classification of Coarse and Fine Air Filters in Accordance with EN 779:2012

Class	Final test Pressure Drop	Average Arrestance of Synthetic Dust ASHRAE 52.1	Average Efficiency of 0,4 μm Particles	Minimum Efficiency of 0,4 μm Particles
			Test Aerosol of DEHS (DiEthylHexylSebacate)	
	Pa	%	%	%
G1	250	$50 \leq A_m < 65$	–	–
G2	250	$65 \leq A_m < 80$	–	–
G3	250	$80 \leq A_m < 90$	–	–
G4	250	$90 \leq A_m$	–	–
M5	450	–	$40 \leq E_m < 60$	–
M6	450	–	$60 \leq E_m < 80$	–
F7	450	–	$80 \leq E_m < 90$	35
F8	450	–	$90 \leq E_m < 95$	55
F9	450	–	$95 \leq E_m$	70

filters with higher efficiency in air-conditioning units of rooms with high air purity requirements.

M and F type fine filters are used in room ventilation systems with high air purity requirements and as pre-filters in room ventilation and air conditioning systems with very high air purity requirements before high efficiency filters.

The International Organization for Standardization (ISO) has prepared a series of new ISO 16890 standards for testing and classification of air filters used in general ventilation. In 2016, EN ISO 16890-1 (Air filters for general ventilation – Part 1: Technical specifications, requirements and classification system based upon particulate matter efficiency (ePM)), EN ISO 16890-2 (Air filters for general ventilation – Part 2: Measurement of fractional efficiency and air flow resistance), EN ISO 16890-3 (Air filters for general ventilation – Part 3: Determination of the gravimetric efficiency and the air flow resistance versus the mass of test dust captured) and EN ISO 16890-4 (Air filters for general ventilation – Part 4: Conditioning method to determine the minimum fractional test efficiency) were introduced in the European Union.

The implementation of the EN ISO 16890 series of standards in the European Union resulted in the withdrawal of the EN 779:2012 standard. Due to the need to adapt the test aerosol to the real atmospheric aerosol and due to health effects related to the inhalation of polluted air by humans, a new way of air filter testing has been included in the standards. It was referred to three different size ranges of particulate matter (PM):

- PM_{10} for particles between 0.3 and 10 μm,
- $PM_{2.5}$ for particles between 0.3 and 2.5 μm,
- PM_1 for particles between 0.3 and 1 μm.

The classification of air filters in accordance with EN ISO 16890-1:2016 is shown in Table 6.3. Air filters are classified based on their initial gravimetric stop, filtration efficiency of ePM_{10}, $ePM_{2.5}$, ePM_1 and a minimum efficiency of $ePM_{1,min}$ and $ePM_{2.5,min}$. Filters with low filtration efficiency obtained as a result of the L2 synthetic dust retention test with a composition that is in accordance with ISO 15957:2015 (Test dusts for evaluating air cleaning equipment) are not assigned the ePM $class_x$.

TABLE 6.3
Classification of Coarse and Fine Air Filters According to EN ISO 16890-1:2016

	Requirement			
	$ePM_{1,min}$	$ePM_{2,5,min}$	ePM_{10}	Class Reporting Value
Group Designation	%	%	%	–
ISO coarse	–	–	< 50	Initial gravimetric arrestance
ISO ePM_{10}	–	–	≥ 50	ePM_{10}
ISO $ePM_{2,5}$	–	≥ 50	–	$ePM_{2,5}$
ISO ePM_1	≥ 50	–	–	ePM_1

Specialists, including Sikonczyk [2018], noted that the new classification of filters makes it easy to determine the minimum required filtration efficiency for a given PM_x fraction ($X = PM_1, PM_{2.5}, PM_{10}$) with knowledge of the concentration of this PM_x fraction before the filter and the required concentrations C (PM_x) behind the filter:

$$ePM_x = 1 - C_{outlet}(PM_x)/C_{inlet}(PM_x) \tag{6.1}$$

Similarities in the requirements of the previous and current standards were described by Wojtas [2016] – e.g. concerning the tested dimension of the filter module 610x610 mm or the nominal air flow V = 3400 m^3/h.

The authors of the publication point out that it is not possible to easily compare or calculate the filter class classified according to the requirements of the previous and new standards [Wojtas 2016; Sikonczyk 2018]. A detailed comparison of research results is required (Brunner 2017). Such studies were conducted by the Association for Indoor Climate, Process Cooling and Food Cold Chain Technologies (Eurovent) and published as recommendations available on the website [Eurovent 4/23 2017]. On the other hand, in another article by Wojtas, there is a filter assessment indicator proposal, combining both the efficiency of dirt removal and operating costs, aimed at presenting the characteristics of the filters in a form understandable for each recipient [Wojtas 2017].

The CIOP-PIB laboratories conduct tests of the effectiveness of initial (Table 6.4) and accurate (Table 6.5) air filters with respect to electrically neutralized solid particles of potassium chloride (KCl) and test dust A2 compliant with ISO 12103-1:2016 (Road vehicles – Test contaminants for filter evaluation – Part 1: Arizona test dust) (Figure 6.4).

The test determines the fractional efficiency of the air filter in the least (24-hour isopropanol conditioning) and the most favorable conditions (initial state of filtration), and then determines the average value of the fractional efficiency, which is the indicator classifying the air filter. This testing procedure allows the air filter to be tested under variable operating conditions.

At the end of 2019, new standards from the ISO 29463 group (high efficiency filters and filter media for removing particles from air) came into force in Europe. The ISO 29463 standard consists of five parts: Part 1 provides the classification, performance, and labeling of filters. Parts 2–5 are implementation standards that describe individual test procedures in the field of aerosol production, measuring equipment and statistical calculations (Part 2), flat material samples (Part 3), filter leakage tests (Part 4) and filtration efficiency tests filters (Part 5).

Parts 2–5 were adopted as Polish standards PN-EN ISO 29463 and have been formally applicable since April 2019. Thus, they were adopted for use without changes. However, European ISO members, instead of part one of ISO 29463-1 (High efficiency filters and filter media for removing particles from air – Part 1: Classification, performance, testing and marking), were adopted as valid for use in Europe, EN 1822-1:2019, which replaced EN 1822-1:2009.

PN-EN 1822-1:2009 was withdrawn on May 17, 2019; however, according to the CEN decision, it was valid within conformity assessment until October 31, 2019.

However, from October 31, 2019, the following standards apply: PN-EN 1822-1:2019 and PN-EN ISO 29463-2,3,4,5:2018.

TABLE 6.4
Classification Parameters of the Coarse Filter (ISO 16890)

Initial resistance to air flow (Pa): 49	Initial arrestance: 67.44%	$ePM_{1, min}$ 4.08%	$ePM_{2,5, min}$ 7.51%		
Final resistance to air flow [Pa]: 200	**Test dust capacity: 2725.00 g**	**ePM_1 3.38%**	**$ePM_{2,5}$ 7.34%**	**ePM_{10} 24.49%**	**ISO rating ISO Coarse 65%**

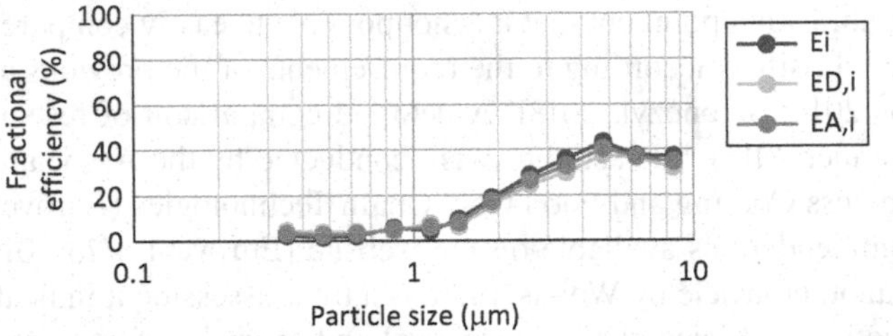

Fractional efficiency (%)
Particle size (μm)
Ei
ED,i
EA,i

E_i – initial fractional efficiency (ISO 16890-2)
$E_{D,i}$ –conditioned fractional efficiency (ISO 16890-4)
$E_{A,i}$ – average fractional efficiency (ISO 16890-1)

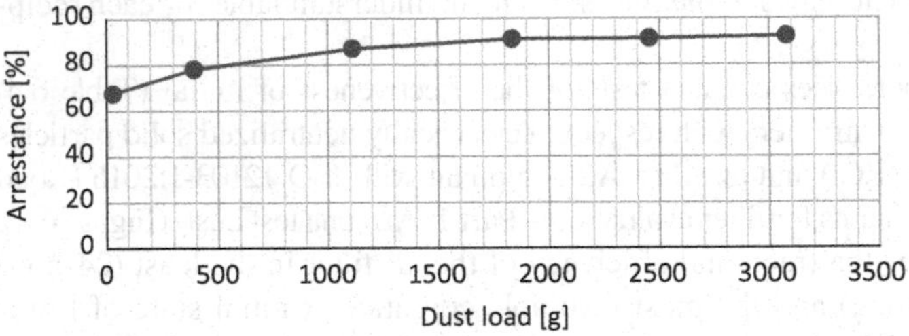

Arrestance [%]
Dust load [g]

Dependence of flow resistance and dust arrestance on the mass of dust given (ISO 16890-3)

The EN 1822-1:2019 standard introduces changes in relation to ISO 29463-1. Wherever ISO 29463-1 is mentioned in ISO 29463-2 to 5, EN 1822-1:2019 should be used. Classification, labeling of filters: the previous classification and labeling are kept, i.e. classes E10 to U17 (instead of classes ISO 15E to ISO 75U). The testing methodology does not change in relation to the EN 1822-2 to 5:2009 standards; however, there are differences in the scope of application:

- the ISO standard allows using a photometer for testing filter leaks; the EN standard does not,
- the ISO standard requires testing the effectiveness of each filter item from ISO 35H (H13) and upwards; the EN standard allows testing HEPA filters, classes H13 and H14, on a statistical basis, provided that the leakage test is performed for each item in accordance with EN ISO 29463-4:2018, Annex A.

Consequently, the EN standard recognizes the possibility of using this method equally with the scanning method, while the ISO standard requires agreement with the recipient. Compared to the previous standard, EN 1822:2009, in terms of testing

TABLE 6.5
Classification Parameters of the Fine Filter (ISO 16890)

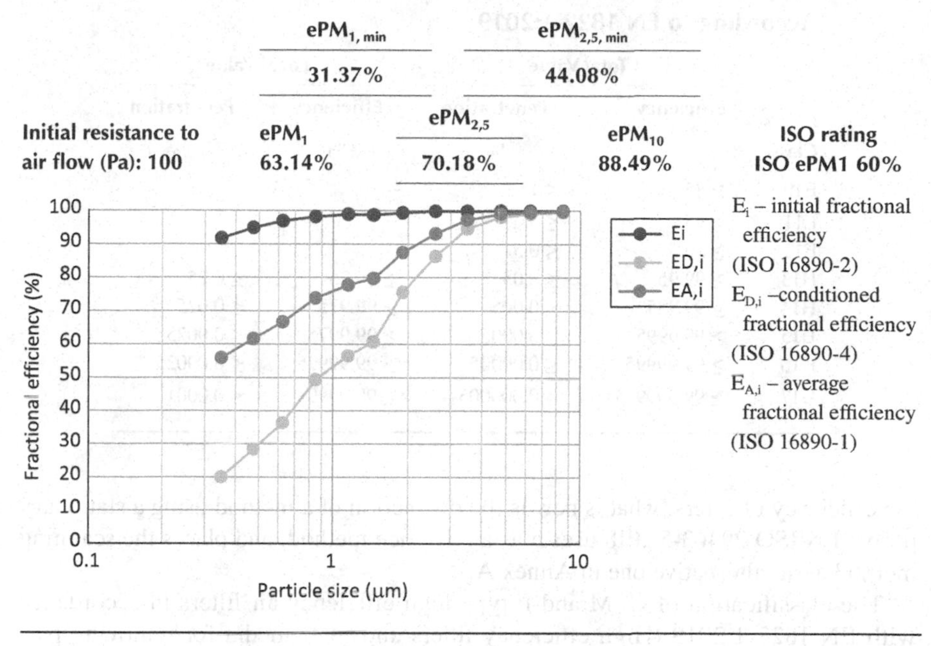

$ePM_{1,\,min}$	$ePM_{2,5,\,min}$
31.37%	44.08%

Initial resistance to air flow (Pa): 100	ePM_1	$ePM_{2,5}$	ePM_{10}	ISO rating
	63.14%	70.18%	88.49%	ISO ePM1 60%

FIGURE 6.4 Stand for testing air filter parameters in accordance with EN ISO 16890.

TABLE 6.6
Classification of EPA, HEPA and ULPA Type Air Filters According to EN 1822-1:2019

	Total Value		Local Value	
	Efficiency	Penetration	Efficiency	Penetration
Class	%	%	%	%
E10	≥ 85	≤ 15	–	–
E11	≥ 95	≤ 5	–	–
E12	≥ 99.5	≤ 0.5	–	–
H13	≥ 99.95	≤ 0.05	≥ 99.75	≤ 0.25
H14	≥ 99.995	≤ 0.005	≥ 99.975	≤ 0.025
U15	≥ 99.9995	≤ 0.0005	≥ 99.9975	≤ 0.0025
U16	≥ 99.99995	≤ 0.00005	≥ 99.99975	≤ 0.00025
U17	≥ 99.999995	≤ 0.000005	≥ 99.9999	≤ 0.0001

the efficiency of filters, what is new is the distinction of a method using a stationary probe: EN ISO 29463-5:2018 uses it as a reference method, and places the scanning method as an alternative one in Annex A.

The classification of G, M and F type high efficiency air filters in accordance with EN 1822-1:2019 (High efficiency filters and filter media for removing particles from air – Part 1: Classification, performance, testing and marking) is shown in Table 6.6. The filter class is determined based on the total and local values of filtration efficiency and penetration. High efficiency filters are used as the last stage in layered fiber composites of ventilation systems in rooms with very high air purity requirements. The overall efficiency determined with reference to the E, H and U type filters is the average efficiency in relation to the total filter face in given operating conditions. Whereas, local efficiency is the effectiveness at a given filter point, under given operating conditions. Filtration efficiency and penetration tests through high efficiency filters is conducted with the bis(2-ethylhexyl)ester of sebacic acid (DEHS) aerosol test or the bis(2-ethylhexyl)ester of phthalic acid (DOP) aerosol test.

High efficiency air filters, such as EPA (classes E10-E12), HEPA (classes H13-H14) and ULPA (classes U15-U17), are used as the last filtration stage in clean room ventilation installations with purity classes higher than ISO 7 (e.g. sterile operating rooms, drug and serum production, film and magnetic tape production, microelectronics production rooms). For very high requirements for air purity, multistage filtration systems are used.

The classification of air purity in terms of concentration of solid particles in the air of clean rooms, clean zones and separation devices, according to EN ISO 14644-1:2015 (Cleanrooms and associated controlled environments – Part 1: Classification of air cleanliness) is presented in Table 6.7.

TABLE 6.7
Room Cleanliness Classes According to EN ISO 14644-1:2015

ISO Classification Number (N)	Maximum Concentration Limits (particles/m³ of air) for Particles Equal to and Larger than the Considered Sizes Shown Below					
	0.1 μm	0.2 μm	0.3 μm	0.5 μm	1 μm	5 μm
ISO 1	10	2	–	–	–	–
ISO 2	100	24	10	4	–	–
ISO 3	1,000	237	102	35	8	–
ISO 4	10,000	2,370	1,020	352	83	–
ISO 5	100,000	23,700	10,200	3,520	832	29
ISO 6	1,000,000	237,000	102,000	35,200	8,320	293
ISO 7	–	–	–	352,000	83,200	2,930
ISO 8	–	–	–	3,520,000	832,000	29,300
ISO 9	–	–	–	35,200,000	8,320,000	293,000

The distribution of air flow in a clean room can be classified as (EN ISO 14644-4:2001 Cleanrooms and associated controlled environments – Part 4: Design, construction and start-up):

- unidirectional air flow – controlled air flow through the entire cross section of the clean zone, with a constant velocity and approximately parallel current lines,
- multidirectional air flow – the air blown into the clean zone mixes with the indoor air by induction.

If a type of flow such as a combination of the above is used, the resulting system is often called a mixed system. For clean rooms of ISO class 5 and above, one-way air flow distribution is usually used at the routine use stage. The multidirectional air flow and mixed distribution is usually used in rooms with ISO 6 purity class and lower purity (EN ISO 14644-4:2001).

When processing nanomaterials, an additional test to which filter materials should be subjected is the measurement of nano-aerosol concentration in accordance with the guidelines of ISO 21083-1:2018 (Test method to measure the efficiency of air filtration media against spherical nanomaterials – Part 1: Size range from 20 to 500 nm). The CIOP-PIB conducts research on nano-aerosol retention on materials used in high efficiency filters [Jakubiak 2019]. For the E11 class filter, the filtration efficiency curve has a characteristic minimum associated with various retention mechanisms for ultrafine and submicrometric particles (Figure 6.5). For H13 and H14 high efficiency filters, the average efficiency was obtained above 99.99% and above 99.999%, respectively.

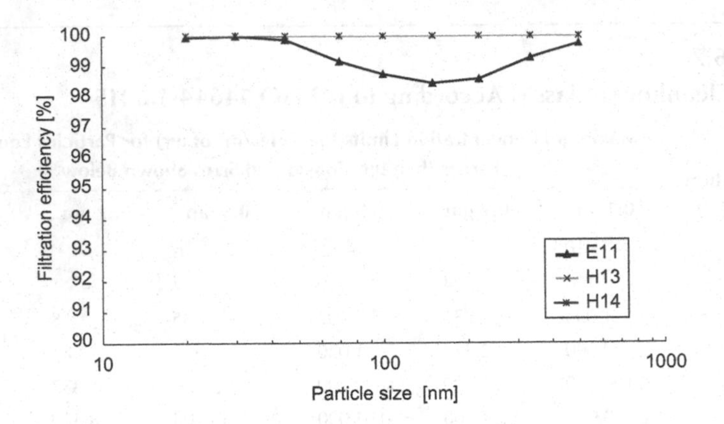

FIGURE 6.5 Results of the efficiency of nanoparticle filtration tests on high efficiency filters.

6.3 ORGANIZATIONAL MEASURES

The risk associated with newly introduced chemical substances and their mixtures, and those substances that cause an increase in hazards in the production, processing and use processes, should also be controlled through the use of appropriate organizational measures in the premises.

Organizational measures may include information to workers about risk, preventive measures, safe work procedures to be used, and minimization of the number of workers exposed to chemicals, the number of hours worked, or areas where exposure may occur.

The National Institute for Occupational Safety and Health (NIOSH) recommends that companies implement the following work practices as part of an overall strategy to control workers' exposure to nanomaterials [NIOSH 2016]:

- Educate workers on safe handling of newly developed nanomaterials to minimize the risk of inhalation and skin contact.
- Provide information on hazardous properties of materials used by workers, together with instructions on how to prevent exposure.
- Encourage workers to use hand washing equipment before eating, smoking or leaving the workplace.
- Provide additional protection measures (e.g. using a buffer zone, devices for decontaminating workers) to ensure that the developed nanomaterials will not be transported outside the working area.
- Where there is a possibility of surface or personnel contamination, showering and uniform changing facilities should be provided to prevent inadvertent contamination of other areas (including at home) due to the transfer of nanomaterials to clothing and skin.
- Avoid processing of nanomaterials in open spaces.
- If possible, store dispersible nanomaterials in the form of liquids or solids in tightly closed containers.

- Make sure that workplaces and designated equipment (e.g. weight) are cleaned at least at the end of each work shift, using collective protection equipment or wet wiping methods (if the use of fluid does not pose additional hazards).
- Cleaning should take place in a way that prevents workers from coming into contact with waste.
- All waste should be disposed of in accordance with all applicable federal, state, and local regulations.
- Avoid storing and consuming food or beverages in work areas where nanomaterials are processed.

The subject of assumptions concerning the requirements of the effectiveness of limiting air pollution threats is the definition of activities that – when introduced through changes in the organization of work and changes in technical parameters of machines and technological processes – will allow reducing workers' occupational risk [CIOP-PIB 2020].

Assumptions about actions concerning organizational changes in the workplace can be directed to:

- limiting workers' threat to emitted air pollution,
- ensuring proper operation of machine and technological processes,
- ensuring an effective operation of technical collective protection equipment.

Assumptions about organizational measures include the following areas:

- operation of machine and technological processes,
- operation of technical collective protection equipment,
- Servicing activities of workers working with machines and technological processes.

Assumptions regarding changes in the organization of work *in the area of machine operation* include:

- phase of selection and use of appropriate production materials, processed etc.,
- phase of preparation for the technological process,
- phase of production, processing and operation,
- maintenance phase.

The selection and use of appropriate materials for processing on a machine should be preceded by:

- obtaining information on the harmfulness of substances contained in the materials, especially carcinogenic or mutagenic substances,
- determining physicochemical properties, at least in terms of their chemical composition and morphology,

- making workers aware of the fact that the use of appropriate raw materials and ensuring their optimal technological parameters (e.g. humidity of dusty materials) will reduce the risk of air pollution at the source of emission.

During the machine preparation phase it is advisable to:

- design, manufacture and equip it so that its operation does not cause the risk of air pollution,
- optimize machine work parameters,
- automate production,
- use tight pipelines and tanks for transporting materials to machines,
- equip the machine with complete or partial housing or local ventilation systems,
- plan its location in the room taking into account environmental conditions of the workplace.

Work organization during the machine's operation phase can be determined by the following actions:

- limitation of use only by workers authorized to do so,
- elimination or limitation of access of people visiting the company, laboratory etc. (e.g. clients, customers),
- location of control systems outside the air pollution hazard zones,
- equipping the workplace with a switch in situations of threat of air pollution emissions,
- application of a control system signaling the state of threat of air pollution in case of failure of the local ventilation cooperating with the system.

To eliminate the phenomenon of secondary air pollution, the following operating methods should be observed during modernization or maintenance of machinery:

- planning and designing tasks to be performed as part of modernization or maintenance,
- creating teams among workers responsible for the performance of planned activities in an efficient and effective manner,
- hiring specialized employers in the field of modernization or maintenance (in the absence of trained personnel),
- marking the area of modernization or maintenance works in a proper way,
- stopping or limiting other work stations in the room that may disturb the course of modernization or maintenance works,
- selecting tools and devices taking into account the type, type of material, size of the work surface, as well as the type of room and its development.

Technical parameters of material and machine have a significant effect on the risk of air pollution emissions resulting from machine operation and the technological process. Technical assumptions should describe the parameters taking into account the type of technological process, material properties and the characteristics of the source of air pollution emissions.

Depending on the conditions of production or material processing, the air pollution hazard zone changes and may include both the work area around the machine and other workstations.

Basic material parameters in the production process or its treatment that should be taken into account are [Kuhlbush et al. 2011]:

- material grade and origin,
- material density and humidity affecting mechanical strength,
- machinability (degree of machining difficulty),
- unusual features of material,
- grain size,
- hardness,
- morphology.

It is also necessary to determine the parameters that characterize the emission source, i.e.

- shape of the tool used (e.g. saws, milling cutters, knife shafts, discs, tapes),
- type of treatment (e.g. concurrent, backward) related to the direction of particle ejection,
- machining parameters (e.g. cutting velocity, feed velocity, depth, cutting width, working direction in relation to the fiber system in material),
- characteristics, circumferential velocity, pressure of the work piece to the abrasive wheel,
- degree of mobility of the source of air pollution emissions during material treatment.

One of the conditions for obtaining effective functioning of *technical collective protection equipment* indoors is to take appropriate organizational measures in the phases of:

- design of enclosures,
- operation of emission source enclosures,
- modernization or maintenance of individual enclosing elements.

Design and selection of equipment elements for enclosures or local ventilation should be focused on:

- type of material used,
- easy cleaning of materials from which the enclosure is made,
- eliminating or limiting the deposition of dust on the internal surfaces of the enclosure,
- determining the method of connecting the housing to the machine,
- size and arrangement of working and installation openings,
- method of opening the enclosure and its working openings,

- location of places for removing polluted air from the housing (connecting ventilation ducts),
- designing filtration systems adapted to the required efficiency of air purification,
- method of sealing connections between housing elements and machines,
- method of preventing air pollution from escaping outside of the enclosure when opening it.

Organizational activities related to the operation of the enclosure or local ventilation system should ensure:

- synchronization of time of turning the enclosing system on and off depending on the time of machine operation,
- tightness of all elements of the enclosure,
- efficiency of dust extraction in accordance with the designed values,
- use of enclosure in accordance with the intended use,
- the housing during machine operation is not opened.

A condition that limits the threat of air pollution emitted during production and processing is to make periodic inspections and modernization of housing or local ventilation systems, taking into account the following assumptions:

- planning and designing tasks to be performed as part of modernization or maintenance,
- creating teams among workers responsible for the performance of planned activities in an efficient and effective manner,
- in the absence of trained personnel, employing specialized employers in the field of modernization or maintenance of technical collective protection equipment,
- marking the area of modernization or maintenance works of enclosures or local ventilation installations in a proper way,
- stopping or limiting other workstations in the room that may disturb the course of modernization or maintenance works,
- selecting tools and devices taking into account the type, size and shape of housing or local ventilation systems.

Activities in the field of requirements related to workers' maintenance work near the source of air pollution emissions, contributing to the reduction of occupational risk for workers, should include:

- making workers aware of the threat of harmful emissions in the workplace,
- providing occupational health and safety instructions concerning the operation of machines, technical collective protection equipment and works related to the threats of air pollution,
- operating machinery and technical collective protection equipment in accordance with the operating instructions, and maintaining occupational health and safety procedures,

- making workers aware of the effects of newly introduced chemical substances and mixtures thereof,
- undertaking actions by workers to prevent air pollutants from getting into the workplace by appropriate handling of technical collective protection equipment,
- making workers aware of the need to prioritize the use of technical collective protection equipment before personal protection equipment.

6.4 PERSONAL PROTECTION EQUIPMENT

If the technical and organizational solutions presented in the above sections cannot be implemented, the final solution is to use appropriate *personal protective equipment* (PPE). PPE should be designed in a way as to protect the workers from exposure and ensure an optimal level of protection. PPE should also be ergonomic.

The Council Directive of November 30, 1989 [EC 1989] on minimum requirements in the field of safety and health protection of workers using protective equipment (third individual directive within the meaning of Article 16 (1) of Directive 89/391/EEC) and Regulation of the EU Parliament and Council 2016/425 of March 9, 2016, on personal protective equipment and repealing Council Directive 89/686/EEC) [EU 2016] treat PPE as well as any additional equipment used for this purpose as individual equipment intended for use or wear by any workers in the purpose of its protection against threats which may affect its health and safety at work.

Information on recommended PPE should be provided in the material safety data sheets and mixtures.

A necessary condition for a proper selection of PPE is to assess the occupational risk at workplaces connected with production, processing and use of newly introduced chemical substances and their mixtures.

To assess occupational risk, information should be used regarding, among others:

- identified threats and sources of emission,
- location of the workplace and technological processes,
- used measures, materials and technological processes,
- activities performed and the manner and time of their performance,
- required legal regulations,
- organizational and technical protection measures applied,
- accidents and potential accidents.

The procedure for selecting PPE should include the following activities:

- determining the types of threats and workers' body parts exposed to these threats,
- determining the necessary types of PPE,
- determining the properties that PPE must possess to provide protection against identified threats,

- based on available knowledge – determining the optimal parameters of personal protective equipment, ensuring effective protection against identified threats,
- selection of PPEs with the required protective parameters.

The EU Regulation 2016/425 [EU 2016] sets out the obligations of individual economic operators: manufacturer, authorized representative, distributor and importer, as well as cases in which manufacturers' obligations apply to importers and distributors.

The regulation introduces the division of PPE into three groups in terms of belonging to the type of threats against which they constitute sufficient protection, and sets out various conformity assessment procedures for individual groups of these measures. The regulation defines the following categories of threats against which PPE is to protect:

- Category I *includes minimum hazards*, such as surface mechanical injuries, contact with weaker cleaning agents or prolonged contact with water, contact with hot surfaces not exceeding 50°C, eye damage due to exposure to sunlight (other than when observing the sun) and atmospheric indicators that are not extreme.
- Category III *includes life hazards or hazards that can cause serious and irreversible damage to health or death* (e.g. hazardous substances and mixtures, harmful biological agents).
- Category II includes hazards not listed in categories I and III.

Depending on the category of PPE, the regulation provides for different conformity assessment procedures. Conformity assessment procedures that apply to PPE for protection against specific hazard categories are:

- Category I: internal production control (module A)
- Category II: EU type examination (module B) followed by a type-conformity test based on internal production control (module C)
- Category III: EU type examination (module B) and one of the following modules:
    - conformity to type based on internal production control and supervised product checks at random intervals (module C2)
    - conformity to type based on quality assurance of the production process (module D).

Individual protection of workers when working with newly introduced chemical substances and mixtures thereof comes down mainly to reducing body penetration by inhalation and through the skin.

The research conducted both in CIOP-PIB and in other scientific centers shows that to protect the respiratory system against nanoparticles, half-masks or complementary masks with class P3 filters should be used. In other cases and during short-term exposure, masks and half-masks with filters class P2 or FFP2 are allowed [Zapór 2012].

Due to work with nano-objects and the nature of the process, it is required that workers, in addition to respiratory protection, are equipped with safety glasses, gloves and footwear, as well as work clothing. For short-term exposure, laboratory aprons (not cotton) can be used as protective clothing, while for long-term work or with a high risk of dusting, it is advisable to use plastic overalls with barrier properties. Workers' hands should be protected with disposable gloves made of nitrile, latex, neoprene or vinyl with the lowest possible permeability for nano-objects and high resistance to chemical substances characteristic for a given technological process [Zapór 2012].

6.5 STORAGE, TRANSPORT AND DISPOSAL

Enterprises which carry out production, processing or use newly introduced chemical substances and mixtures, a plan for storage and utilization of harmful agents should be developed, depending on their quantity and degree of danger to people and the environment. Waste management, according to guides [NSRC 2008; EA 2005], should mainly concern: uncontaminated chemicals, including NOAA, objects contaminated with nanomaterials (storage containers, disposable personal protective equipment etc.), NOAA containing liquid suspensions and products containing fragile and poorly bound nanomaterials. Any other material in contact with NOAA, which has not been subjected to the decontamination procedure, should also be treated as waste [Jankowska 2015].

Nanomaterial waste should be stored in plastic bags (laboratory waste) or airtight containers (industrial waste), labeled with basic information about the content. Waste nanomaterials should not be disposed of with other harmless waste. The waste management system, including nanomaterials, should be operated in accordance with national regulations.

Waste management systems are organized differently in European countries and include different technologies, e.g. recycling, waste water, incineration and storage. The latter involves various processing steps, such as comminution (size reduction). For this reason, solid waste will be processed using different technologies, depending on local waste management systems [Gressler et al. 2014; Caballero-Guzman et al. 2015]. Outside the European Union, waste disposal is the dominant solution [Keller et al. 2013]. Partially, nanomaterials can also get into the water and be further processed under controlled conditions in wastewater treatment plants. The most common example is titanium dioxide nanoparticles, contained in cosmetics for protection against sunlight, which are removed from the skin during bathing.

Currently, European legislation on waste containing nanomaterials does not provide for any specific treatment regulations. Solid wastes containing nanomaterials enter national waste management systems as part of waste material.

During various stages of processing, nanomaterials can be released from raw waste [EA 2005; 2008]. During recycling, waste size is reduced mechanically by crushing, grinding or milling, which can lead to the release of nanomaterial from the waste. During the combustion process, waste material is almost completely burned at very high temperatures, and therefore nanomaterials can be released into the environment. During waste storage, the material also degrades and the particles can also be released.

Nanomaterials contained in waste can be released into the environment by means of typical treatment processes. Thus, nanomaterials get into different areas of the environment, including air, water and soil.

Nanomaterials can float in the air when they are mechanically treated during recycling or during combustion.

Release into water can also occur during the recycling process or during long-term storage of waste, where rainwater is not collected and completely retained. At landfills, released nanomaterials can also penetrate the soil.

Basically, waste treatment processes include systems that collect all materials emitted from waste, e.g., by filtration systems, gas treatment systems or closed landfills. The results of research on the effectiveness of selected protective systems indicate the presence of small or medium risk associated with the release of nanomaterials into the environment [Walser and Gottschalk 2014; Andersen et al. 2014; Part et al. 2015; FOEN 2017].

Pursuant to the Regulation of the Minister of the Environment of October 7, 2016, on detailed requirements for waste transport [Dz. U. 2016, poz. 1742], non-hazardous waste may be transported simultaneously with hazardous waste, but only provided that mutual contact is impossible. In addition, waste transport must be made in a manner that prevents mixing of individual types of waste, except where the flow of mixed types of waste is entirely directed to processing in the same process.

Waste transport must take place in a way that minimizes odor nuisance and prevents it from spreading outside the means of transport, in particular dumping, dusting and leakage. It should also be conducted so as to minimize the impact of atmospheric indicators on waste, if it can cause a negative effect of the transported waste on the environment or human life and health.

A carrier must also arrange or secure transported goods (in particular those in containers or bags) in the means of transport in a way so as to prevent them from moving and tipping over. This regulation does not apply to waste transported in bulk or in tanks.

If there is a change in the type of waste transported, before transporting waste the carrier must ensure that the means of transport does not contain residues from the previous transport of waste, unless the residues from the previous transport of waste do not affect the properties of the transported waste in a way that causes a threat to the environment or human life and health.

According to guidelines, waste should be transported along with a document confirming the type and details of the ordering party. This document may be a waste transfer card, waste sales invoice, basic waste characteristics, a document regarding transboundary waste shipment or other documents confirming the type of waste transported and data of party ordering the waste transport.

The regulation also introduces a decision regarding the marking of waste transport vehicles. Means of transport of waste (constituting a vehicle or combination of vehicles) should be marked with an appropriate plate placed in a visible place in front of the means of transport on its external surface. The marking should be legible and durable, as well as weatherproof.

6.6 CONCLUSION

This chapter discussed the general recommendations for methods and measures for protecting human health in rooms with newly introduced chemical substances and their mixtures, and those substances that cause increased risks in the processes of their production, processing and use.

Occupational risks related to chemical harmful agents should be controlled by introducing appropriate actions – in accordance with the priority order presented in ISO/TS 12901-1:2012 (Nanotechnologies – Occupational risk management applied to engineered nanomaterials – Part 1: Principles and approaches).

The methods and measures to protect human health against the hazard of chemical emissions and particulate matter emissions presented in the following subsections have been characterized in relation to the design, operation and modernization or maintenance phases of individual elements of production, transport and storage process.

The spread of air pollutants emitted at workstations can be reduced by using various types of protective measures. Substitution is priority. The harmful indicator replacement method reduces investment and operating costs associated with technical protection equipment. In most cases, substitution is impossible. Treatment rooms of newly introduced chemical substances and their mixtures, and those substances that cause increased risks in their production, processing and use should be equipped with appropriate collective protection equipment, the use of which is a priority in relation to the use of personal protective equipment. Collective protection equipment against dust and chemical harmful indicators are mainly general mechanical ventilation systems and local mechanical ventilation devices. Mechanical ventilation is an organized air exchange that is ensured by technical measures such as fans. When it is necessary to create a microclimate in a room that constantly meets specific requirements, regardless of external weather conditions, air conditioning must be used, which not only provides the right amount of air, but also its required air quality (cleanless, humidity, temperature etc.).

The purpose of ventilation that consists in constant or periodic exchange of air in this type of room is therefore to:

- improve the condition and composition of air at workstations in accordance with hygiene and technological requirements,
- regulate such parameters of the air environment in rooms as concentration of pollution, temperature, humidity, velocity and direction of air movement.

The lack or improper design of collective protection equipment and inadequate further control of them may cause irrational air exchange, which if:

- too low – does not ensure a volume air flow supplied to the room in relation to the number of people,
- too high – irrational use of energy [EC 2010].

The issue of rational design of collective protection equipment occurs during various technological processes related to the release of air pollution.

The greatest challenge is to design a collective protection equipment that takes into account the work zones of persons and machines together with the air pollution they generate, and at the same time are economical in terms of operating costs. Actions aimed at reducing the cost of collective protection equipment can be implemented both as part of an existing installation, and by designing a completely new protection measure.

A well-designed protective measure should only work with maximum efficiency if there are people in the workplace or the exposure to air pollution is particularly high.

REFERENCES

ACGIH [American Conference of Governmental Industrial Hygienists]. 2019. *Industrial ventilation: A manual of recommended practice for design*. 30th ed. Cincinnati, OH: ACGIH.

ACS [American Chemical Society]. 2011. American Chemical Society of Green Chemistry Institute. Green nanotechnology challenges and opportunities. http://greennano.org/sites/greennano1.uoregon.edu/files/GCI_WP_GN10.pdf (accessed January 30, 2020).

Ahn, K., S. Woskie, L. DiBerardinis, and M. Ellenbecker. 2008. A review of published quantitative experimental studies on factors affecting laboratory fume hood performance. *J Occup Environ Hyg* 5(11):735–753. doi: 10.1080/15459620802399989.

Albuquerque, P. C., J. F. Gomes, C. A. Pereira, and R. M. Miranda. 2015. Assessment and control of nanoparticles exposure in welding operations by use of a Control Banding Tool. *J Clean Prod* 89:296–300. doi: 10.1016/j.jclepro.2014.11.010.

Andersen, L., F. M. Christensen, and J. M. Nielsen. 2014. *Environmental Project No. 1608*. Copenhagen, Denmark: Environmental Protection Agency, Danish Ministry of the Environment.

Bell, G., D. Sartor, and E. Mills. 2002. The Berkeley hood: Development and commercialization of an innovative high-performance laboratory fume hood. Lawrence Berkeley National Laboratory Report LBNL-48983. https://www.researchgate.net/publication/237639221_The_Berkeley_Hood_Development_and_Commercialization_of_an_Innovative_High-Performance_Laboratory_Fume_Hood (accessed January 30, 2020).

Bell, G. C. 2009. Optimizing laboratory ventilation rates: Process and strategies. *J Chem Health Saf* 16(5):14–19.

Bémer, D., M. T. Lecler, R. Régnier, G. Hecht, and J. M. Gerber. 2002. Measuring the emission rate of an aerosol source placed in a ventilated room using a tracer gas: Influence of particle wall deposition. *Ann Occup Hyg* 46(3):347–354.

Brunner, A. 2017. Feinstaub: Die Klassifizierungsgröße der neuen ISO-Luftfilternorm. HLH Bd. 68, No 3. https://www.hlh.de/2017/Ausgabe-03/Editorial2/Feinstaub-Die-Klassifizierungsgroesse-der-neuen-ISO-Luftfilternorm (accessed January 30, 2020).

Caballero-Guzman, A., T. Sun, and B. Nowack. 2015. Flows of engineered nanomaterials through the recycling process in Switzerland. *Waste Manag* 36:33–43. doi: 10.1016/j.wasman.2014.11.006.

Chen, J. K., and R. F. Huang. 2014. Flow characteristics and robustness of an inclined quad-vortex range hood. *Ind Health* 52:248–255. doi: 10.2486/indhealth.2013-0138.

Chen, J. K., R. F. Huang, P. Y. Hsin, C. M. Hsu, and C. W. Chen. 2012. Flow and containment characteristics of an air-curtain fume hood operated at high temperatures. *Ind Health* 50:103–114. doi: 10.2486/indhealth.ms1326.

Chosewood, L. C., and D. E. Wilson. 2009. *Biosafety in microbiological and biomedical laboratories.* 5th ed. Washington, DC: U.S. Department of Health and Human Services, Public Health Service, Centers for Disease Control and Prevention, National Institutes of Health.

CIOP-PIB 2020. Knowledge base on chemical and dust hazards. https://www.ciop.pl/CIOPPortalWAR/appmanager/ciop/pl?_nfpb=true&_pageLabel=P13800141641345795944292 (accessed January 30, 2020).

EA [Environment Agency]. 2005. What is a hazardous waste? A guide to the hazardous waste regulations and the list of waste regulations in England and Wales (HWR01). Almondbury, Bristol: Environment Agency. https://www.hazwasteonline.com/marketing/media/Regulations/wm2_what_is_hazardous_waste.pdf (accessed January 30, 2020).

EA [Environment Agency]. 2008. Interim advice on wastes containing unbound carbon nanotubes. https://nanotech.law.asu.edu/Documents/2009/08/nano-waste_199_5154.pdf (accessed January 30, 2020).

EC [European Commission]. 1989. Council Directive 89/656/EEC of 30 November 1989 on the minimum health and safety requirements for the use by workers of personal protective equipment at the workplace (third individual directive within the meaning of Article 16 (1) of Directive 89/391/EEC). https://eur-lex.europa.eu/legal-content/en/TXT/?uri=CELEX:31989L0656 (accessed January 30, 2020).

EC [European Commission]. 2010. Directive 2010/31/EU of the European Parliament and of the Council of 19 May 2010 on the energy performance of buildings. https://eur-lex.europa.eu/eli/dir/2010/31/oj (accessed January 30, 2020).

EPA [US Environmental Protection Agency]. 2015. Design for the Environment Alternatives Assessments. http://www2.epa.gov/saferchoice/design-environment-alternatives-assessments (accessed January 30, 2020).

EU [European Union]. 2016. Parliament and Council Regulation 2016/425 of 9 March 2016 on personal protective equipment and repealing Council Directive 89/686/EEC. *OJ L 81, 31.3.2016, 51–98.* http://data.europa.eu/eli/reg/2016/425/oj (accessed January 30, 2020).

Eurovent 4/23. 2017. Selection of EN ISO 16890 rated air filter class general ventilation applications. https://eurovent.eu/sites/default/files/field/file/Eurovent%20REC%204-23%20-%20Selection%20of%20EN%20ISO%2016890%20rated%20air%20filter%20classes%20-%202017.pdf (accessed January 30, 2020).

FOEN [Federal Office for the Environment]. 2017. Nanowaste. https://www.bafu.admin.ch/bafu/en/home/topics/waste/guide-to-waste-a-z/nanowaste.html (accessed January 30, 2020).

Gressler, S., F. Part, and A. Gazsó. 2014. Nanowaste: Nanomaterial-containing products at the end of their life cycle. *NanoTrust Dossier* 040en:1–5. doi: 10.1553/ita-nt-040en.

Hampl, V., and R. Niemelä. 1986. Use of tracer gas technique for industrial exhaust hood efficiency evaluation: Where to sample? *Am Ind Hyg Assoc J* 47(5):281–287.

Hodson, L., and M. Hull. 2016. *Building a safety program to protect the nanotechnology workforce: A guide for small to medium-sized enterprises.* Cincinnati, OH: U.S. Department of Health and Human Services, Centers for Disease Control and Prevention, National Institute for Occupational Safety and Health. https://www.cdc.gov/niosh/docs/2016-102/pdfs/2016-102.pdf (accessed January 30, 2020).

Huang, R. F., J. K. Chen, and W. L. Hung. 2013. Flow and containment characteristics of sashless, variable-height inclined air-curtain fume hood. *Ann Occup Hyg* 57(7):934–952.

Jakubiak, S. 2019. ISO 2183: Nowa norma międzynarodowa do określania skuteczności filtracji nanocząstek. (ISO 21083: New international standard for determination of nanoparticles filtration efficiency). *Podstawy i Metody Oceny Środowiska Pracy* 2(100):7–11.

Jankowska, E. 2015. Zasady zarządzania ryzykiem zawodowym związanym z narażeniem na nanoobiekty, ich aglomeraty i agregaty (NOAA) [Principles of occupational risk management related to exposure to nano-objects, their agglomerates and aggregates (NOAA)]. *Podstawy i Metody Oceny Środowiska Pracy* 2(84):17–36.

Jankowski, T. 2011. Impact of air distribution on efficiency of dust capture from metal grinding: Bench Test Method. *Ind Health* 49:735–745. doi: 10.2486/indhealth.ms1293.

Jankowski, T. 2019. Narażenia zawodowe na dymy spawalnicze i wentylacja na stanowiskach spawalniczych. *Bezpieczeństwo Pracy. Nauka i Praktyka* 4:10–14. doi: 10.5604/01.3001.0013.1575.

Jung, A., and M. Zeller. 1994. An analysis of different tracer gas techniques to determine the air exchange efficiency in a mechanically ventilated room. In *Air Distribution in Rooms, Roomvent '94, Fourth International Conference, Krakow, June 15-17, 1994.* Vol. 2, 315–332.

Keller, A. A., S. McFerran, A. Lazareva, and S. Suh. 2013. Global life cycle releases of engineered nanomaterials. *J Nanopart Res* 15(6):1692. doi: 10.1007/s11051-013-1692-4.

Kuhlbusch, T. A. J., C. Asbach, H. Fissan, D. Göhler, and M. Stintz. 2011. Nanoparticle exposure at nanotechnology workplaces: A review. *Part Fibre Toxicol* 8:22. doi: 10.1186/1743-8977-8-22.

Lieckfield Jr., R. G., and R. C. Poore. 2001. Health and safety factors in designing an industrial hygiene laboratory. In *Patty's industrial hygiene*, eds. R. L. Harris, and F. A. Patty. New York, NY: Wiley. doi: 10.1002/0471435139.hyg063.

Mierzwiński, S. 2015. *Aerodynamika wentylacji.* Gliwice: Wydawnictwo Politechniki Śląskiej.

Moore, R. W. 2004 Tracer gas testing applications for industrial hygiene evaluations. *Occupational Hazards* 66(5):39–43.

NIOSH [National Institute for Occupational Safety and Health]. 2016. Building a Safety Program to Protect the Nanotechnology Workforce: A Guide for Small to Medium-Sized Enterprises. https://www.cdc.gov/niosh/docs/2016-102/pdfs/2016-102.pdf (accessed January 30, 2020).

NSRC [Nanoscale Science Research Centres]. 2008. Department of Energy Nanoscale Science Research Centres. Approach to nanomaterial ES&H. Revision 3a, May 12, 2008.

Ojima, J. 2002. Worker exposure due to reverse flow in push-pull ventilation and development of a reverse flow preventing system. *J Occup Health* 44:391–397.

Ojima, J. 2007. Tracer gas evaluation of local exhaust hood performance. *San Ei Shi* 49(5):209–215.

Ojima, J. 2009. Tracer gas evaluations of push-pull ventilation system performance. *Industrial Health* 47(1):94–96.

Part, F., G. Zecha, T. Causon, E. K. Sinner, and M. Huber-Humer. 2015. Current limitations and challenges in nanowaste detection, characterisation and monitoring. *Waste Man* 43:407–420. doi: 10.1016/j.wasman.2015.05.035.

Salasinska, K., M. Borucka, M. Leszczyńska et al. 2017. Analysis of flammability and smoke emission of rigid polyurethane foams modified with nanoparticles and halogen-free fire retardants. *J Therm Anal Calorim* 130:131–141. doi: 10.1007/s10973-017-6294-4.

Saunders, G. T. 1993 *Laboratory fume hoods: A user's manual.* New York, NY: John Wiley & Sons.

Sikończyk, I. 2018. Dobór filtrów sklasyfikowanych wg PN-EN ISO 16890 w systemach wentylacji ogólnej [Selection of filters classified according to EN ISO 16890 in general ventilation systems]. *Chłodnictwo i Klimatyzacja* 2:76–80.

Szymański, M., R. Gorzeński, and K. Szkarlat. 2015. HVAC installations for chemical laboratories: Design. *Rynek Instalacyjny* 11:72–75.

Tsai, S. J., R. F. Huang, and M. J. Ellenbecker. 2010. Airborne nanoparticle exposures while using constant-flow, constant-velocity and air-curtain-isolated fume hood. *Ann Occup Hyg* 54(1):78–87.

Tschudi, W. T., G. Bell, and D. Sartor. 2004a. Side-by-side fume hood testing: ASHRAE 110 Containment Report: Comparison of a conventional and a Berkeley fume hood. LBID-2560. https://hightech.lbl.gov/sites/default/files/documents/SidebySide%20Berkeley%20Hood%20HAM_2004.pdf (accessed January 30, 2020).

Tschudi, W. T., G. Bell, and D. Sartor. 2004b. Side-by-side fume hood testing: Human-as-Mannequin report: Comparison of a conventional and a Berkeley fume hood. LBID-2561. http://hightech.lbl.gov/Documents/HOOD/LBNL_HAM_TestECT.pdf (accessed January 30, 2020).

Tseng, L. C., R. F. Huang, and C. Chen. 2010. Significance of face velocity fluctuation in relation to laboratory fume hood performance. *Ind Health* 48:43–51. doi: 10.2486/indhealth.48.43.

Tseng, L., R. F. Huang, C. Chen, and C. Chang. 2006. Correlation between airflow patterns and performance of a laboratory fume hood. *J Occup Environ Hyg* 3(12):694–706.

Tseng, L., R. F. Huang, C. Chen, and C. Chang. 2007. Effects of sash movement and walk-bys on aerodynamics and contaminant leakage of laboratory fume cupboards. *Ind Health* 45:199–208. doi: 10.2486/indhealth.45.199.

Varley, J. O. 1993. Measuring fume hood diversity in an industrial laboratory. *ASHRAE Trans* 99(2):429–439.

Walser, T., and F. Gottschalk. 2014. Stochastic fate analysis of engineered nanoparticles in incineration plants. *J Clean Prod* 80:241–251. doi: 10.1016/j.jclepro.2014.05.085.

Wang, H. Q., C. H. Huang, D. Liu et al. 2012. Fume transports in a high rise industrial welding hall with displacement ventilation system and individual ventilation units. *Building and Environment* 52:119–128. doi: 10.1016/j.buildenv.2011.11.004.

Weale, J., P. H. Rumsey, D. Sartor, and L. E. Lee. 2007. Low-pressure-drop HVAC design for laboratories. https://www.aivc.org/sites/default/files/members_area/medias/pdf/CR/CR06_Low-pressure-drop.pdf (accessed January 30, 2020).

Wojtas, K. 2016. Wymagania i zasady nowej klasyfikacji filtrów w systemach wentylacji budynków [Requirements and rules for the new classification of filters in building ventilation systems]. *Rynek Instalacyjny* 12:1–7.

Wojtas, K. 2017. Konsekwencje wprowadzenia nowej klasyfikacji filtrów dla wentylacji wg normy EN-ISO 16890. (Practical aspects of the new classification of filters for ventilation according to EN-ISO 16890 standard). *Rynek Instalacyjny* 1/2:22–28.

Worrell, E., J. A. Laitner, M. Ruth, and H. Finman. 2003. Productivity benefits of industrial energy efficiency measures. *Energy* 28(11):1081–1098. doi: 10.1016/S0360-5442(03)00091-4.

Zapór, L. 2012. Toksyczność nanocząstek metali: Wybrane zagadnienia [Selected aspects of the toxicity of metals nanoparticles]. *Przemysł Chemiczny* 91(6):1237–1240.

Zatorski, W., and K. Salasinska. 2016. Analiza palności nienasyconych żywic poliestrowych modyfikowanych nanocząstkami [Combustibility studies of unsaturated polyester resins modified by nanoparticles]. *Polimery* 61(11/12):815–823. doi: 10.14314/polimery.2016.815.

Index

Printed in the United States
by Baker & Taylor Publisher Services

Printed in the United States
by Baker & Taylor Publisher Services